Models of Particles and Moving Media

MODELS OF PARTICLES AND MOVING MEDIA

DONALD A. DUNN

DEPARTMENT OF ENGINEERING-ECONOMIC SYSTEMS AND
INSTITUTE FOR PLASMA RESEARCH
STANFORD UNIVERSITY
STANFORD, CALIFORNIA

1971

ACADEMIC PRESS New York and London

ACADEMIC PRESS, INC.
111 Fifth Avenue, New York, New York 10003

United Kingdom Edition published by
ACADEMIC PRESS, INC. (LONDON) LTD.
24/28 Oval Road, London NW1 7DD

LIBRARY OF CONGRESS CATALOG CARD NUMBER: 70-154391

PRINTED IN THE UNITED STATES OF AMERICA

Contents

Chapter 4 Forces and Wave Interaction with Uniformly Moving Circuits and Continua

Chapter 5 Nonuniform Motion of Charged Particles in Prescribed Electric and Magnetic Fields

Chapter 6 **Motion of Many Interacting Particles—The Lagrangian Model**

Chapter 7 **Motion of Charged and Conducting Fluids—The Eulerian Model**

Preface

This book is designed to provide a foundation for an understanding of the literature in a number of related fields of electrodynamics in which directed motion of either free charges or solid or fluid conducting media occurs. Each is a major field of research and a potential or established industry:

1. Electric motors and generators;
2. Electron tubes;
3. Particle accelerators;
4. Plasma devices, including switches, gas tubes, controlled thermonuclear reactors, and magnetohydrodynamic generators;
5. Space physics.

The treatment of electrodynamics that is presented here builds on elementary electromagnetic theory by proceeding step by step from (1) moving frames of reference in Newtonian mechanics to (2) the Lorentz transformation and waves in bulk media in motion at a constant velocity to (3) accelerated motion of single particles and finally to (4) many-particle and continuous fluid models. The emphasis throughout is on the precise definition of mathematical models of physical systems and on the process of model building. Where possible a connection between the analysis of these models and physical experiments is presented.

Controlled thermonuclear fusion for stationary electric power generation is a field of research of long-term interest that requires an understanding of electrodynamics. The operation of a thermonuclear reactor necessarily involves relativistic effects and nuclear chemistry of considerable complexity. However, there is a broad class of problems associated with heating a plasma to the necessary temperature, and with containing it once it has this temperature for a sufficient time for a reaction to occur, that involves only Newton's laws and Maxwell's equations. The problems are those of obtaining self-consistent field solutions to the motion of 10^{20} or more interacting and highly energetic particles. If there is more than one charged particle in a given region of space, the electric and magnetic fields acting at the position of each particle depend on the positions and velocities of the other particles in the space. The "self-consistent field" is the field in the region of interest, as a function of time and position, that is consistent with the motion of all particles in the space. It requires very little imagination to see that this type of problem can become very complex.

Intuitively, it appears that there is a class of solutions to this problem that is sufficiently stable and that contains enough particles at a high enough temperature to make a thermonuclear reaction go on a steady-state basis. A large number of carefully thought-through experiments aimed at containing a thermonuclear plasma have been performed. Although none have been successful, many new types of instabilities have been discovered, and the richness and complexity of the behavior of plasmas is now much better appreciated than it was before this thermonuclear research was begun.

Many of the stability problems that have arisen in plasma physics research have their counterparts in various types of microwave electron tubes that make use of these instabilities to provide amplification. Similarly, many of the mechanisms of solar phenomena and of the motion of charged particles in the extraterrestrial plasma in the earth's magnetic field are closely related to laboratory plasma phenomena.

There has also been a developing interest in recent years in "direct" energy conversion systems in which electricity is generated directly from heat energy without the interposition of mechanical motion of a turbine or other macroscopic elements. In the magnetohydrodynamic (MHD) generator, heat energy is converted first to directed motion of the combustion gases which are, in one way or another, made electrically conducting. This moving conductor is then passed through a region of electric and magnetic fields in which it is slowed down and the energy extracted in the form of electric current. Similarly, the thermionic energy converter converts heat to electricity by producing thermionic electron emission from a cathode and extracting the kinetic energy of these electrons by passing them through an appropriate electric field.

This book presents a unified treatment of the electrodynamics common to all of these devices and research areas. The technique that enables us to examine the foundations of a series of rather deep research problems of this level of complexity is simply that of taking a very simple model for each device that we examine. In many cases the analysis of a highly simplified model of a device can yield useful information about the nature of the mechanism of interest, and the art of selecting the right model is the art that needs to be learned in this field. Of course, in many cases the simplest model simply does not act like a real device. In this case, however, it is extremely important to be able to identify the level of complexity of the model at which the real world behavior of interest sets in. If this identification can be made, the mechanism of interest can often be understood with much greater precision than is the case when one is simply confronted with similar behavior in a single complex model and in a real world system. In other words, it can be very informative to simplify the model to a point at which the desired behavior is lost. As an example, in many plasma problems the mechanism of interest involves an interplay between the electrons and the ions of the plasma. In a mathematical model of the plasma it is often easy to assign an infinite mass to the ions. If doing so eliminates the behavior of interest, it is possible to identify the mechanism as one in which ion motion is essential.

The general plan of study consists of three basic steps. First, an understanding of both particle motion and Maxwell's equations as viewed from moving frames of reference is sought. Learning to understand these equations is accomplished by using them to solve problems of interaction with moving solid media in which motion takes place only at a constant velocity. The next step is the study of nonuniform particle motion in prescribed fields. The third step is to learn to do self-consistent field problems in which the field is influenced by nonuniform particle or material motion.

In the study of problems of moving solid media, energy transfer between fields and bulk material is of most significance. We examine several basic energy conversion devices from a somewhat higher level of analytic sophistication than is ordinarily used. These devices include dc motors and generators, synchronous and induction ac machines, and electrohydrodynamic and magnetohydrodynamic devices in which the details of the fluid motion can be omitted. In each case the emphasis is on obtaining an understanding of how the device works using analysis of a simplified model.

Two types of models are studied that can be used in the solution of self-consistent field problems involving nonuniform motion of particles or fluids. First, the Lagrangian discrete particle model with many interacting particles is developed and used to solve some simple one-dimensional problems. With the advent of fast digital computers it has been possible to do problems in this field that had been outstanding for 30 years, but that were never touched because of the tedious nature of the computations required. An effort is made here to formulate these problems in a way that is adapted to solution on the computer. It is the author's belief that, with the next generations of computers, whole new areas of understanding will become possible using extensions of these analytical techniques.

Second, some problems for which analytic solutions to the Eulerian fluid model can be obtained are solved and the results are compared with those obtained from the Lagrangian discrete particle model. Two types of problems are studied, those appropriately solved by the use of static theory and those for which small-signal perturbation theory is relevant. Examples from the fields of electron and plasma diodes and ion propulsion systems naturally fall into the category of static Eulerian theory. Examples from the fields of microwave electron tubes and plasma waves are natural illustrations of small-signal perturbation theory in which perturbations of a static state are studied.

Acknowledgments

I am indebted to the many students who have worked through the problems and examples in this book with me and who have helped me to see the right way to view them. I am particularly indebted to Dr. Christopher Barnes for his assistance in producing many of the previously unpublished computer diagrams that appear in Chapters 5 and 6 and to Dr. Heinrich Derfler for his helpful comments on the manuscript. I am grateful to my wife, Jane, for her encouragement and to my secretary, Elaine Christensen, for her help in typing and preparing the manuscript in its many drafts.

Models of Particles and Moving Media

CHAPTER 1

Space-Time and the Galilean Transformation

According to our viewpoint, the concept of time is determined by the operations by which it is measured. We have to distinguish two sorts of time; the time of events taking place near each other in space, or local time, and the time of events taking place at considerably separated points in space, or extended time. As we now know, the concept of extended time is inextricably mixed up with that of space. This is not primarily a statement about nature at all, and might have been made simply by the observation that the operations by which extended time is measured involve those for measuring space. Of course historically the doctrine of relativity was responsible for the critical attitude which led to an examination of the operations of measuring time, but relativity was not necessary for a realization of the spatial implications of time.

P. W. Bridgman, "The Logic of Modern Physics." Macmillan, New York, 1927. Reprinted by permission.

1.0 Introduction

We begin the study of electrical interactions with moving particles and moving media by examining some of the ways in which motion can be described. A central idea in the study of motion is that of being able to adopt the point of view of an observer riding with whatever it is that is moving. This idea leads to the description of motion from two or more frames of reference, and, in turn, this involves us with the mathematics of coordinate transformations involving time as well as space. In this chapter we examine some of these concepts as they arise in classical Newtonian mechanics.

1.1 Trajectories in Various Spaces

If a projectile is fired from a gun, its vertical height will vary with both horizontal position and time. The path the projectile follows in space is commonly referred to as the trajectory of the projectile. We will be studying the motion of particles and bulk matter and will have occasion to use this term trajectory in a somewhat broader sense. We will typically obtain, by one means or another, a differential equation that describes the motion of a particle in one or more space dimensions as a function of time. The solution will be expressible in a variety of forms, including the position of the particle as a function of time, one space coordinate as a function of another space coordinate, and velocity or momentum as a function of time or position. In each of these situations, the trajectory is defined as the path followed by a particle in the space defined by the diagram in which we plot one or more of these variables versus another. We are familiar with the idea of the trajectory of a particle in configuration space, say in the form x as a function of z, as in Fig. 1.1-1, which is further explained in Example I.

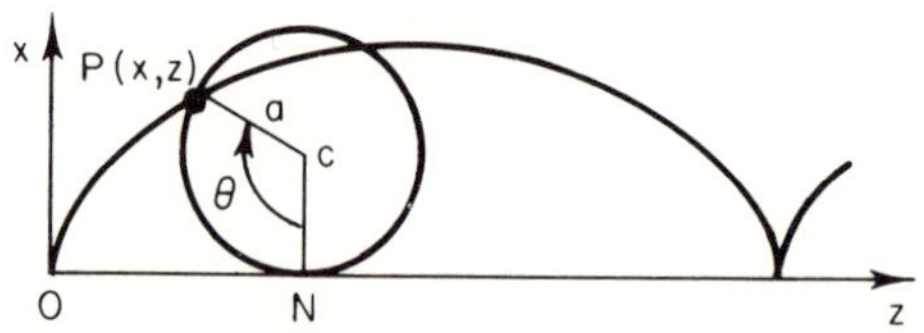

Fig. 1.1-1. The trajectory in configuration space, $P(x, z)$, of a point on the rim of a wheel rotating at a uniform angular velocity and moving along the z-axis with one point of the wheel N always in contact with the z-axis. The curve traced out by $P(x, z)$ is a cycloid (see Example I).

A second type of space that we will use extensively is the "space" in which the dimensions are one or more distance coordinates and time. In relativity theory this type of space is often referred to as Minkowski space. There is no standard nomenclature in the nonrelativistic literature to refer to this type of space and we will refer to it, therefore, either as Minkowski space or as distance-time space. When motion is restricted to one distance coordinate and time the resulting diagram can be drawn in two dimensions

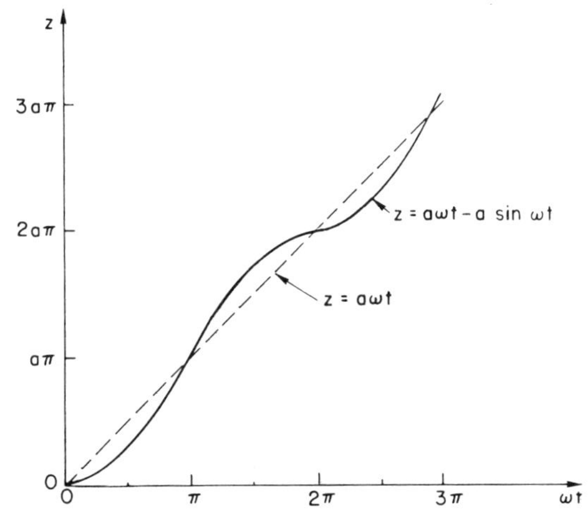

Fig. 1.1-2. The trajectory in Minkowski space, $P(z, t)$, the position along the z-axis as a function of time of the point P on the wheel rim of Fig. 1.1-1.

as in Fig. 1.1–2. The trajectory of a particle in such a space is referred to in relativity theory as a world line. The slope of such a trajectory in Minkowski space is evidently the velocity.

If the velocity is plotted as a function of position, we have what is usually

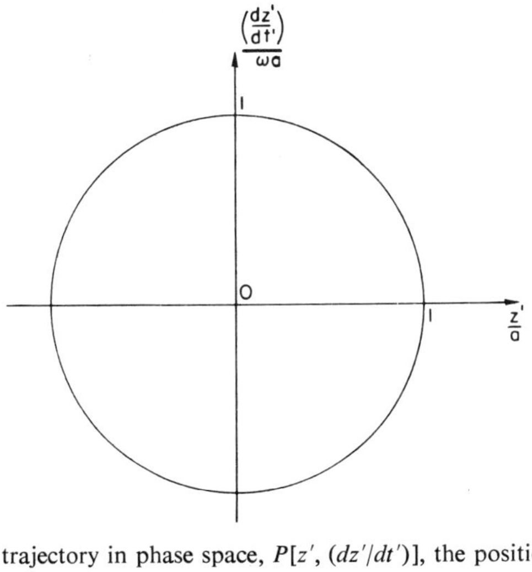

Fig. 1.1-3. The trajectory in phase space, $P[z', (dz'/dt')]$, the position along the z-axis as a function of velocity in the z-direction of the point P on the wheel rim of Figs. 1.1-1 and 1.1-2, as viewed by an observer at rest with respect to a frame of reference moving at a uniform velocity in the z-direction with the wheel axis. Viewed from the rest frame of this moving observer the point on the rim executes simple harmonic motion. Viewed from the laboratory frame the motion is more complex (see Example I).

referred to as a phase diagram. The six-dimensional space having as dimensions the three position coordinates and the three corresponding velocity coordinates is referred to as phase space. If only one distance and one velocity coordinate are involved in the motion, the phase diagram can also be plotted in a plane which is the phase plane. Figure 1.1-3 is an example of this method of plotting the particle trajectory. Different initial conditions result in different trajectories in all diagrams and it is often of interest to plot a set of trajectories corresponding to many possible initial conditions.

A great deal can be learned about particles and their motion by studying problems that involve only one space dimension. Then, as in Figs. 1.1-2 and 1.1-3 the motion can be visualized in two-dimensional diagrams. Evidently such problems are two-dimensional, but as a matter of convention they are usually referred to as one-dimensional, referring to the fact that there is only one position coordinate involved plus time.

Example I Simple Harmonic Motion in a Moving Frame of Reference

A particularly simple example of motion in various spaces is worked out here to illustrate the various types of trajectories of interest. A wheel of radius a is pictured as rolling along the z-axis as in Fig. 1.1-1 without slipping, so the point N on the rim that is touching the z-axis is always at rest. The point $P(x, z)$ is the point on the rim of the wheel that started out at the origin at $t = 0$. The wheel is rotating at a uniform angular velocity ω and the angle measured from the line CN to CP is θ, so that

$$\omega = d\theta/dt \tag{I.1}$$

and the distance ON is then

$$ON = a\theta = a\omega t \tag{I.2}$$

The distances (x, z) which define $P(x, z)$ are then by inspection

$$z = a\theta - a \sin\theta, \qquad x = a - a\cos\theta \tag{I.3}$$

Equations (I.3) are the equations of a cycloid and constitute the equations for the trajectory in configuration space, as in Fig. 1.1-1.

As a function of time the position coordinates vary as

$$z = a\omega t - a \sin \omega t, \qquad x = a - a\cos\omega t \tag{I.4}$$

and the first of these equations provides the diagram of $P(z, t)$ of Fig. 1.1-2 which is a plot of the z-coordinate of the point on the wheel rim as a function of time. The velocity of this point in the z-direction is obtained by differentiating Eq. (I.4)

$$dz/dt = a\omega - a\omega \cos \omega t \tag{I.5}$$

It is not obvious how to get a relation for the velocity as a function of z, and, if it were obtained, it would be an involved expression, as may be seen by comparing Eqs. (I.4) and (I.5). However, if we transform to a moving coordinate system, moving in the positive z-direction at the velocity of the wheel axis $a\omega$ we find that the rim point is executing simple harmonic motion in this frame of reference. In order to transform to this frame of reference using a Galilean transformation we use Eqs. (1.3.9) and (1.3.11)

$$\begin{gathered} z' = z - a\omega t, \qquad x' = x, \qquad t' = t \\ dz'/dt' = (dz/dt) - a\omega \end{gathered} \tag{I.6}$$

and obtain

$$\begin{gathered} z' = -a \sin \omega t, \qquad x' = a - a \cos \omega t, \qquad t' = t \\ dz'/dt' = -a\omega \cos \omega t \end{gathered} \tag{I.7}$$

It is easy to obtain a relation between velocity and position in the z-direction in this frame of reference. Square and add the first and last relations of Eq. (I.7) to get

$$a^2\omega^2 \sin^2 \omega t + a^2\omega^2 \cos^2 \omega t = (\omega z')^2 + (dz'/dt')^2 \tag{I.8}$$

which is

$$\frac{z'^2}{a^2} + \frac{(dz'/dt')^2}{\omega^2 a^2} = 1 \tag{I.9}$$

This is the equation of a circle, plotted in Fig. 1.1-3, which is the trajectory of the rim point in the phase plane as viewed by an observer at rest in the primed frame. If a phase plane trajectory in the laboratory rest frame were desired, it could be obtained readily from a back transformation of Eq. (I.9) using Eqs. (I.6).

If we wish to view the transformation to the rest frame of the moving observer graphically, a simple approach due to Minkowski is illustrated in Fig. 1.3-2 which allows us to use the same line for the trajectory in both frames of reference. A plot of distance versus time for the z-coordinate of the point on the wheel rim as viewed from the laboratory frame is plotted in Fig. 1.3-2 from Eq. (I.4). This is exactly the same curve that was plotted in Fig. 1.1-2 and the z- and t-coordinates are plotted in the same way. In addition, Fig. 1.3-2 has the z'- and t'-axes present with the z'-axis aligned with the z-axis and the t'-axis at an angle $\phi = \tan^{-1} a$ to the t-axis. The value of $z'(t')$ from Eq. (I.7) is evidently correctly given by the same curve in the z'–t' plane that defines $z(t)$ in the z–t plane.

1.2 Frames of Reference

A frame of reference is simply the coordinate system that describes position in configuration space as seen by an observer at rest with respect to the coordinate system. Obviously things look different to observers who are moving with respect to each other, and it often simplifies a problem in particle or wave motion to be able to use a frame of reference moving at the velocity of the wave or the particle. The most familiar example is perhaps the Doppler effect for sound, in which the frequency of a sound wave is altered by the relative motion of source and observer. Since sound is carried by a medium, such as air, that may be moving relative to both the source of sound and the observer, it is useful in this case to choose a frame of reference in which the air is at rest. Because a sound wave is carried by a material medium, there is an inherent asymmetry in the equations relating the observed frequency to the source frequency of the sound for cases in which the observer is in motion relative to the medium or in which the source is in motion. The Doppler effect for electromagnetic waves in free space has no corresponding asymmetry, as we shall see, because electromagnetic waves in free space behave as if there were no medium present.

Transformation equations can be used to transform to any other frame of reference, once the physical effect has been properly described in the frame of reference most convenient for analysis. The mathematics of such transformations, when there are three or more dimensions involved, can most elegantly be handled with matrix algebra. The transformation to moving frames is no different mathematically from that of a simple rotation from one frame at rest to another also at rest but rotated by some angle with respect to the original frame.

However, if we stick to one-dimensional problems, it is usually easier to keep track of what we are doing without using matrices and so we shall not have occasion here to use much of the extensive mathematical machinery of this type that has been developed for use in special and general relativity problems.

Of great importance in choosing frames of reference and in stating physical laws mathematically is the concept of invariance. We would like to be able to mathematically describe physical events as they are viewed by observers in various different frames of reference. An event, we assume, occurs in exactly the same way physically, regardless of how we observe it, in classical physics. Nevertheless, the equations describing the event may take different forms in different frames of reference. If the physical law governing an event can be expressed in the same form in two frames of reference, we say the law is invariant to the mathematical transformation that takes us between the two frames of reference. This statement is explained in the form of an example in the next section.

The terminology and notation that we shall use in describing frames of reference is as follows:

1. The "laboratory frame" is the frame of reference of an observer at rest in the laboratory where measurements can be made with the usual instrumentation. Quantities such as distance and time in this frame are unprimed.
2. The "rest frame" is the frame of reference of an observer at rest with respect to a coordinate system moving at constant velocity v in the positive z-direction. Quantities in this frame are primed.
3. The "proper frame" is the frame of reference of an observer at rest with respect to a moving body such as a clock that may (or may not) be accelerated. Quantities in this frame are double primed if necessary to distinguish from primed frame quantities.

In a proper frame an observer can read time on the same clock at different times, because he is at rest with respect to the clock. Similarly, in a proper frame an observer can make a measurement of length using a meter stick, because he is at rest with respect to the meter stick. In most cases we will be concerned with unaccelerated motion and so will be able to choose the rest frame of the moving observer in such a way that it is a proper frame. For example, an unaccelerated but drifting beam of ions emits light at frequencies determined by the atomic properties of the drifting material and thus can be thought of as a large number of moving atomic clocks. A frame of reference moving with the drift velocity of the ions is thus both a proper frame and the rest frame of an observer moving with the ions [1–3].

Example II Doppler Effect for Acoustic Waves

The velocity of sound in air at normal temperature and pressure is about 333 m sec^{-1} and it is not difficult to produce mechanical motion at velocities exceeding this. When we discuss the effects of motion of a source of sound and an observer listening to the frequency of the sound wave, we must take into account the possible motion of the air as well.

Let s be the speed of sound and let v be the relative velocity of source and observer. Let us take a frame of reference at rest with respect to the medium of propagation and consider a case in which both source and observer are also at rest in this frame of reference. The sound is emitted by the source with constant frequency f. The wavelength λ_0 of the wave is given by $\lambda_0 = s/f$. Now consider the effect of the motion of the observer. If he moves toward the source at velocity v, he will see a higher frequency f_1 because he passes an additional number of wavelengths per unit time, thus raising the observed frequency by an amount v/λ_0. Thus

$$f_1 = f + (v/\lambda_0) = f[1 + (v/s)] \tag{II.1}$$

For motion away from the source evidently the sign is changed.

Now consider again that we are at rest with respect to the medium and let the source move toward the observer. In this case the wavelength of the sound is altered. During the course of a cycle the source moves a distance v/f thus shortening the wavelength of the emitted sound by this amount. Thus the frequency of the sound as observed is

$$f_1 = \frac{s}{\lambda_0 - (v/f)} = f\frac{1}{1 - (v/s)} \tag{II.2}$$

Again the sign is changed if the direction of motion is reversed. The frequency shift is thus in the same direction for motion of either source or observer, but is only the same in value to first order in (v/s). For (v/s) equal to unity the observed frequency is twice the source frequency for the case of a moving observer and is equal to infinity for the case of the moving source. Here motion of source and observer can be distinguished, and radically different results are obtained, because motion of either can be referred to the special frame of reference attached to the medium carrying the wave.

Problem 1.2-1 Wind Velocity Measurement Apparatus

Devise an experiment to measure wind velocity by sending sound waves of constant known frequency in various directions and measuring the received frequency at various pickup points.

1.3 Newton's Laws and Galilean Invariance

We shall first state Newton's laws in a more or less conventional form.

I. Bodies not subject to external forces continue in a state of rest or straight-line uniform motion.

II. Bodies subject to forces undergo accelerations. The acceleration is in the direction of the force and its magnitude is the ratio of the force to the inertial mass of the body.

III. If two bodies exert forces on each other, these forces are equal in magnitude and opposite in direction.

A few comments on these laws are worth making. First of all, it is evident that the first two laws can be regarded simply as definitions of the concepts of force, inertial mass, and acceleration, and contain no physics. The third law tells us the physical law governing the interaction between two bodies or particles. A force causes acceleration in an amount dependent on the inertial mass of the body as specified by the second law. Thus, we may write the second law in the familiar form

$$\mathbf{f} = m\mathbf{a} \tag{1.3.1}$$

where $\mathbf{a}$ is the vector

$$\mathbf{a} = \mathbf{1}_x(d^2x/dt^2) + \mathbf{1}_y(d^2y/dt^2) + \mathbf{1}_z(d^2z/dt^2) \tag{1.3.2}$$

and where $\mathbf{1}_x$, $\mathbf{1}_y$, and $\mathbf{1}_z$ are unit vectors along their respective axes. If we have two bodies in an isolated system in which no forces act other than forces between the bodies, we may write, from the third law,

$$\mathbf{f}_1 = -\mathbf{f}_2 \tag{1.3.3}$$

where $\mathbf{f}_1$ and $\mathbf{f}_2$ are the forces acting on bodies 1 and 2, respectively. Then

$$m_1\mathbf{a}_1 = -m_2\mathbf{a}_2 \tag{1.3.4}$$

where $m_{1,2}$ and $\mathbf{a}_{1,2}$ are the inertial masses and accelerations of the bodies respectively. The ratio of inertial masses is then, evidently,

$$m_1/m_2 = |\,\mathbf{a}_2\,|/|\,\mathbf{a}_1\,| \tag{1.3.5}$$

as required by the second law. If we are dealing with gravitational force, we can determine the relative masses of two objects by simply weighing them. The question then arises as to whether the mass so determined, which is termed the gravitational mass, will be the same as the inertial mass of the same bodies that might be measured if the bodies were subjected to another type of force such as centrifugal force. Recent experiments by Dicke [4] have confirmed early results by Eötvös *et al.* [5] indicating an equivalence to a part in 10^{10}. The principal of equivalence, as the term is used in general relativity theory, is the assumption of exact equality of gravitational and inertial mass.

Mathematically, we may state the first law in the form

$$x = a_1 + b_1 t, \qquad y = a_2 + b_2 t, \qquad z = a_3 + b_3 t \tag{1.3.6}$$

where x, y, z are space coordinates in a Cartesian coordinate system, t is time, and the a's and b's are constants that characterize the location of the body and its velocity at $t = 0$. We may also write the first law in the form

$$d^2x/dt^2 = d^2y/dt^2 = d^2z/dt^2 = 0 \tag{1.3.7}$$

obtained by differentiating Eq. (1.3.6) twice. The verbal statement of the first law corresponding to Eq. (1.3.7) is: If a body is not subject to forces, it is possible to find a coordinate system with respect to which the motion of the body is unaccelerated.

The term inertial systems is used to refer to all frames of reference from which the body appears unaccelerated. Obviously, a body at rest or in uniform motion in an inertial system will appear to be accelerated if viewed from any frame of reference that is accelerated with respect to the inertial system. The corollary to this statement is still another statement of the first law: If no forces are acting on a body in one frame of reference, there will be no forces acting in any frame of reference in uniform motion with respect to the first frame. Even though this statement is obvious, let us prove it, in order to get used to the concepts involved in moving frames.

Figure 1.3-1 shows two coordinate systems S and S'. The primed system

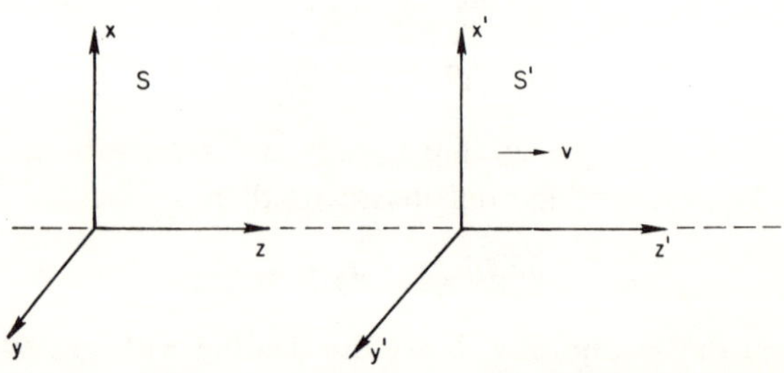

Fig. 1.3-1. Two Cartesian coordinate systems. The origin of system S' is moving along the z-axis of the S system at velocity v and the z and z' axes are colinear. The S system is the laboratory frame and the S' system is the rest frame (of a moving observer).

is moving in the positive z-direction at the velocity v with respect to the unprimed system. To locate a position in the primed system we first locate the origin of the primed system. It has coordinates in the unprimed system as follows, if the origins were coincident at $t = 0$:

$$z = vt, \qquad x = y = 0 \tag{1.3.8}$$

It is then clear that if we set

$$z' = z - vt, \qquad x' = x, \qquad y' = y, \qquad t' = t \tag{1.3.9}$$

we have expressed the relative motion of the two systems correctly and we can use this (Galilean) transformation to convert the position of a particle as expressed in one system to a position in the other. The same transformation can of course, be expressed in reverse.

$$z = z' + vt', \qquad x = x', \qquad y = y', \qquad t = t' \tag{1.3.10}$$

We may transform not only positions, but also velocities and accelerations. The transformation for velocity is obtained by differentiating Eqs. (1.3.9) and using $dt = dt'$:

$$\begin{aligned} dz'/dt' &= (dz/dt') - v\,dt/dt' = (dz/dt) - v \\ dx'/dt' &= dx/dt \\ dy'/dt' &= dy/dt \end{aligned} \tag{1.3.11}$$

The transformation for acceleration is obtained by differentiating Eqs. (1.3.11):

$$\begin{aligned} d^2z'/dt'^2 &= d^2z/dt'^2 = d^2z/dt^2 \\ d^2x'/dt'^2 &= d^2x/dt^2 \\ d^2y'/dt'^2 &= d^2y/dt^2 \end{aligned} \tag{1.3.12}$$

From Eqs. (1.3.12) it is seen that acceleration has the same form in both systems, i.e., it is invariant to a coordinate transformation of this type. Thus, our statement of Newton's first law is correct because the force involves only the second derivative of position and is thus unaffected by a coordinate transformation involving a constant velocity.

We also see that the second law, Eq. (1.3.1), would be written in the same form by observers in both the primed and unprimed frames. We therefore say that, in general, Newton's laws are invariant under a Galilean transformation. Invariant means simply that the equations of motion have the same form. For example, if the force between two particles is proportional to the distance between them, and if the particles lie on the z-axis at z_1 and z_2, their equation of motion in the unprimed system is

$$d^2(z_2 - z_1)/dt^2 = -\omega^2(z_2 - z_1) \tag{1.3.13}$$

where ω^2 is a constant of proportionality. In the primed system, from Eqs. (1.3.9)

$$(z_2' - z_1') = (z_2 - z_1) \tag{1.3.14}$$

and from Eqs. (1.3.12)

$$d^2(z_2 - z_1)/dt^2 = d^2(z_2' - z_1')/dt'^2 \tag{1.3.15}$$

and because the coordinate difference and the acceleration transform unchanged in form, the equation of motion in the primed system has exactly the same form that it had in the unprimed system:

$$d^2(z_2' - z_1')/dt'^2 = -\omega^2(z_2' - z_1') \tag{1.3.16}$$

It is helpful to look at transformations graphically and if there is only motion in one direction, it is possible to express the interesting features of the transformation in a graph showing distance and time in both coordinate systems simultaneously [1–3]. Figure 1.3-2 is a Minkowski (or space–time) diagram of this type that shows the trajectories of a point P in both the z–t plane and the z'–t' plane. The z'-axis has been taken coincident with the z-axis and with the same scale as the z-axis. As indicated, $z = z' + vt'$, as required by Eqs. (1.3.10). The t'-axis has been drawn at an angle ϕ to the t-axis determined by the velocity v, according to the relation $v = \tan\phi$. The scales along the t- and t'-axes have been chosen to make lines of constant time be lines parallel to the z-axis, which is of course a line of constant time equal to zero in both coordinate systems. Similarly it is helpful in using this diagram to remember that the t- and t'-axes are the lines of constant z and z' equal to zero. The case shown in Fig. 1.3-2 is the same one that was shown in the z–t plane in Fig. 1.1-2 and represents a case of simple harmonic motion in the moving z'–t' frame as described by Eq. (1.3.16) and by a differential equation of exactly the same form in the z–t plane, Eq. (1.3.13). It is customary in special relativity theory to refer to the trajectory in Minkowski space as a world line and to a point on the trajectory as a world point or event, emphasizing that a particular time and position are being specified.

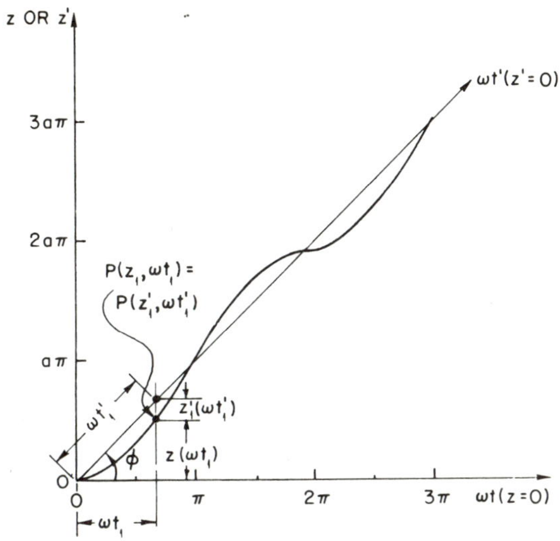

Fig. 1.3-2. Trajectories in two superimposed Minkowski spaces of the point *P* on the wheel rim of Figs. 1.1-1–1.1-3. Here $P(z, t)$, the trajectory as viewed from the laboratory frame, as in Fig. 1.1-2, is plotted along with the trajectory $P(z', t')$ of the same point as viewed from the rest frame (of a moving observer). Because the axes have been drawn in a particular way, the two trajectories are identical (see Example I).

REFERENCES

1, J. B. Marion, "Classical Dynamics of Particles and Systems," 2nd ed. Academic Press, New York, 1970.
2. E. P. Ney, "Electromagnetism and Relativity." Harper, New York, 1962.
3. P. W. Bridgman, "A Sophisticate's Primer of Relativity." Wesleyan Univ. Press, Middletown, Connecticut, 1962.
4. R. H. Dicke, The Eötvös experiment, *Sci. Amer.* **205**, 84–94 (Dec. 1961).
5. R. V. Eötvös, D. Pekar, and E. Fekete, Beitrage zum Gesetze der Proportionalität von Tragheit und Gravitat, *Ann. Phys.* (*Leipzig*) **68**, 11 (1922).

CHAPTER 2

The Lorentz Transformation of Time and Space

The example I saw before me was thermodynamics. The general principle was there given in the theorem: the laws of nature are such that it is impossible to construct a perpetuum mobile (of the first and second kind). How, then, could such a universal principle be found? After ten years of reflection such a principle resulted from a paradox upon which I had already hit at the age of sixteen: If I pursue a beam of light with the velocity c *(velocity of light in a vacuum), I should observe such a beam of light as a spatially oscillatory electromagnetic field at rest. However, there seems to be no such thing, whether on the basis of experience or according to Maxwell's equations. From the very beginning it appeared to me intuitively clear that, judged from the standpoint of such an observer, everything would have to happen according to the same laws as for an observer who, relative to the earth, was at rest. For how, otherwise, should the first observer know, i.e., be able to determine, that he is in a state of fast uniform motion?*

One sees that in this paradox the germ of the special relativity theory is already contained. Today everyone knows, of course, that all attempts to clarify this paradox satisfactorily were condemned to failure as long as the axiom of the absolute character of time, viz., of simultaneity, unrecognizedly was anchored in the unconscious. Clearly to recognize this axiom and its arbitrary character really implies already the solution of the problem.

A. Einstein, Autobiographical Notes, in "Albert Einstein: Philosopher-Scientist" (The Library of Living Philosophers, P. A. Schilpp, ed.). Open Court, La Salle, Illinois, 1949, Reprinted by permission.

2.0 Introduction

We have seen that Newton's laws are invariant under a Galilean transformation and that one of the conditions of a Galilean transformation is the condition that time runs at the same rate in all inertial (unaccelerated) systems, i.e., that $t = t'$. This so-called universality of time condition never has had any experimental evidence in favor of it. In fact, it was not widely noticed that this assumption underlay all physical theory during the nineteenth century.

It was evident from the time of Maxwell and Hertz that the velocity of light in vacuum is a maximum propagation velocity for electrical effects. Maxwell's equations yield this velocity as a natural constant and they are invariant under a transformation due to Lorentz that incorporates the condition that the velocity of light is the same, when viewed from any frame of reference. According to the Galilean transformation, the velocity of light is different for observers moving at different velocities, just as any other velocity is relative. This fact implies that one can catch up with a light wave, and for some reason this idea was not acceptable to Einstein when he first encountered it. He did accept the Lorentz transformation and the idea that the velocity of light should be a constant, but this left the worlds of mechanics and light operating under quite different rules.

The bold solution proposed by Einstein in 1905 was to assume that the velocity of light represented a limitation on mechanical motion as well as electrical effects and to propound the special theory of relativity as a universal description of space and time [1]. In the special theory of relativity, the velocity of light is regarded as a maximum velocity of mechanical motion. Since Einstein's statement of this theory, all of the experimental evidence that has been obtained has confirmed the correctness of the theory. In this chapter we will explore the meaning of this transformation in relation to time and space and such concepts as simultaneity. In the next chapter we examine the Lorentz transformation of Maxwell's equations.

2.1 Transformation of Coordinates

The special theory of relativity is based on two postulates: (1) The velocity of light is the same when viewed from any frame of reference. (2) it is impossible to detect unaccelerated translatory motion, because the fundamental laws of physics are the same in all unaccelerated frames of reference.

The second postulate is the same for both Newtonian mechanics and the special theory. The first postulate replaces the postulate of universality of time in Newtonian mechanics.

If the first postulate is to be correct, it should be possible to sit at the origin of a coordinate system and watch a spherical wavefront traveling outward at the velocity of light and, during the same time, but perhaps with time running at a different rate, have an observer in a moving coordinate system see the same spherical wavefront also traveling spherically outward from him at the same velocity from his frame of reference. This picture is a little difficult to visualize; however, mathematically there is no difficulty. The key to visualization is the fact that time does run at different rates in the two frames.

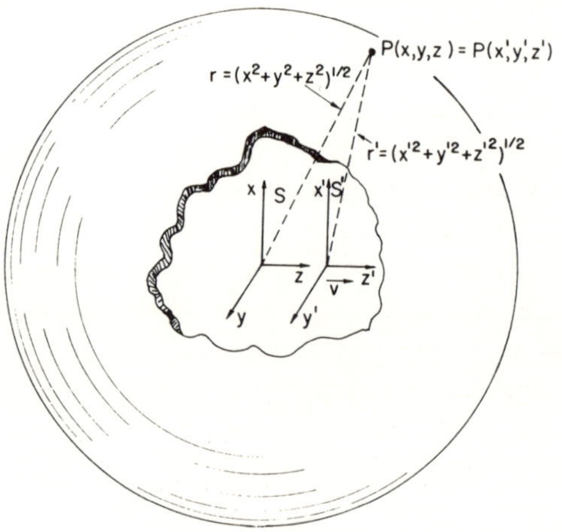

Fig. 2.1-1. A light pulse emitted at the time the origins of the S and S' systems were coincident has a spherical wavefront. The point $P(x, y, z)$ is a point on the sphere.

Picture two frames of reference S and S' with S' moving at a velocity v with respect to S along the z-axis of S as in Fig. 2.1-1. A light pulse is emitted at time $t = t' = 0$ when the two coordinate systems are coincident in space. The equation for the wavefront in the unprimed system S is

$$x^2 + y^2 + z^2 = c^2t^2 \tag{2.1.1}$$

In the primed system of coordinates S' the equation has the same form

$$x'^2 + y'^2 + z'^2 = c^2t'^2 \tag{2.1.2}$$

From symmetry, with motion in the z-direction only, it can be shown that

$$x' = x \tag{2.1.3}$$

$$y' = y \tag{2.1.4}$$

Rotational symmetry allows us to write $x' = \alpha(v)x$ and $y' = \alpha(v)y$, where $\alpha(v)$ is a function of velocity that must be the same for both x and y. From axial symmetry $\alpha(v) = \alpha(-v)$ from which it follows that $\alpha(v) = 1$, if Eqs. (2.1.1) and (2.1.2) are to be satisfied [2]. We may thus set the z and t and z' and t' terms of Eqs. (2.1.1) and (2.1.2) equal, using Eqs. (2.1.3) and (2.1.4):

$$z'^2 - c^2t'^2 = z^2 - c^2t^2 \tag{2.1.5}$$

Now, if we take the simplest linear transformation we can think of that properly expresses the positions of the origins O ($z = 0$) and O' ($z' = 0$), we may write

$$z' = \gamma(v)(z - vt), \qquad z = \gamma'(v)(z' + vt') \tag{2.1.6}$$

where $\gamma(v)$ and $\gamma'(v)$ are functions of velocity which we will attempt to solve for. As viewed from S, at the time t, O' is at the coordinate $z = vt$ giving z' equal to zero in the first equation in Eq. (2.1.6). As viewed from S', at the time t', O is at $z' = -vt'$, giving the correct value of z equal to zero for all values of t' in the second equation in Eq. (2.1.6), just as in a Galilean transformation. If we eliminate z' from Eq. (2.1.6) we find

$$t' = \gamma\left[t - \frac{z}{v}\left(1 - \frac{1}{\gamma\gamma'}\right)\right] \tag{2.1.7}$$

In a Galilean transformation, $\gamma = \gamma' = 1$ which clearly yields $t = t'$. Now, if z' from Eq. (2.1.6) and t' from Eq. (2.1.7) are substituted in Eq. (2.1.5), we obtain an equation in the unprimed coordinates only:

$$\gamma'^2(z - vt)^2 - c^2\gamma^2\left[t - \frac{z}{v}\left(1 - \frac{1}{\gamma\gamma'}\right)\right]^2 = z^2 - c^2t^2 \tag{2.1.8}$$

We can rearrange Eq. (2.1.8) in a form in which we separate out and solve for the coefficients of z^2, t^2, and zt. If our transformation is to be effective, this equation must be satisfied for any positive t or z, and this can only be true if each of the coefficients of z^2, zt, and t^2 is separately equal to zero. Solving for γ and γ' from these equations yields the form of γ and γ'

$$\gamma = \gamma' = [1 - (v^2/c^2)]^{-1/2} = (1 - \beta^2)^{-1/2} \tag{2.1.9}$$

where

$$\beta = v/c \tag{2.1.10}$$

The transformation then becomes, from Eqs. (2.1.6) and (2.1.7)

$$\begin{aligned} x' &= x, \qquad y' = y \\ z' &= \gamma(z - vt) = \gamma(z - \beta ct) \\ t' &= \gamma[t - (zv/c^2)] = \gamma[t - (\beta z/c)] \end{aligned} \tag{2.1.11}$$

$$\begin{aligned} x &= x', \qquad y = y' \\ z &= \gamma(z' + vt') = \gamma(z' + \beta ct') \\ t &= \gamma[t' + (z'v/c^2)] = \gamma[t' + (\beta z'/c)] \end{aligned} \tag{2.1.12}$$

and Eqs. (2.1.9)–(2.1.12) define the Lorentz transformation. It is seen that if β approaches zero the Lorentz transformation approaches the Galilean transformation and that t is approximately equal to t'. Thus for low velocities in comparison with c, Newtonian mechanics and the universality of time become good approximations. In the Lorentz transformation both z and t are functions of both z' and t' and vice versa, so that space and time are intertwined in a more intimate manner than is the case in Newtonian mechanics where time runs at the same rate in all frames of reference.

We now examine the Lorentz transformation in a Minkowski diagram in Fig. 2.1-2 in which z-, ct-, z', and ct'-axes are indicated [3, 4]. The time axes are multiplied by c to give both axes the same dimensions. To find the angle ϕ_1, we set $z' = 0$ in Eq. (2.1.11), as it must be along the ct'-axis, and we find $z = \beta ct$ along the ct'-axis. Hence $\tan \phi_1 = \beta$. Similarly ϕ_2 is found by

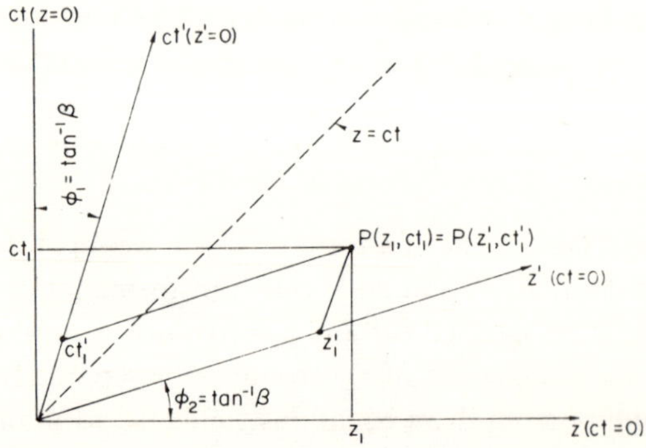

Fig. 2.1-2. A Minkowski diagram for the Lorentz transformation.

setting $t' = 0$ in Eq. (2.1.11) and we find $\beta z = ct$ along the z'-axis. Hence, $\tan \phi_2 = \beta$ and the diagram is completely symmetrical. A rectangular grid with lines at right angles to the z- and ct-axes is used to determine position in the z–ct plane as usual. Similarly lines of constant z' are parallel to the ct'-axis and lines of constant ct' are parallel to the z'-axis. This system is illustrated in Fig. 2.1-2 in which P is located by means of lines at right angles to the z- and ct-axes from z_1 and ct_1 and also by lines in the z'–ct' plane parallel to the primed axes from z_1' and ct_1'.

It remains to set the scales on the axes. It was seen in Example I in connection with Fig. 1.3-2 that if we set the scales on the t- and t'-axes in the right way, a point located by a given value of t also was located by the same value of t', even though the t- and t' axes were not coincident. In such a case any trajectory that is drawn in one coordinate system is automatically the same line in the other, and distance versus time can be read off from either set of axes. In the Minkowski diagram of the Lorentz transformation we can accomplish this same result by using Eq. (2.1.5) and letting it have the values $\pm 1, \pm 2, \pm 3, \ldots$. For example, if we let the value be unity

$$z^2 - c^2t^2 = z'^2 - c^2t'^2 = \pm 1 \tag{2.1.13}$$

which is equivalent to

$$(z + ct)(z - ct) = (z' + ct')(z' - ct') = \pm 1 \tag{2.1.14}$$

which defines a set of hyperbolas asymptotic to $z = \pm ct$ as shown in Fig. 2.1-3 and marked unity. A similar set of hyperbolas are defined for ± 2, etc., but these are not needed, because the scales are linear on all axes. So once we have determined the positions for unity on all scales we can mark off the rest of the scales as in Fig. 2.1-3. The equation such as Eq. (2.1.14) that defines the relative scales of a graph is called the metric. In the present case we have a hyperbolic metric. In the Minkowski diagram of the Galilean transformation we had a simple rectangular grid metric.

We are now in a position to understand a number of interesting relativistic effects. First let us consider the basic consequence of not being able to transmit matter or information any faster than c. In Fig. 2.1-3 the lines $z = \pm ct$ define the velocity of light, and if we picture any trajectory on this diagram it must not contain a slope greater than unity. This restriction has the consequence that you cannot start out from the origin and travel at a physically realizable velocity and get to the parts of the diagram marked "elsewhere." We are thus restricted to the regions within the dashed triangles marked past and future. In two or more space dimensions this would

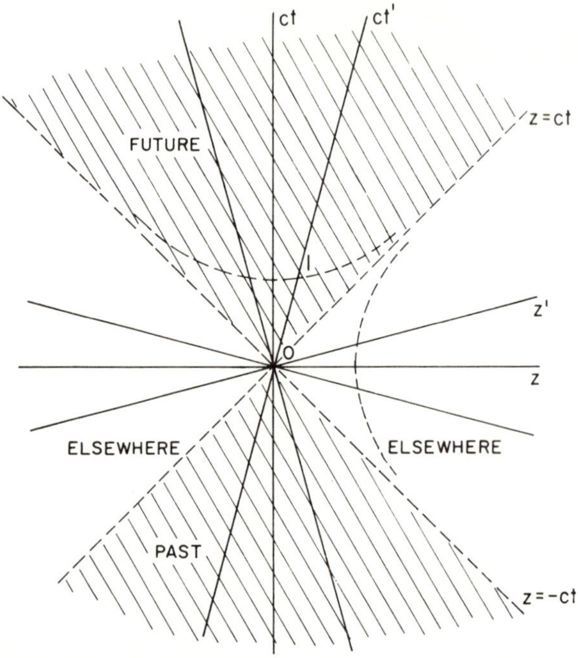

Fig. 2.1-3. A four quadrant Minkowski diagram of the Lorentz transformation. The light cone (triangle) divides space–time into past, future, and elsewhere.

be a cone, and it is usually referred to as the light cone. The present is the origin and all possible pasts meet here and pass through to the possible futures.

The concept of causality is preserved in the Lorentz transformation, just as in the Galilean transformation. Events taking place in the past can be thought of as determining or causing the present which is unique and which is represented by the origin in Fig. 2.1-3. There are many possible futures but only one present. Note that the diagram corresponding to Fig. 2.1-3 for the Galilean transformation would have the light cone as horizontal lines, corresponding to an infinite velocity of light. Under the Galilean transformation there is no elsewhere region, i.e., events occurring throughout all space can affect the present, as long as they occur at earlier times than the present. Under the Lorentz transformation only events occurring within the light cone can be said to have caused the present. Events taking place elsewhere can only affect events in the future.

Let us now consider two events (x_1, y_1, z_1, t_1) and (x_2, y_2, z_2, t_2). The "interval" between these two events in Newtonian mechanics we would think of as having a space part equal to

$$[(x_1 - x_2)^2 + (y_1 - y_2)^2 + (z_1 - z_2)^2]^{1/2}$$

and a time part equal to $(t_1 - t_2)$. We have seen that in a Galilean transformation these intervals transform independently as invariants, i.e.,

$$(x_1 - x_2)^2 + (y_1 - y_2)^2 + (z_1 - z_2)^2 = (x_1' - x_2')^2 + (y_1' - y_2')^2 + (z_1' - z_2')^2 \qquad (2.1.15)$$

and

$$(t_1 - t_2) = (t_1' - t_2') \qquad (2.1.16)$$

In special relativity it is the combined space–time interval

$$[(x_1 - x_2)^2 + (y_1 - y_2)^2 + (z_1 - z_2)^2 - c^2(t_1 - t_2)^2]$$

that is invariant. As a differential it is the quantity

$$ds^2 = c^2\, dt^2 - dx^2 - dy^2 - dz^2 \qquad (2.1.17)$$

that is invariant. The differential ds involves both space and time, but not symmetrically. We can have timelike intervals for which $ds^2 > 0$ or spacelike intervals for which $ds^2 < 0$. In Fig. 2.1.3 we can thus regard the regions marked past and future as temporal regions within which events have a timelike interval with relation to the origin. Similarly the elsewhere regions are connected with the origin by spacelike intervals.

Let us now consider a specific case in which two events take place at the same point in space. Then the space part of the interval is zero and the time part records the time as it would be measured by a clock sitting at this point in space. This time is called proper time; in the frame of reference of a proper system one can read the time between two events on the same clock.

There need not always be such a frame of reference. If one event occurs elsewhere in Fig. 2.1-3 and another event occurs at the origin, then no frame of reference can be found within which the two events occur at the same point in space. However, in this case a frame of reference can be found within which the two events appear to be simultaneous. In Newtonian mechanics there is no conceptual difficulty with the idea that two events are simultaneous, because time runs at the same rate in all frames of reference and simultaneous means that the two events occur at the same time. In our present context, two events are simultaneous if a frame of reference can be found within which the time part of the interval between the two events is zero. The space part of the interval in this case tells us the separation in configuration space of the two events.

Let us see what simultaneity looks like on a Minkowski diagram [4]. Let A_1 and B_1 be the positions of two receivers at rest in the S frame as shown in Fig. 2.1-4a. A transmitter at C sends out signals at the velocity of light

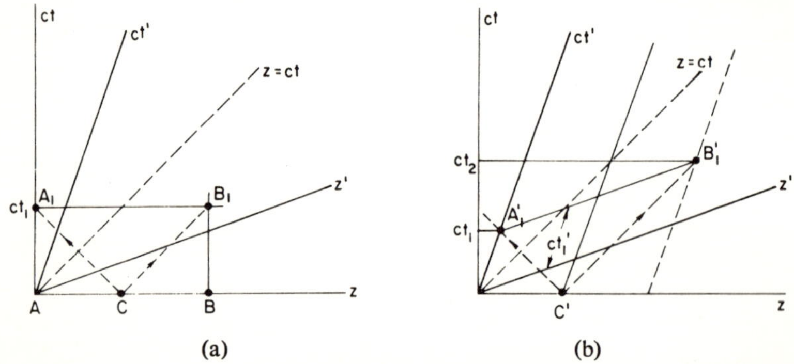

Fig. 2.1-4.(a) An observer at rest in S would say signals from C simultaneously arrived at A_1 and B_1 at time t_1. (b) An observer at rest in S' would say signals from C' simultaneously arrived at A_1' and B_1' at t_1'.

which are received at A_1 and B_1 at time ct_1. An observer in S would say that these signals were received simultaneously. An observer in S' who receives a signal from C' at equally spaced receivers at rest in S' at A_1' and B_1' in Fig. 2.1-4b would say these signals arrived simultaneously. Obviously, to an observer in S, these signals arrived at ct_1 and ct_2 and their arrivals were not simultaneous.

Problem 2.1-1 Trajectories in Minkowski Space

A particle executes simple harmonic motion as viewed from the S' frame of reference, such that

$$z' = -a \sin \omega t'$$

a. Using a Lorentz transformation, transform to a frame of reference (the S frame) moving with a velocity v in the negative z'-direction and find $z(t)$.

b. Sketch the trajectory in a Minkowski diagram, indicating how the trajectory could be plotted in detail, if desired.

2.2 Time Dilation and the Lorentz Contraction

Consider a clock at rest at the origin of S' that is set to zero as it passes the origin in Fig. 2.2-1. As time passes, time advances in the S' frame along the ct'-axis to the point A at which an observer in S' says that a time t_1' has elapsed. For convenience we take $ct_1' = OA$ equal to unity. Now, at the time that the observer in S' states that time ct_1' has elapsed, an observer in S claims that the event consisting of the arrival of the S' observer at position z_1 occurred at time ct_1. To find out if these times are the same or different, we must mark off the relative scales with our hyperbolic metric, as shown in Fig. 2.2-1 for this case. It is seen that ct_1 is longer than ct_1',

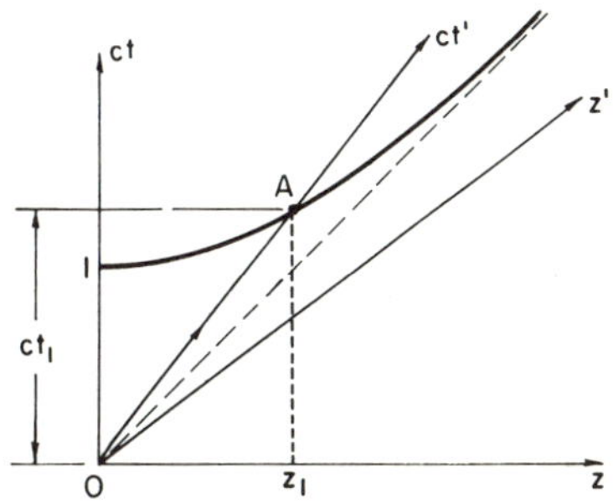

Fig. 2.2-1. Minkowski diagram illustrating time dilation. $OA = ct_1' = 1$.

when correctly measured, although the opposite appears to be true at first glance. This is a general result and the proper time is always shorter than the time observed from any other frame of reference. The difference in readings is obtained from Eq. (2.1.12) for a case with $z' = 0$. From this equation

$$t = \gamma t' \tag{2.2.1}$$

and a time dilation by a factor γ is observed, i.e., time in a proper system appears dilated by γ when viewed from another system moving at velocity v with respect to the proper system [3, 5–8].

The Lorentz contraction is a related effect. Let the proper system again be S' and consider a unit length measuring rod extending from the origin along the z'-axis in Fig. 2.2-2 to A. An observer in S will observe the trajectories or world lines of the ends of the rod as the ct'-axis and the parallel line ABD. To measure the length of the rod, the distance between these trajectories must be measured at a time when the ends of the rod are simultaneously observed in S. In Fig. 2.2-2 the length OB is the length that would be measured by an observer in S, if there were a way of measuring the

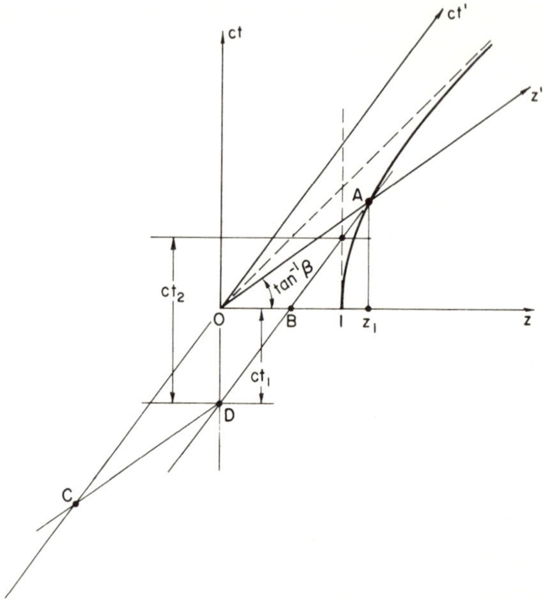

Fig. 2.2-2. Minkowski diagram illustrating the Lorentz contraction.

length of a moving rod, because *OB* lies along a line of constant time in *S* ($ct = 0$). *OB* is clearly shorter than *OA* when the effect of the hyperbolic metric is taken into account. The observed effect is thus a contraction when the rod is viewed from a nonproper frame. The amount of the contraction can be found from Eq. (2.1.11) by setting $t = 0$, in which case

$$z = z'/\gamma \tag{2.2.2}$$

and γ is of course greater than unity.

It is of interest in any problem in relativity to be precise about the experiment or measurement that is to be performed. It is difficult to prescribe a direct method for measuring the length of a moving rod, as required to measure *OB* in Fig. 2.2-2. Therefore, it is worth considering a measurement of time that would yield the Lorentz contraction. Consider an observer in Fig. 2.2-2 in *S* at *B* with a clock that is started when the meter stick occupies the position *CD* with the end marked *D* at $z = 0$. When the end of the meter stick marked *D* arrives at *B*, the observer at *B* will find that his clock indicates a time t_1. Let us also suppose that the observer in *S* records the time t_2 that it takes for an end of the meter stick to travel a unit distance in S. We now have a way of comparing the apparent length of the moving meter stick l_0 with a known unit length, because both lengths are proportional to vt. The observed ratio is

$$l_0/1 = t_1/t_2 = 1/\gamma \tag{2.2.3}$$

which may be verified by the geometrical construction shown in Fig. 2.2-2.

Problem 2.2-1 Time Dilation Measurement

Show that time dilation can be calculated by measuring the distance in the unprimed laboratory frame between two events occurring at the origin of the primed rest frame and dividing this distance by v to get time in the unprimed laboratory frame.

Problem 2.2-2 Time Measured in Separated Locations

Observers in S and S' synchronize clocks at the origin. At the time that the clock in S reads 10 seconds, the observer in S looks at the clock in S' through a telescope. What will he see as the reading of the clock in S', if $\gamma = 2$? Consider the same problem from the point of view of an observer in S' who looks at the clock in S when the clock in S' reads 10 seconds. Show that the reading on the distant clocks are the same when the roles of the observers are reversed.

Problem 2.2-3 The Twin Paradox

One twin A stays on earth. The other twin B travels away from earth at a constant velocity $v = 0.6c$ for one year as measured on earth by A (neglect the problem of how he accelerates from zero speed to this velocity). At the end of the year B undergoes an elastic collision which reverses his direction. He returns to earth in one year of earth time. The twins set their clocks to zero when they parted. A's clock reads 2 years when B returns. What does B's clock read? Note that in this problem a reversal of roles causes a reversal of clock readings. Assuming that acceleration is not detectable, which result is correct and how can we predict the clock readings? See Arzelies [11].

2.3 Transformation of Velocity and Acceleration

Velocities in the S system can be defined as derivatives of unprimed quantities as follows:

$$u_x = dx/dt, \qquad u_y = dy/dt, \qquad u_z = dz/dt \tag{2.3.1}$$

and velocities in the S' system defined as derivatives of primed quantities

$$u_x' = dx'/dt', \qquad u_y' = dy'/dt', \qquad u_z' = dz'/dt' \tag{2.3.2}$$

Taking derivatives with respect to t and t' in Eqs. (2.1.11) and (2.1.12) we obtain equations involving only dt/dt' and dt'/dt and the desired velocities:

$$\begin{aligned}
&\frac{dx'}{dt'} = \frac{dx}{dt}\frac{dt}{dt'}, && \frac{dx}{dt} = \frac{dx'}{dt'}\frac{dt'}{dt} \\
&\frac{dy'}{dt'} = \frac{dy}{dt}\frac{dt}{dt'}, && \frac{dy}{dt} = \frac{dy'}{dt'}\frac{dt'}{dt} \\
&\frac{dz'}{dt'} = \gamma\left(\frac{dz}{dt}\frac{dt}{dt'} - \beta c\frac{dt}{dt'}\right), && \frac{dz}{dt} = \gamma\left(\frac{dz'}{dt}\frac{dt'}{dt} + \beta c\frac{dt'}{dt}\right) \\
&\frac{dt'}{dt} = \gamma\left(1 - \frac{\beta}{c}\frac{dz}{dt}\right), && \frac{dt}{dt'} = \gamma\left(1 + \frac{\beta}{c}\frac{dz'}{dt'}\right)
\end{aligned} \tag{2.3.3}$$

Now dt'/dt and dt/dt' can be eliminated from Eq. (2.3.3) to yield the transformation equations for velocity

$$\begin{aligned}
u_x' &= \frac{u_x}{\gamma[1 - (\beta u_z/c)]}, & u_x &= \frac{u_x'}{\gamma[1 + (\beta u_z'/c)]} \\
u_y' &= \frac{u_y}{\gamma[1 - (\beta u_z/c)]}, & u_y &= \frac{u_y'}{\gamma[1 + (\beta u_z'/c)]} \\
u_z' &= \frac{u_z - \beta c}{[1 - (\beta u_z/c)]}, & u_z &= \frac{u_z' + \beta c}{[1 + (\beta u_z'/c)]}
\end{aligned} \tag{2.3.4}$$

When the two coordinate systems have the relative velocity c, $\beta = 1$ and the primed system is moving to the right at velocity c. Under these conditions, if a particle has a velocity u_z' equal to c to the right in the primed system, the velocity observed in the unprimed system is not $2c$, but is only c. Thus this transformation accomplishes the purpose of limiting the maximum value of velocity to c.

It is useful at this point to also obtain an expression for the transformation of the Lorentz contraction factor γ_0 for a particle with velocity u defined as

$$\gamma_0 = [1 - (u^2/c^2)]^{-1/2} \tag{2.3.5}$$

where

$$u^2 = u_x^2 + u_y^2 + u_z^2 \tag{2.3.6}$$

Similarly we define for the primed frame

$$\gamma_0' = [1 - (u'^2/c^2)]^{-1/2} \tag{2.3.7}$$

where

$$u'^2 = u_x'^2 + u_y'^2 + u_z'^2 \tag{2.3.8}$$

Using Eqs. (2.3.4) it can be shown that

$$\gamma_0 = \gamma\gamma_0'[1 + (\beta u_z'/c)], \qquad \gamma_0' = \gamma\gamma_0[1 - (\beta u_z/c)] \tag{2.3.9}$$

The transformation equations for acceleration can be obtained by differentiating Eqs. (2.3.4) and eliminating time derivatives of t and t' as before. The result [9] is

$$\begin{aligned}
\frac{du_x'}{dt'} &= \left[\gamma\left(1 - \beta\frac{u_z}{c}\right)\right]^{-2}\frac{du_x}{dt} + \frac{\beta}{\gamma^2}\frac{u_x}{c}\left(1 + \beta\frac{u_z}{c}\right)^{-3}\frac{du_z}{dt}\\
\frac{du_x}{dt} &= \left[\gamma\left(1 + \beta\frac{u_z'}{c}\right)\right]^{-2}\frac{du_x'}{dt'} - \frac{\beta}{\gamma^2}\frac{u_x'}{c}\left(1 + \beta\frac{u_z'}{c}\right)^{-3}\frac{du_z'}{dt'}\\
\frac{du_y'}{dt'} &= \left[\gamma\left(1 - \beta\frac{u_z}{c}\right)\right]^{-2}\frac{du_y}{dt} + \frac{\beta}{\gamma^2}\frac{u_y}{c}\left(1 - \beta\frac{u_z}{c}\right)^{-3}\frac{du_z}{dt}\\
\frac{du_y}{dt} &= \left[\gamma\left(1 + \beta\frac{u_z'}{c}\right)\right]^{-2}\frac{du_y'}{dt'} - \frac{\beta}{\gamma^2}\frac{u_y'}{c}\left(1 - \beta\frac{u_z'}{c}\right)\frac{du_z'}{dt}\\
\frac{du_z'}{dt'} &= \left[\gamma\left(1 - \beta\frac{u_z}{c}\right)\right]^{-3}\frac{du_z}{dt}\\
\frac{du_z}{dt} &= \left[\gamma\left(1 + \beta\frac{u_z'}{c}\right)\right]^{-3}\frac{du_z'}{dt'}
\end{aligned} \tag{2.3.10}$$

An obvious change from the results of the Galilean transformation is the fact that the acceleration in the primed or moving frame is dependent on the relative velocity of the frames, β, as well as the velocity of the particle, $u_{x,y,z}$. For example, a constant acceleration in the z-direction, independent of time in the unprimed frame, will not transform to a constant acceleration in the primed frame, because u_z is evidently linearly proportional to time, and it appears in the factor multiplying (du_z/dt) in the fifth equation of Eq. (2.3.10) which gives (du_z'/dt').

Note that although there is no formal difficulty in arriving at expressions for the transformation of acceleration that are required by the Lorentz transformation, the justification for using them to solve problems does not arise from the special theory of relativity. We are simply able to state that, if you know the acceleration in one frame of reference, these equations can be used to describe the acceleration as it will be seen from another frame of reference moving at a constant velocity with respect to the original frame [7, 8, 10, 11].

2.4 Transformation of Mass

We here consider some of the consequences of the second postulate of special relativity, i.e., as in Newtonian mechanics, conservation of mass, energy, and linear momentum apply in all frames of reference. We have required that the velocity of light be the same for all observers. We now, in addition, require conservation of mass and momentum and calculate the results of a simple elastic collision of two particles.

In the S' frame let us picture two particles traveling toward each other along the z'-axis. The particles have equal mass and are traveling at velocities $+u'$ and $-u'$ before the collision. After the collision they will have rebounded and be traveling at $+u'$ and $-u'$ as well. In this primed coordinate system conservation of mass and momentum are obviously satisfied by arguments of symmetry.

Now, in the unprimed S system moving at velocity $-v$ with respect to the S' system, let u_1 and u_2 and m_1 and m_2 refer to the velocities and masses before collision. At some time during the moment of collision the two particles will come to rest in the S' frame and both will be moving at velocity v with respect to the S frame. Let the total mass of the two particles at this instant be m. The conservation of mass and linear momentum require that

$$m_1 + m_2 = m, \qquad m_1u_1 + m_2u_2 = mv = (m_1 + m_2)v \qquad (2.4.1)$$

If we had taken the case of an inelastic collision in which the particles came to rest in the primed system and dissipated all their kinetic energy in heat, we would have written exactly the same equations. From our equations for the transformation of velocity, Eqs. (2.3.4),

$$u_1 = \frac{u' + \beta c}{1 + (\beta u'/c)}, \qquad u_2 = \frac{-u' + \beta c}{1 - (\beta u'/c)} \tag{2.4.2}$$

From Eqs. (2.4.1) and (2.4.2) we can solve for the mass ratio:

$$\frac{m_1}{m_2} = \frac{1 + (\beta u'/c)}{1 - (\beta u'/c)} \tag{2.4.3}$$

Using Eq. (2.3.9) we can then rewrite Eq. (2.4.3) as

$$\frac{m_1}{m_2} = \frac{[1 - (u_2{}^2/c^2)]^{1/2}}{[1 - (u_1{}^2/c^2)]^{1/2}} \tag{2.4.4}$$

We assume that both masses have the same mass at rest m_0. According to Eq. (2.4.4) the masses are inversely proportional to $[1 - (u^2/c^2)]^{1/2}$ when moving with velocity u. We therefore write

$$m = \frac{m_0}{[1 - (u^2/c^2)]^{1/2}} = \gamma_0 m_0 \tag{2.4.5}$$

Although this expression for the variation in mass with velocity is derived from a simple head-on collision, Lewis and Tolman [12] have shown that the same equation applies in any coordinates with any type of collision, if u is taken as the total velocity. Since the mass becomes infinite at the velocity of light, so does the energy, indicating the velocity of light is an upper bound for particle velocity.

Obviously Eq. (2.4.5) requires that mass be different in different coordinate systems, because velocity is different. The transformation for mass may be written from Eqs. (2.4.5) and (2.3.9) as

$$m = m'\gamma[1 + (\beta u_z'/c)] \tag{2.4.6}$$

where $m' = m_0\gamma_0'$.

There is only one type of mass and Eq. (2.4.6) gives the transformation for it. However, as we will see in the next section, the apparent inertia in the sense of the ratio of force to acceleration for an object with mass m, depends on the direction in which the force is applied.

As in the case of acceleration, special relativity does not provide any justification for using the concept of mass, since mass is a concept that only arises when acceleration occurs. The equation for the transformation of mass is simply a statement about the mass seen by an observer who is moving with a constant velocity with respect to a frame in which the mass is known.

PROBLEM 2.4-1 A RELATIVISTIC COLLISION

Consider an elastic collision between two particles which have equal masses and equal speeds, and which have initial velocities in the primed frame:

$$u'_{z1} = a, \qquad u'_{z2} = -a$$
$$u'_{y1} = b, \qquad u'_{y2} = -b$$
$$u'_{x1} = 0, \qquad u'_{x2} = 0$$

where subscripts 1, 2 refer to the first and second particles and a and b are constants.

Find the velocities after the collision in the primed frame and the velocities both before and after the collision in the unprimed frame which is moving in the negative z'-direction at velocity a, using the Lorentz transformation. Compare with the results using the Galilean transformation.

Assume that the particles collide in such a way as to exchange momenta in the z'-direction and to leave the y'-momenta unchanged.

PROBLEM 2.4-2 PARTICLES WITH VELOCITIES GREATER THAN c

We have seen that the fact that a particle has infinite energy at the velocity of light would act to prevent us from accelerating matter that starts from rest in the laboratory frame to a velocity greater than c. However, if a particle were somehow traveling faster than light it could see the velocity of light as a lower bound on its velocity rather than an upper bound. Such particles are discussed by G. Feinberg, *Phys. Rev.* **159**, 1089 (1967).

Suppose that you believed in the existence of such particles and wished to observe them. Describe an experiment that might be used to measure their presence in cosmic rays.

While the existence of faster than light particles does not violate the infinite energy condition, the existence of such particles would lead to a different concept of causality than we have discussed in connection with Fig. 2.1-3. Discuss this issue using this figure.

2.5 Transformation of Force

If we take as a definition of force the rate of change of momentum

$$\mathbf{F} = \frac{d}{dt}(m\mathbf{u}) = m\frac{d\mathbf{u}}{dt} + \mathbf{u}\frac{dm}{dt} \tag{2.5.1}$$

from Eqs. (2.3.5) and (2.4.5)

$$\begin{aligned}\mathbf{F} &= \frac{d}{dt}(m_0\gamma_0\mathbf{u}) = m_0\gamma_0\frac{d\mathbf{u}}{dt} + m_0\gamma_0^3\mathbf{u}\frac{u}{c^2}\frac{du}{dt} \\ &= m_0\gamma_0^3\left[\left(1 - \frac{u^2}{c^2}\right)\frac{d\mathbf{u}}{dt} + \frac{u\mathbf{u}}{c^2}\frac{du}{dt}\right]\end{aligned} \tag{2.5.2}$$

Let the unit vector in the direction of $\mathbf{u}$ be $\mathbf{1}_u$ so that $\mathbf{u} = \mathbf{1}_u u$. Consider first a force $\mathbf{F}_l$ applied in the $\mathbf{u}$ direction, so that both terms of Eq. (2.5.2) are in the same direction, and

$$d\mathbf{u}/dt = \mathbf{1}_u(du/dt) \tag{2.5.3}$$

For this case

$$\mathbf{F}_l = m_0\gamma_0^3(d\mathbf{u}/dt) = \gamma_0^2 m(d\mathbf{u}/dt) \tag{2.5.4}$$

If the force $\mathbf{F}_t$ is applied transversely, du/dt is zero and the acceleration is in the same direction as the force which is at right angles to $\mathbf{u}$. For this case

$$\mathbf{F}_t = m_0\gamma_0(d\mathbf{u}/dt) = m(d\mathbf{u}/dt) \tag{2.5.5}$$

Note that a particle has a greater inertia with respect to acceleration along the direction of motion than with respect to transverse acceleration.

The force observed in S' is related to the force in S through transformation equations that can be obtained by requiring the same form for the equation relating force and momentum in both frames. We have already used this technique in obtaining the transformations for velocity and acceleration. For example, we required that $u_x = dx/dt$ and $u_x' = dx'/dt'$ in Eqs. (2.3.1) and (2.3.2) and then proceeded to use the Lorentz transformation of distance and time of Eqs. (2.1.11) and (2.1.12) to relate u_x and u_x'. A key step in the logic was the requirement that velocity have the form that we are accustomed to in both frames. In a sense, all that we have said is that the Lorentz transformation of time and space requires that u_x' and u_x be related in the manner described by Eq. (2.3.4), if and only if we mean the usual thing by velocity in both frames, as defined in Eqs. (2.3.1) and (2.3.2). We then make various physical arguments about the significance of the velocity transformation in interpreting Eq. (2.3.4). We are now about to do the same thing for force that we have already done for velocity and acceleration, and it is important to understand that the logic is based entirely on the idea that force should be defined by the derivative of momentum with respect to time in both frames. If we accept this definition of force, then we may write $\mathbf{F}$ and $\mathbf{F}'$ in the form

$$\begin{aligned} \mathbf{F}' &= \frac{d}{dt'}(m_0\gamma_0'\mathbf{u}') = \mathbf{1}_x'F_x' + \mathbf{1}_y'F_y' + \mathbf{1}_z'F_z' \\ \mathbf{F} &= \frac{d}{dt}(m_0\gamma_0\mathbf{u}) = \mathbf{1}_xF_x + \mathbf{1}_yF_y + \mathbf{1}_zF_z \end{aligned} \tag{2.5.6}$$

Now, in order to obtain equations that properly relate the components of force in the two frames, it is simply necessary to use the Lorentz transformation equations for time and space and the velocity transformations we have already derived. We start by expressing F_z in unprimed quantities and then substitute for γ_0 and u_z from Eqs. (2.3.4) and (2.3.9) and for (dt/dt') from Eq. (2.3.3), as follows [9]:

$$F_z = m_0\frac{d}{dt}(\gamma_0 u_z) = m_0\left(1 + \frac{\beta u_z'}{c}\right)^{-1}\left[\frac{d}{dt'}\gamma_0'(u_z' + v)\right] \tag{2.5.7}$$

Eq. (2.5.7) can then be expressed in terms of F_z' in the form

$$\begin{aligned} F_z &= F_z' + m_0 v\left(1 + \frac{\beta u_z'}{c}\right)^{-1}\left[\frac{d\gamma_0'}{dt'} - \frac{u_z'}{c^2}\frac{d}{dt'}(\gamma_0'u_z')\right] \\ &= F_z' + \frac{m_0 v}{c^2 + u_z'v}\left[u_x'\left(\gamma_0'\frac{du_x'}{dt'} + u_x'\frac{d\gamma_0'}{dt'}\right)\right. \\ &\quad \left. + u_y'\left(\gamma_0'\frac{du_y'}{dt'} + u_y'\frac{d\gamma_0'}{dt'}\right)\right] \\ &= F_z' + \frac{u_x'v}{c^2 + u_z'v}F_x' + \frac{u_y'v}{c^2 + u_z'v}F_y' \end{aligned} \tag{2.5.8}$$

For F_x and F_y we obtain by a similar process

$$F_x = \frac{c^2}{\gamma(c^2 + u_z'v)}F_x', \qquad F_y = \frac{c^2}{\gamma(c^2 + u_z'v)}F_y' \tag{2.5.9}$$

As in the preceding two sections, it is important to emphasize that we have not derived a "relativistically correct" force equation. In fact, we must rely on some other theory to provide any information on the correct form of the force equation. All that we get out of the Lorentz transformation is a transformation that yields consistent results in frames that are in motion with respect to each other.

2.6 Conservation of Mass and Energy

As we have seen, the Lorentz transformation intertwines space and time in a way that makes it necessary to consider a combination of both quantities in any consideration of motion, regardless of the frame of reference. This intermixture of quantities that we have come to think of as independent in Newtonian mechanics also occurs with respect to mass and energy in the Lorentz transformation.

Consider the mass m that we have shown in Section 2.4 to be a function of velocity u:

$$m = \gamma_0 m_0 = \frac{m_0}{[1 - (u^2/c^2)]^{1/2}} \tag{2.6.1}$$

Now consider a change in the kinetic energy of the particle. Since the velocity will change as the kinetic energy of the particle changes, it is obvious from Eq. (2.6.1) that the mass m will also change. The relativistic interrelationship of mass and energy is thus qualitatively clear. Let us now obtain the quantitative relation between them, $\mathscr{E} = mc^2$, which is perhaps the best known equation that comes out of special relativity. If we differentiate Eq. (2.6.1) we can obtain a relation between the rate of change of mass with time and the corresponding rate of change of energy:

$$\begin{aligned}
\frac{dm}{dt} &= \frac{d}{dt}\left\{\frac{m_0}{[1 - (u^2/c^2)]^{1/2}}\right\} = \frac{m_0 u}{c^2[1 - (u^2/c^2)]^{3/2}} \frac{du}{dt} \\
&= \frac{\mathbf{u}}{c^2} \cdot \frac{d}{dt}\left\{\frac{m_0 \mathbf{u}}{[1 - (u^2/c^2)]^{1/2}}\right\} \\
&= \frac{\mathbf{u}}{c^2} \cdot \frac{d}{dt}(m\mathbf{u}) \\
&= \frac{\mathbf{u}}{c^2} \cdot \mathbf{F}
\end{aligned} \tag{2.6.2}$$

where $\mathbf{F}$ is the force defined as usual in Section 2.5 by Eq. (2.5.2). The rate of doing mechanical work is, in terms of energy $\mathscr{E}$

$$d\mathscr{E}/dt = \mathbf{F} \cdot \mathbf{u} \tag{2.6.3}$$

Therefore, from Eq. (2.6.2)

$$d\mathscr{E}/dt = d(mc^2)/dt \tag{2.6.4}$$

This equation tells us that the rate of change in energy is proportional to the rate of change in mass. If conservation of energy is to be valid in a given frame, the integral of Eq. (2.6.4) must also be correct, i.e.,

$$\mathscr{E} = mc^2 + \text{constant} \tag{2.6.5}$$

The evaluation of the constant in Eq. (2.6.5) is not obvious. That it is correctly equal to zero has been established experimentally with considerable accuracy in experiments in which the fractional mass change is large. It has also been established experimentally that the mechanism of energy change need not be due to work by mechanical forces as implied in Eq. (2.6.3), but applies to energy changes generally. It may also be shown [2] that mathematically the transformations of energy and momentum require that the constant in Eq. (2.6.5) be equal to zero. So the Einstein equation is obtained

$$\mathscr{E} = mc^2 \tag{2.6.6}$$

The total energy of a particle can be written in terms of the sum of its "rest energy," m_0c^2 and its kinetic energy V, i.e.,

$$\mathscr{E} = V + m_0c^2 \tag{2.6.7}$$

At velocities small in comparison with c

$$\begin{aligned}\mathscr{E} &= m_0c^2(1 + \tfrac{1}{2}(u/c)^2 + \cdots) \\ &= m_0c^2 + \tfrac{1}{2}m_0u^2 + \cdots\end{aligned}$$

the first two terms being the rest energy and the Newtonian kinetic energy.

We now have a measure of when a particle is "relativistic" in terms of its energy as well as its velocity in comparison with c. In order to establish some orders of magnitude it is interesting to make a few numerical calculations of energy and velocity for various particles and energies, as in Table 2.6-1. Some of the data from Table 2.6-1 has been plotted in Fig. 2.6-1 on a log scale. In order to calculate particle velocity from the energy

TABLE 2.6-1

PARTICLE VELOCITY AS A FUNCTION OF ENERGY

Particle	Rest mass (m_0) (kg $\times 10^{-27}$)	Rest energy (m_0c^2/e) (GeV)	u/c ($V = 1$ eV)	u/c (V=100 keV)	u/c (V=100 GeV)
Electron	9.11×10^{-4}	0.511×10^{-3}	1.97×10^{-3}	0.548	0.999
Proton	1.67	0.939	45.9×10^{-6}	14.5×10^{-3}	0.999
Lithium ion	9.98	5.60	18.8×10^{-6}	5.95×10^{-3}	0.999
Sodium ion	38.4	21.55	9.56×10^{-6}	3.02×10^{-3}	0.984
Cesium ion	222	124.5	3.98×10^{-6}	1.26×10^{-3}	0.833
Mercury ion	332	186.0	3.26×10^{-6}	1.03×10^{-3}	0.759

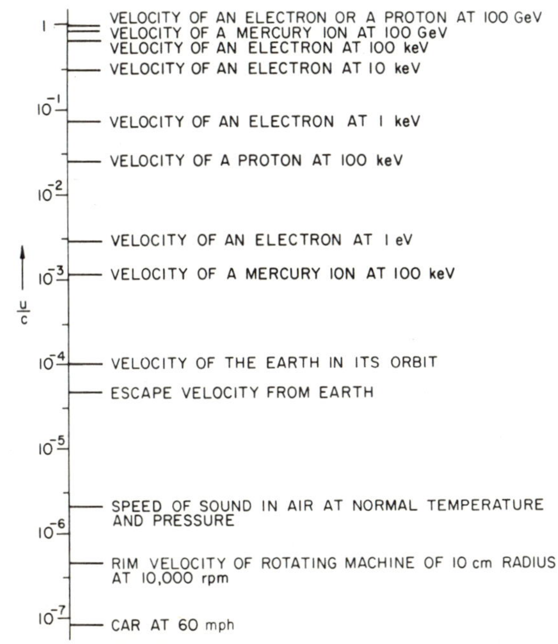

Fig. 2.6-1. Some typical velocities of motion of physical objects compared with the velocity of light.

it is convenient to note that

$$\gamma_0 = \mathscr{E}/m_0c^2 \tag{2.6.8}$$

and that

$$u/c = [1 - (1/\gamma_0^2)]^{1/2} \tag{2.6.9}$$

so we can solve for values of (u/c) as in Table 2.6-1 from $\mathscr{E}$ and m_0c^2. For small β it is convenient to use the approximate relation

$$u/c = (\sqrt{V}/506)(m_e/m_i)^{1/2} \tag{2.6.10}$$

where u is the velocity of a particle of mass m_i; V is the kinetic energy in electron volts, and m_e is the electron rest mass.

In Table 2.6-1, three energies were considered, 1 eV which is a typical thermal energy for particles in a low pressure gas discharge, 100 keV which is a typical energy for particles accelerated by a high voltage dc power supply, and 100 GeV which is a typical energy for a moderately large particle accelerator, such as a 300-ft linear electron accelerator. It is seen that electrons become at least slightly relativistic at 100 keV, a proton becomes highly relativistic at 100 GeV, and a mercury ion only moderately so at this energy. An electron in a particle accelerator at 100 GeV energy has nearly all of its energy in kinetic energy while a cesium ion at the same energy has a kinetic energy about equal to its rest energy.

Problem 2.6-1 Particle Beams

A particle beam has a particle velocity of 0.5c and a dc current of 1 μamp. Calculate the power carried by this beam, if the particles are (1) electrons, (2) protons, or (3) liquid particles, each with the charge of an electron and a mass of 1 μgm. [See D. S. Swatik and C. D. Hendricks, *AIAA J.* **6**, 1596–1597 (1968).] Calculate the energy necessary to accelerate one particle with a mass of 1 mgm to a velocity of 0.5c. If energy is delivered to this particle uniformly in time over a period of 10 μsec, what peak power is required? Winterberg has suggested a form of pulsed thermonuclear fusion reactor in which small pellets of tritium or deuterium ice would be caused to collide at velocities of the order of 0.5c. Such a device avoids the necessity of confining charged particles during the prefusion stage. [See F. Winterberg, *J. Nucl. Energy Pt. C* **8**, 541–553 (1966).]

2.7 Particle Accelerators

In this section a few elementary features of particle accelerators are described in a qualitative manner. Problems are given in various sections throughout the remainder of this book that deal with some of the aspects of accelerator design that can be treated using one-dimensional models.

The simplest type of particle accelerator is the electron or ion gun discussed in Section 5.7 in which a dc voltage is applied between two electrodes, one of which emits particles to be accelerated and the other of which has a hole in it to allow the particles to pass through it into the region where the particles are to be utilized. A basic limitation on such accelerators is the maximum value of dc field that can be tolerated between electrodes in a vacuum without breakdown as a result of various effects such as field emission and residual gases. A large voltage can be tolerated if the spacing is made large so the field limitation is not exceeded, but then power supply limitations arise at voltages of around 1 MeV. Van de Graaf machines offer one escape from this dilemma and allow the use of very high dc voltages with the attendant convenience of a dc beam [13].

The other route to high energy is the rf accelerator. Slater [14] describes a number of types of machines in a unified presentation that goes beyond the ideas presented here. Perhaps the simplest rf accelerator is the cyclotron shown in Fig. 2.7-1, invented by Lawrence [15] in the early 30s. In this

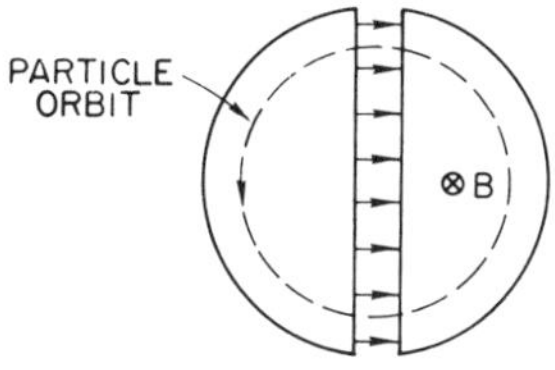

Fig. 2.7-1. Cyclotron dees and lines of force between them. A dc magnetic field into the paper fills the space. The electric field varies sinusoidally in time.

machine the particles are synchronized with the rf electric field in such a way that each time a particle passes between the dees (each dee is half of a cylindrical metal box within which particles travel in circular orbits) the electric field accelerates it further. Thus the same electric field can, in a sense, be reused to produce particles with energies many times the voltage appearing between the dees.

A variety of circular machines have been built to overcome second-order difficulties with the simple cyclotron. The synchro-cyclotron uses a variable frequency to keep the rf field in step with the particles when they acquire enough energy to experience a relativistic mass increase and a consequent shift in their resonant frequency of rotation in the magnetic field, qB/m. The synchrotron changes the magnetic field rather than the rf frequency to keep synchronization. The betatron delivers all of its energy by means of a change in magnetic field rather than through an rf field across a gap.

All of these machines encounter difficulties with very high velocity particles which begin to radiate energy as they become relativistic due to the acceleration resulting from the circular orbit. Consequently, circular machines are most useful for ions rather than electrons.

If we wish to deliver a lot of energy to an electron a linear accelerator is necessary. In a linear electron accelerator such as the Stanford 2-mile machine a waveguide is used that is designed to carry a wave with a strong axial electric field in the hole where the electron beam travels at exactly the velocity of light [16]. Figure 2.7-2 is a sketch of the type of structure

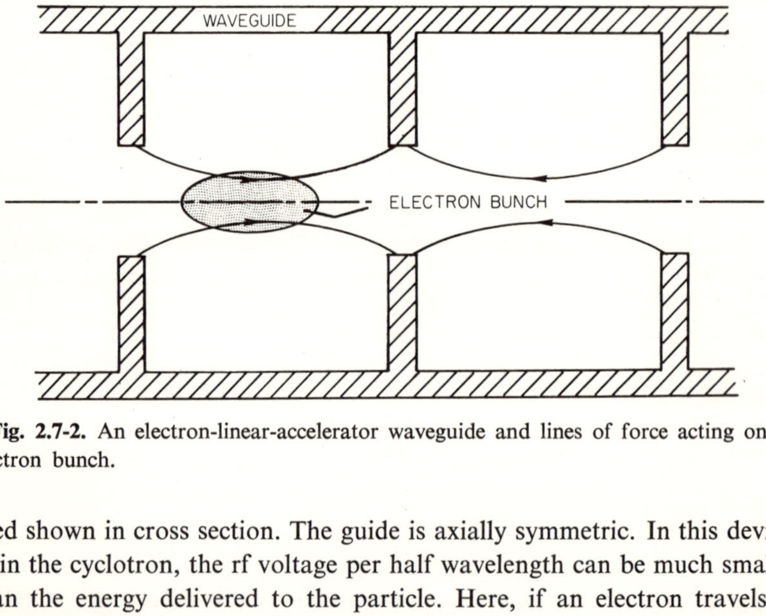

Fig. 2.7-2. An electron-linear-accelerator waveguide and lines of force acting on an electron bunch.

used shown in cross section. The guide is axially symmetric. In this device, as in the cyclotron, the rf voltage per half wavelength can be much smaller than the energy delivered to the particle. Here, if an electron travels in exact synchronism with the wave at the peak of the electric field, as shown in Fig. 2.7-2, it will see a total voltage of EL, where E is the peak field and L is the total length of the accelerator (such as 2 miles). Of course, we cannot expect to do quite this well for a variety of reasons. The particles will not all be at the peak of the field and many will move about within the field pattern, so they will see a field with a time-varying amplitude during their trip down the accelerator.

If we wish to use a linear machine for ions rather than electrons, we have to vary the velocity of the wave with distance along the length of the machine. This is also true of the electron accelerator in the startup or buncher section. This section which is typically 10-ft long receives a dc beam from an electron gun at a few hundred kiloelectron volts and emits a bunched beam at a few megaelectron volts suitable for injection into the constant velocity section, which is typically a few hundred feet long. In order to maintain the electric fields in the rf field structure, energy is supplied periodically by high power klystron amplifiers which deliver pulses a few microseconds long at power levels of the order of 10 MW per tube. Most of the power is delivered to losses in the walls of the waveguide and only a very small fraction is delivered to the beam in most accelerator designs. The long guide is actually composed of a number of sections about 10-ft long, each with its own source of power. In order to synchronize the fields in all these sections it is necessary to use a common rf source for all of the high-power amplifiers. It is evident that an amplifier rather than an oscillator approach is necessary, in order to keep synchronism. Historically, high-power magnetron oscillators were developed before high-power klystron amplifiers and linear electron accelerators had to await the development of suitable amplifiers of this type. Figs. 2.7-3 and 2.7-4 are photographs of the Stanford two-mile linear accelerator.

Problem 2.7-1 Electron Accelerator

A linear electron accelerator can be designed to subject a bunch of particles to an approximately constant axial electric field during the entire time that the bunch is in the accelerator. The bunch travels at approximately the velocity of light down the length of the accelerator and a TM electromagnetic wave with a strong axial electric field travels down the accelerator structure at the same velocity as the particle bunch. In the Stanford two-mile accelerator a short variable velocity bunching section (10-ft long) is used to bring the bunch up nearly to the velocity of light. The main guide which is two miles long operates at a single velocity over its entire length. Assume that the length is exactly two miles and that the injection energy of electrons is 10 MeV. If the exit energy is 2000 GeV, what must the electric field have been (assumed constant over the entire length) as seen by the electrons in their frame of reference? What is the corresponding electric field in the laboratory frame? The limiting factor in the design of such an accelerator is the value of the maximum allowable

Fig. 2.7-3. An aerial photograph of the Stanford two-mile linear electron accelerator. The beam switchyard and experimental area is in the foreground. [Courtesy Stanford University and U. S. Atomic Energy Commission.]

electric field, as determined by vacuum breakdown. A rule of thumb for the maximum dc field that can be used is 200 kV per inch to 400 kV per inch between two electrodes. In an rf accelerator there is only one electrode, but the rf voltage between adjacent portions of the accelerator structure must not exceed a limit which is comparable to this dc field value. Calculate the rf voltage per half wavelength in this type of accelerator at 2950 MHz for the electric field you have computed above.

Fig. 2.7-4. A photograph of the accelerator and support structure. The big pipe is the support and alignment pipe which is aligned by means of a laser beam which passes along its axis. The smaller pipe above is the accelerator. [Courtesy Stanford University and U. S. Atomic Energy Commission.]

REFERENCES

1. A. Einstein, Zur electrodynamik bewegter Körper, *Ann. Phys.* (*Leipzig*), **17**, 891–921 (1905).
2. W. Pauli, "Theory of Relativity." Pergamon Press, Oxford, 1958.
3. E. F. Taylor and J. A. Wheeler, "Spacetime Physics." Freeman, San Francisco, California, 1963.
4. E. P. Ney, "Electromagnetism and Relativity." Harper, New York, 1962.
5. R. P. Feynman, R. B. Leighton, and M. Sands, "The Feynman Lectures on Physics," Vols. I and II. Addison-Wesley, Reading, Massachusetts, 1964.
6. D. Bohm, "The Special Theory of Relativity." Benjamin, New York, 1965.
7. P. W. Bridgman, "A Sophisticate's Primer of Relativity." Wesleyan Univ. Press, Middletown, Connecticut, 1962.
8. J. B. Marion, "Classical Dynamics of Particles and Systems, 2nd ed. Academic Press, New York, 1970.
9. W. R. Smythe, "Static and Dynamic Electricity," 3rd ed. McGraw-Hill, New York, 1968.

10. P. G. Bergmann, "Introduction to the Theory of Relativity." Prentice-Hall, Englewood Cliffs, New Jersey, 1942.
11. H. Arzelies, "Relativistic Kinematics." Pergamon Press, Oxford, 1966.
12. G. H. Lewis and R. C. Tolman, *Phil. Mag.* **18**, 510 (1909).
13. P. H. Rose and A. B. Wittkower, Tandem Van de Graaff accelerators, *Sci. Amer.* **233**, 24–33 (1970).
14. J. C. Slater, "Microwave Electronics." Van Nostrand-Reinhold, Princeton, New Jersey, 1950.
15. E. O. Lawrence and M. S. Livingston, *Phys. Rev.* **40**, 19 (1932).
16. R. B. Neal, ed., "The Stanford Two Mile Accelerator." Benjamin, New York, 1968.

CHAPTER 3

The Lorentz Transformation of Maxwell's Equations

This kind of one-sidedness appears over and over again in the history of science. I am not saying this as a reproach: on the contrary, we must be glad that there are people who are courageous enough to be immoderate and one-sided. It is to them that we owe our discoveries. What is regrettable is that each should defend his one-sidedness so passionately. Scientific theories are merely suggestions as to how things might be observed.

G. Adler, M. Fordham, and H. Read, eds., "The Collected Works of C. G. Jung" (translated by R. F. C. Hull), Bollingen Series XX, Volume 4, Freud and Psychoanalysis, p. 107. Princeton Univ. Press, Princeton, New Jersey, 1961. Reprinted by permission.

3.0 Introduction

The basic idea that space and time are so designed that there is a maximum realizable velocity and that this is the velocity of light stems directly from Maxwell's equations. The homogeneous wave equation which can be derived directly from Maxwell's equations has wave solutions in which this velocity appears as a natural constant of free space. The derivation of the Lorentz transformation at the beginning of Chapter 2 was based on the requirement that a spherical wave of this type should appear to travel at the same velocity when viewed from two frames of reference moving with respect to each other.

In this chapter we examine the transformation of Maxwell's equations including charge and current, and we find that the Lorentz force equation is a direct consequence of the fact that Maxwell's equations are invariant to a Lorentz transformation. Often the Lorentz force equation is treated as if it were a separate law that must be obeyed in addition to Maxwell's equations. In fact, it is contained in Maxwell's equations as an implicit consequence of the Lorentz transformability of these equations. Thus, the definitions of the electric and magnetic fields that appear in Maxwell's equations are given by the Lorentz force equation

$$\mathbf{F} = q(\mathbf{E} + \mathbf{v} \times \mathbf{B}) \tag{3.0.1}$$

where **F** is the force, **E** and **B** are electric and magnetic fields, q is charge and **v** is the velocity of the charge relative to an observer in the frame of reference within which the electric and magnetic field are defined. The electric field is thus the force per unit charge in a frame of reference in which the charge is at rest, while $(\mathbf{E} + \mathbf{v} \times \mathbf{B})$ is the force per unit charge on a moving but unaccelerated charge. In following chapters we will discuss the motion of charges and currents in prescribed fields, according to this equation. Here we will develop the basic equations to be used for such calculations. We will also, in this chapter, examine the effects of motion of dielectric media on an electromagnetic wave that is propagating through it and compare the Doppler effect for electromagnetic waves with the corresponding effect for acoustic waves. Many practical electromechanical energy conversion devices such as induction motors and traveling-wave tubes involve the motion of charged or conducting materials through electromagnetic waves at velocities much less than the velocity of light. We will examine here the approximate forms of Maxwell's equations that are suitable for the study of such devices.

3.1 The Lorentz Transformation of Maxwell's Equations in Free Space

The usual formulation of Maxwell's equations in differential form is as follows:

$$\begin{aligned}
\nabla \cdot \mathbf{D} &= \varrho \\
\nabla \cdot \mathbf{B} &= 0 \\
\nabla \times \mathbf{E} &= -\partial \mathbf{B}/\partial t \\
\nabla \times \mathbf{H} &= \mathbf{J} + (\partial \mathbf{D}/\partial t)
\end{aligned} \tag{3.1.1}$$

The corresponding integral form of Maxwell's equations is as follows, where $d\mathbf{l}$ is a length element, $d\mathbf{S}$ is a surface element, and dV a volume element:

$$\begin{aligned}
&\int_S \mathbf{D} \cdot d\mathbf{S} = \int_V \varrho \, dV \\
&\int_S \mathbf{B} \cdot d\mathbf{S} = 0 \\
&\oint \mathbf{E} \cdot d\mathbf{l} = -\int_S \frac{\partial \mathbf{B}}{\partial t} \cdot d\mathbf{S} \\
&\oint \mathbf{H} \cdot d\mathbf{l} = \int_S \mathbf{J} \cdot d\mathbf{S} + \int_S \frac{\partial \mathbf{D}}{\partial t} \cdot d\mathbf{S}
\end{aligned} \tag{3.1.2}$$

We follow the usual convention that $d\mathbf{l}$ encircles $d\mathbf{S}$ and is positive in the direction of the angular coordinate in a right-handed cylindrical coordinate system and $d\mathbf{S}$ is outward from a closed surface. We restrict our analysis to simple media in which $\mathbf{B} = \mu\mathbf{H}$ and $\mathbf{D} = \varepsilon\mathbf{E}$.

Defining ∇^2 of a vector in the usual way

$$\nabla(\nabla \cdot \mathbf{A}) - \nabla^2\mathbf{A} = \nabla \times \nabla \times \mathbf{A} \tag{3.1.3}$$

it is possible to obtain immediately the homogeneous wave equations for $\mathbf{E}$ and $\mathbf{B}$ in uniform lossless isotropic media containing neither charges nor currents by taking the curl of the last two equations of Eq. (3.1.1).

$$\nabla^2\mathbf{E} - \mu\varepsilon(\partial^2\mathbf{E}/\partial t^2) = 0 \tag{3.1.4}$$

$$\nabla^2\mathbf{B} - \mu\varepsilon(\partial^2\mathbf{B}/\partial t^2) = 0 \tag{3.1.5}$$

Equations (3.1.4) and (3.1.5) may be written as follows for variations in a single spatial dimension, if we align the x-axis with the electric field and consider variations only in the z-direction:

$$(\partial^2 E_x/\partial z^2) - \mu\varepsilon(\partial^2 E_x/\partial t^2) = 0 \tag{3.1.6}$$

$$(\partial^2 B_y/\partial z^2) - \mu\varepsilon(\partial^2 B_y/\partial t^2) = 0 \tag{3.1.7}$$

Any function whatever of the variable $[t \pm z(\mu\varepsilon)^{1/2}]$ may be shown to be a solution to an equation of this form. A solution of Eq. (3.1.6) is, thus,

$$E_x = f[t \pm (z/w)] \tag{3.1.8}$$

where $w = 1/(\mu\varepsilon)^{1/2}$ is the velocity of the wave, as seen by an observer in the frame of reference of the xyz-coordinate system in which the fields

E_x and B_y are defined and f is any function. It is called a wave because, if z is varied at the rate defined by the equation $z = wt$, a constant value of f is obtained corresponding to motion of a wave pattern at the wave velocity. If $\mu = \mu_0$ and $\varepsilon = \varepsilon_0$ corresponding to free space, $w = c \cong 3 \times 10^8$ m sec^{-1}.

In the preceding chapter, in requiring that spherical waves traveled at the same velocity when viewed by observers in the S and S' frames that were moving with respect to each other, we assured that the homogeneous wave equations were Lorentz transformable. We now examine the more complete form of Maxwell's equations including charges and currents. We will specifically treat the case of the motion of charges in free space and put

$$\mathbf{B} = \mu_0\mathbf{H}, \qquad \mathbf{D} = \varepsilon_0\mathbf{E}, \qquad \mathbf{J} = \varrho\mathbf{u} \tag{3.1.9}$$

Then we may write Maxwell's equations in the S and S' frames in the same form

$$\begin{aligned} \nabla \cdot \mathbf{B} &= 0, & \nabla' \cdot \mathbf{B}' &= 0 \\ \nabla \cdot \mathbf{E} &= \varrho/\varepsilon_0, & \nabla' \cdot \mathbf{E}' &= \varrho'/\varepsilon_0 \\ \nabla \times \mathbf{E} &= -\partial\mathbf{B}/\partial t, & \nabla' \times \mathbf{E}' &= -\partial\mathbf{B}'/\partial t \\ \nabla \times \mathbf{B}/\mu_0 &= \varrho\mathbf{u} + \varepsilon_0\, \partial\mathbf{E}/\partial t, & \nabla' \times \mathbf{B}'/\mu_0 &= \varrho'\mathbf{u}' + \varepsilon_0\, \partial\mathbf{E}/\partial t' \end{aligned} \tag{3.1.10}$$

if, in fact, Maxwell's equations in free space are invariant to a Lorentz transformation. The charge density ϱ is moving at the velocity $\mathbf{u}$ in S and has a value that may be different in the two coordinate systems, i.e., ϱ' is not assumed to be equal to ϱ. We make use of Eqs. (2.3.4) and obtain the following transformation due to Lorentz, the correctness of which may be demonstrated by substituting in Eq. (3.1.10):

$$\begin{aligned} E_z &= E_z', & B_z &= B_z' \\ E_x &= \gamma(E_x' + \beta c B_y'), & E_y &= \gamma(E_y' - \beta c B_x') \\ B_x &= \gamma[B_x' - (\beta/c)E_y'], & B_y &= \gamma[B_y' + (\beta/c)E_x'] \\ \varrho &= \gamma\varrho'[1 + (\beta u_z'/c)] \end{aligned} \tag{3.1.11}$$

To get primed quantities in terms of unprimed quantities it is only necessary to interchange unprimed and primed quantities and to change the sign of β.

The way an electromagnetic field is divided into electric field and magnetic field thus depends on the velocity of the observer. For example, if an observer sees a pure electric field with nonzero x and y components in the primed frame, it is evident from Eq. (3.1.11) that an observer in the unprimed frame will see nonzero magnetic field components, B_x and B_y, even though B_x' and B_y' are zero. In other words, moving through a pure electric field causes it to appear to be a combination of electric and magnetic fields. This statement can be verified qualitatively by picturing an observer with a small test charge that he can use to determine the fields present in his region of space. The test charge is first held at rest with respect to the observer and then released. If the observer is sitting in a pure electric field uniform in space, the test charge will move at an increasing velocity in a straight line in the direction of the electric field, assuming the effect of gravity is negligible. If now a second observer watches the motion of the same test charge from a frame of reference moving at a right angle to the direction of the electric field, he will see the test charge move in a curved path, because it has a time-varying velocity. He will ascribe this curvature in the trajectory to the Lorentz force and say that there is a magnetic field present in his frame of reference. The precise value of this magnetic field is given by Eq. (3.1.11). In Chapter 5 we will examine this trajectory problem in detail. Here we simply note that it illustrates qualitatively the fact that the mix of electric and magnetic fields observed depends on one's frame of reference.

The charge density transformation in Eq. (3.1.11) can be understood most simply if a point charge of value q is considered. We may multiply the charge equation in Eq. (3.1.11) by γ_0' and rewrite it as

$$\varrho\gamma_0' = \varrho'\gamma_0 = \varrho'\gamma\gamma_0'[1 + (\beta u_z'/c)] \tag{3.1.12}$$

where γ_0 and γ_0' are from Eq. (2.3.9), so that

$$\varrho[1 - (u^2/c^2)]^{1/2} = \varrho'[1 - (u'^2/c^2)]^{1/2} \tag{3.1.13}$$

Now consider a volume element $d\tau = dx\, dy\, dz$ and using Eq. (2.3.3) to express $d\tau' = dx'\, dy'\, dz'$ it can be shown that the volume element contraction is expressed by

$$d\tau[1 - (u^2/c^2)^{-1/2} = d\tau'[1 - (u'^2/c^2)]^{-1/2} \tag{3.1.14}$$

so that

$$\varrho\, d\tau = \varrho'\, d\tau' = q = q' \tag{3.1.15}$$

and there is invariance of electric charge in a Lorentz transformation, although not of charge density

If we wish to apply these results to the motion of a continuum of charge rather than a point charge, we need to examine the transformation of current density. Let us picture the situation in a conductor in which an applied electric field produces a flow of electrons through the background of positive charge ion cores in a solid conductor. We assume that the net charge density in the frame of reference in which the ions are at rest is zero. We take this to be the primed frame and thus $\varrho' = -\varrho_i'$, where ϱ' is the charge density of the electrons and ϱ_i' is the charge density of the ion background. The electron current density in the primed frame is then $\mathbf{J}' = \varrho'\mathbf{u}'$ where $\mathbf{u}'$ is the drift velocity of the electrons in this frame. To obtain the net charge density and current density in the unprimed frame moving at velocity v in the negative z'-direction, we use the transformations of velocity and charge density from Eqs. (2.3.4) and (3.1.11), respectively. Evidently

$$\begin{aligned}
\varrho_i &= \gamma\varrho_i' \\
\varrho &= \gamma[\varrho' + (\beta\varrho' u_z'/c)] = \gamma[\varrho' + (\beta/c)J_z'] \\
J_x &= \varrho u_x = \varrho' u_x' = J_x' \\
J_y &= \varrho u_y = \varrho' u_y' = J_y' \\
J_z &= \varrho u_z = \gamma\varrho'(u_z' + v) = \gamma(J_z' + \beta c\varrho')
\end{aligned} \tag{3.1.16}$$

where ϱ and ϱ_i are the charge densities of the electrons and ions in the laboratory frame.

There are several interesting features of this result. First, note that J_z has a nonzero value, even if J_z' is zero, due to the transport of charge ϱ', just as we would expect. However, in addition ϱ is nonzero, even if ϱ' is zero, if there is a z-directed component of current in the primed frame, which is not expected. Nor is this effect canceled out by a change in the apparent charge density by the ions, as is seen if we consider the net charge density ϱ_t for the case in which the medium has zero net charge ϱ_t' in the primed frame in which the ions are at rest. From Eq. (3.1.16)

$$\varrho_t = \varrho_i + \varrho = \gamma\beta J_z'/c = \gamma\beta\varrho' u_z'/c \tag{3.1.17}$$

so that, if there is a z-directed current in the primed frame there is a nonzero net charge in a frame of reference moving with respect to the primed frame, even though the net charge is zero in the primed frame. This effect is a very real one and is an effect of relativity that is observable at very low velocity. Picture a wire carrying current. Outside this wire there is a test charge at rest. It will not move because there is no electric field. Now, if we cause the test charge to move parallel to the axis of the wire, it will experience a force due to the magnetic field surrounding the wire. However, if we leave the test charge alone and move the wire along its axis, at first glance we would say there is no force on the test charge. Yet there is relative motion of the charge and the wire. In fact, the result of moving the wire is the creation of the net charge given by Eq. (3.1.17) and the test charge experiences a force due to this apparent charge density which is exactly equal to the

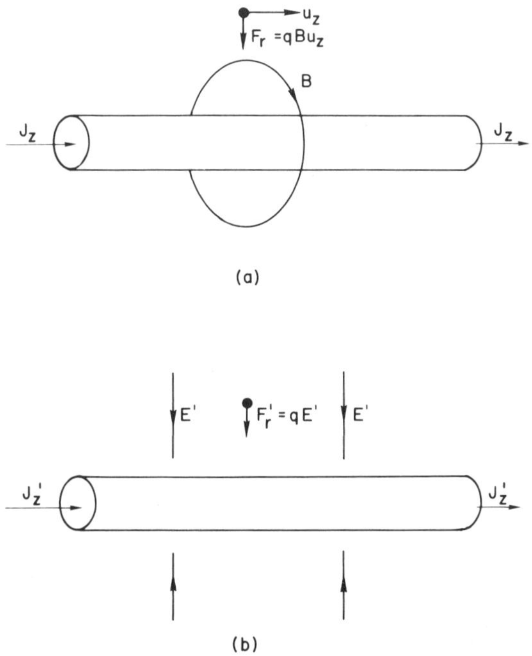

Fig. 3.1-1. (a) A test charge q moving at velocity u_z near a wire carrying a dc current density J_z experiences a force $F_r = qBu_z$. (b) When the same problem is viewed from a frame of reference moving at velocity u_z, there is no magnetic force, but there is an electric force due to the apparent net charge density required by the Lorentz transformation.

magnetic force in the case where the wire is at rest and the test charge moves, assuming the same relative velocity in both cases. We thus have an example here of a magnetic field that transforms to an electric field when viewed from a moving frame. Figure 3.1-1 illustrates these concepts.

We now examine this question of the force on a particle in a general way and find that we can derive the Lorentz force equation from the transformation of fields in Eq. (3.1.11) and the transformation of force in Eqs.

(2.5.8) and (2.5.9). We consider the case of a pure electric field in the primed frame which is the proper frame in which a test charge is at rest, so that in this frame

$$\mathbf{F}' = q'\mathbf{E}' \tag{3.1.18}$$

The unprimed (laboratory) frame is moving at velocity v in the negative z'-direction. From Eq. (3.1.15) we know that the charge is unchanged in going to the unprimed frame and we can use Eq. (3.1.11) to transform $\mathbf{E}'$. We then obtain for $\mathbf{F}'$ in terms of laboratory frame quantities

$$F_x' = q\gamma(E_x - vB_y), \qquad F_y' = q\gamma(E_y - vB_x), \qquad F_z' = qE_z \tag{3.1.19}$$

In order to write the force equation in the laboratory frame that is analogous to Eq. (3.1.18) in the primed frame we must use Eqs. (2.5.8) and (2.5.9) for the force transformation. We find that for the simple case of velocity in the z-direction,

$$\begin{aligned} F_x &= (1/\gamma)F_x' = q(E_x - vB_y) \\ F_y &= (1/\gamma)F_y' = q(E_y + vB_x) \\ F_z &= F_z' = qE_z \end{aligned} \tag{3.1.20}$$

We therefore have in the laboratory frame in vector form, just what we expect,

$$\mathbf{F} = q(\mathbf{E} + \mathbf{v} \times \mathbf{B}) \tag{3.1.21}$$

Problem 3.1-1 Lorentz Transformation of Maxwell's Equations

Show that the transformation of Eq. (3.1.11) does accomplish the transformation of Eq. (3.1.10). Note that by differentiating Eqs. (2.1.11) and (2.1.12) we can show that

$$\frac{\partial z'}{\partial z} = \gamma, \qquad \frac{\partial t'}{\partial z} = -\gamma\frac{\beta}{c}, \qquad \frac{\partial z'}{\partial t} = -\gamma\beta c, \qquad \frac{\partial t'}{\partial t} = \gamma$$

so that

$$\begin{aligned} \frac{\partial}{\partial z} &= \frac{\partial}{\partial z'}\frac{\partial z'}{\partial z} + \frac{\partial}{\partial t'}\frac{\partial t'}{\partial z} & \frac{\partial}{\partial t} &= \frac{\partial}{\partial z'}\frac{\partial z'}{\partial t} + \frac{\partial}{\partial t'}\frac{\partial t'}{\partial t} \\ &= \gamma\left(\frac{\partial}{\partial z'} - \frac{\beta}{c}\frac{\partial}{\partial t'}\right), & &= \gamma\left(\frac{\partial}{\partial t'} - \beta c\frac{\partial}{\partial z'}\right) \\ \frac{\partial}{\partial y} &= \frac{\partial}{\partial y'}, & \frac{\partial}{\partial x} &= \frac{\partial}{\partial x'} \end{aligned}$$

3.2 The Fields of a Moving Point Charge in Free Space

If a point charge is moving at a uniform velocity in free space, there will be no forces acting on it and it will not radiate. The electric field surrounding the charge, when viewed from a frame of reference moving with the charge, will of course be simply the value given by Coulomb's law and the magnetic field will be zero. We here use the Lorentz transformation to find the electric and magnetic field associated with the charge, when viewed from the laboratory frame, where, as usual, the charge and the primed frame are moving at a constant velocity v with respect to the laboratory frame.

Consider a point charge q' at rest in the moving S' frame at the point P on the z'-axis, as shown in Fig. 3.2-1. We take the coordinates of P to be $(0, 0, z_1')$ and we measure the field at Q at $(x_1, 0, 0)$. The vector $\mathbf{r}$ extends from Q to P and is directed toward P. The field as measured in S' is, from Coulomb's law,

$$\mathbf{E}' = -\frac{q'}{4\pi\varepsilon_0}\frac{\mathbf{r}}{r^3} \tag{3.2.1}$$

The value of $\mathbf{r}$ in Fig. 3.2-1 is

$$\mathbf{r} = -\mathbf{1}_x'x_1' + \mathbf{1}_z'z_1' \tag{3.2.2}$$

The components of the field at Q in S' are therefore

$$\begin{aligned} E_x' &= \frac{q'}{4\pi\varepsilon_0}\frac{x_1'}{(x_1'^2 + z_1'^2)^{3/2}} \\ E_z' &= -\frac{q'}{4\pi\varepsilon_0}\frac{z_1'}{(x_1'^2 + z_1'^2)^{3/2}} \\ E_y' &= 0 \end{aligned} \tag{3.2.3}$$

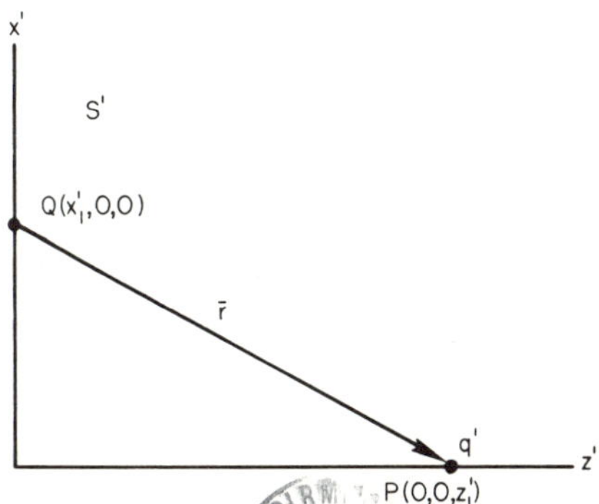

Fig. 3.2-1. A point charge with a charge q' is at rest in the S' frame at point P. The field at Q is found from Coulomb's law.

Evidently, since the field is symmetrical about the z'-axis, we will obtain a complete description of the problem by examining the x'–z' plane.

Now we wish to obtain expressions for the field in the S frame with respect to which the charge is moving at velocity v. We use the transformations of distance and time given in Eq. (2.1.11) and the transformations of field given in Eq. (3.1.11) taking $t = 0$ since this simplifies the equations and t is arbitrary. If $t = 0$, $t' = \beta z'/c$, $z_1 = z_1'/\gamma$, and $x = x_1'$, $q = q'$, so that at Q in S

$$E_x = \gamma E_x' = \frac{q}{4\pi\varepsilon_0} \frac{\gamma x_1}{(x_1^2 + \gamma^2 z_1^2)^{3/2}}$$

$$E_z = E_z' = -\frac{q}{4\pi\varepsilon_0} \frac{\gamma z_1}{(x_1^2 + \gamma^2 z_1^2)^{3/2}} \tag{3.2.4}$$

$$B_y = \frac{\beta\gamma}{c} E_x' = \frac{\mu_0 q v}{4\pi} \frac{\gamma x_1}{(x_1^2 + \gamma^2 z_1^2)^{3/2}}$$

Let us examine these fields. First of all note that there is a circumferential magnetic field encircling the current qv in this frame. Next note that for high velocities, as $\beta \to 1$ and $\gamma \to \infty$, the spherically symmetric field in the S' frame is compressed in the direction of motion into a pancake shaped distribution as indicated in Fig. 3.2-2. We can see this by finding the shape

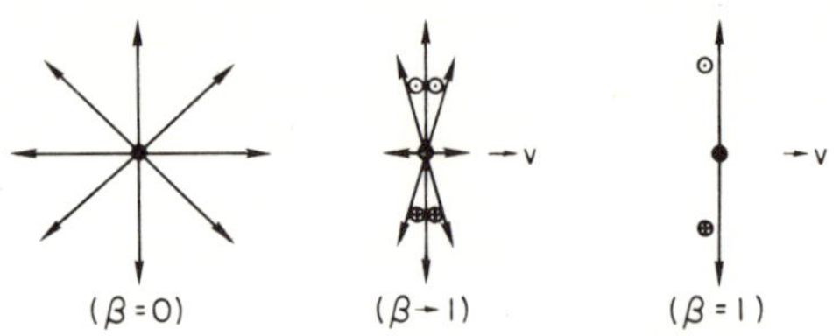

Fig. 3.2-2. The electric fields (arrows) and magnetic fields (circled dots and crosses) of a point charge at three values of velocity. For $\beta = 0$ there is a spherically symmetric pure electric field. For $\beta = 1$ there is a circularly-symmetric electric field concentrated in a plane and an associated magnetic field in the same plane. For an intermediate value of β the value of $|\mathbf{E}|$ has a pancake-shaped distribution in space.

of $|\mathbf{E}|$ = constant lines in the S frame. Finally for $\beta = 1$ the pancake has zero thickness and all of the field is contained in a sheet moving with the charge. Under these conditions we see that E_z is negligible in comparison with E_x and that the ratio of E_x to B_y has the value

$$E_x/B_y = 1/\mu_0\varepsilon_0 c = (1/\mu_0)(\mu_0/\varepsilon_0)^{1/2} \tag{3.2.5}$$

as in a plane wave. The field pattern is moving with a velocity c, but it is obviously not a sinusoidal plane wave that fills all space. Here there is only a delta function of field moving at the velocity of light with the particle. Of some interest is the fact that the electric field lines remain straight for all β. This is clear from the ratio of E_x/E_z which is equal to $(-x_1/z_1)$, just as in the S' frame.

Problem 3.2-1 Fields of a Moving Point Charge

In Eq. (3.2.4) the equations for the fields of a point charge in motion are given. Show that these equations do in fact represent a pancake shaped field distribution for high velocities and find the equation for the shape of the pancake.

Problem 3.2-2 Cerenkov Radiation

The phenomenon of Cerenkov radiation [1] is exhibited when charges move through a medium with a velocity greater than the velocity of light in the medium. A particle with a velocity near the velocity of light will penetrate a considerable distance through glass, water, or other typical materials before it is stopped. Most of the energy loss of such a particle is due to collisions that cause ionization of the atoms of the material, but some loss is due to Cerenkov radiation. Presumably, the same effect is possible with a bunched relativistic beam traveling through a hole many atomic radii (but only a few bunch radii) in diameter, and hence, in principle, suffering almost no losses except those due to Cerenkov radiation. However, the experimental evidence for Cerenkov radiation in this sort of configuration is unclear. Partly this is because the effect is not a large one, except at very high energies.

The problem can be analyzed by treating the case of a single point charge traveling through an infinitesimal hole in a background of dielectric material. If we view this problem from the primed frame of reference in which the charge is at rest, we see a stationary charge and a medium moving past to the left at velocity v. If we view the problem from the frame in which the medium is at rest, the charge is in motion.

Start the analysis of this problem in the frame in which the charge is at rest, using Gauss's law to write the **D** field in the primed frame. We also know that $\mathbf{H} = 0$ in the primed frame because the charge is at rest. In the laboratory frame $\mathbf{D} = \varepsilon\mathbf{E}$ and $\mathbf{B} = \mu\mathbf{H}$. Find all fields in both frames.

The solution is given by B. D. Nag and A. M. Sayied, *Proc. Roy. Soc., Ser. A* **235**, 544 (1956), but they use a different approach to the analysis.

3.3 The Lorentz Transformation of Maxwell's Equations in Material Media

In this section we consider some of the elementary features of a topic that has been a focus of controversy for over fifty years and one that is still at the center of fundamental research in electrodynamics [2]. The fundamental work in this field was done by Minkowski [3], and certain aspects of the subject are not controversial.

We begin by selecting only simple media characterized by scalar permittivity ε, permeability μ, and conductivity σ. In the frame of reference in which the medium is at rest, the primed frame or rest frame

$$\mathbf{D}' = \varepsilon\mathbf{E}', \qquad \mathbf{B}' = \mu\mathbf{H}', \qquad \mathbf{J}' = \sigma\mathbf{E}' \tag{3.3.1}$$

Maxwell's equations in this frame take the standard form

$$\begin{aligned} \nabla' \cdot \mathbf{B}' &= 0, & \nabla' \cdot \mathbf{D}' &= \varrho' \\ \nabla' \times \mathbf{E}' &= -\partial\mathbf{B}'/\partial t', & \nabla' \times \mathbf{H}' &= \varrho'\mathbf{u}' + (\partial\mathbf{D}'/\partial t') \end{aligned} \tag{3.3.2}$$

where we have taken $\mathbf{J}' = \varrho'\mathbf{u}'$. If these equations are transformed to the laboratory frame we will obtain the effective μ and ε observed in the laboratory frame. It turns out that the laboratory frame equations can be written in standard form

$$\begin{aligned} \nabla \cdot \mathbf{B} &= 0, & \nabla \cdot \mathbf{D} &= \varrho \\ \nabla \times \mathbf{E} &= -\partial\mathbf{B}/\partial t, & \nabla \times \mathbf{H} &= \varrho\mathbf{u} + (\partial\mathbf{D}/\partial t) \end{aligned} \tag{3.3.3}$$

provided the constitutive relations between **B**, **H**, **D**, and **E** have the correct form, as specified by

$$\begin{aligned} \mathbf{D} &= \varepsilon\mathbf{E} + \gamma^2\varepsilon[1 - (1/\mu\varepsilon c^2)]\mathbf{v} \times [\mathbf{B} - (\mathbf{v}/c^2) \times \mathbf{E}] \\ \mathbf{H} &= (\mathbf{B}/\mu) + \gamma^2\varepsilon[1 - (1/\mu\varepsilon c^2)]\mathbf{v} \times (\mathbf{E} + \mathbf{v} \times \mathbf{B}) \end{aligned} \tag{3.3.4}$$

first obtained by Minkowski. Implicit in Eqs. (3.3.1) and (3.3.4) are the transformations for the components of **D** and **H**, which may be written in the form

$$\begin{aligned} D_z &= D_z', & H_z &= H_z' \\ D_x &= \gamma[D_x' + (\beta/c)H_y'], & H_x &= \gamma(H_x' - \beta c D_y') \\ D_y &= \gamma[D_y' - (\beta/c)H_x'], & H_y &= \gamma(H_y' + \beta c D_x') \end{aligned} \tag{3.3.5}$$

which turn out to be of the same form as Eq. (3.1.11) for **E** and **B** and must be satisfied in addition to Eq. (3.1.11) in a material medium. As usual, the inverse transformations are obtained by interchanging primed and unprimed quantities and changing the sign of the terms involving β.

Equations (3.1.11) and (3.3.5) can be written somewhat more compactly, in terms of the components of field parallel to or perpendicular to the direction of the velocity $\mathbf{v}$, as follows, where the subscripts $\|$ and $\perp$ refer to these two directions.

$$
\begin{aligned}
E_\perp' &= \gamma(\mathbf{E} + \mathbf{v} \times \mathbf{B})_\perp, & E_\|' &= E_\| \\
D_\perp' &= \gamma\left(\mathbf{D} + \frac{\mathbf{v} \times \mathbf{H}}{c^2}\right)_\perp, & D_\|' &= D_\| \\
H_\perp' &= \gamma(\mathbf{H} - \mathbf{v} \times \mathbf{D})_\perp, & H_\|' &= H_\| \\
B_\perp' &= \gamma\left(\mathbf{B} - \frac{\mathbf{v} \times \mathbf{E}}{c^2}\right)_\perp, & B_\|' &= B_\|
\end{aligned}
\tag{3.3.6}
$$

We next consider several examples of the use of these transformations for the case of static electric and magnetic fields. We first look at a problem of interaction with dc fields uniform in space. Figure 3.3-1 is a diagram of a type of apparatus studied by Rowland [4], Röntgen [5], and Eichenwald [6] before either relativity theory or the existence of the electron were very well established. This device can be operated in either of the modes shown and the amount of current in the θ-direction in mode 2 is adjusted to produce the same needle deflection with the disk stationary as is produced when the disk is rotating in mode 1. By this means the amount of current produced by rotating the disk in mode 1 can be inferred.

The only applied field in mode 1 is an electric field, yet a magnetic field is measured, and hence a current must flow to produce it as a result of the motion of the disk. The amount of this current can be calculated by computing the charge on the disk and multiplying charge by the velocity of rotation to get current. At velocities much less than the velocity of light we need not be concerned with the fact that the particles are in accelerated rather than uniform motion, due to rotation, so we will analyze a model in which motion is uniform, as shown in Fig. 3.3-2, and apply the results to the apparatus of Fig. 3.3-1. In the configuration of Fig. 3.3-2 we first calculate the effects of the applied field in the primed frame moving with the material at velocity v. In this frame

$$D_x' = \varepsilon E_x', \qquad H_y' = B_y' = 0 \tag{3.3.7}$$

From Eqs. (3.1.11) and (3.3.5) we can find the apparent resultant fields

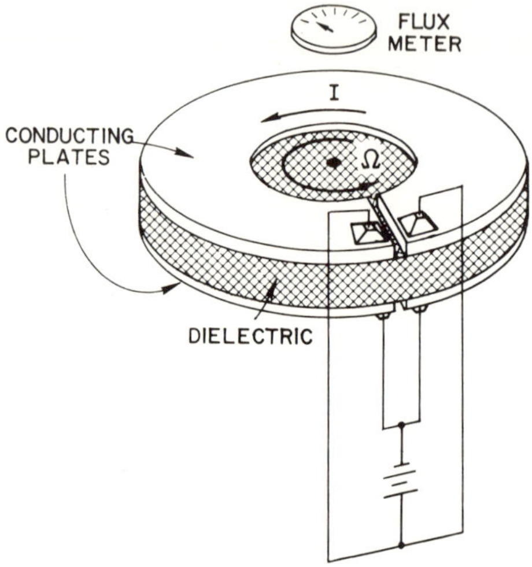

Fig. 3.3-1. An apparatus studied extensively in the nineteenth and early twentieth century by Röntgen, Rowland, and Eichenwald. A dielectric disc can be rotated at angular velocity Ω between a pair of conducting plates that have the shape of split rings.

Contact to the conducting plates is made through carbon brushes that can slide on the plates. The plates can either be held at rest or rotated with the dielectric. A flux meter measures the magnetic field present near the axis of the disc, if any. A voltage can be applied between the plates in mode 1 as shown. In mode 2, not shown, a small voltage is applied across the gap in the split rings causing a current to flow around the circumference of the rings allowing the flux meter to be calibrated. In mode 2 the dielectric and the rings are at rest.

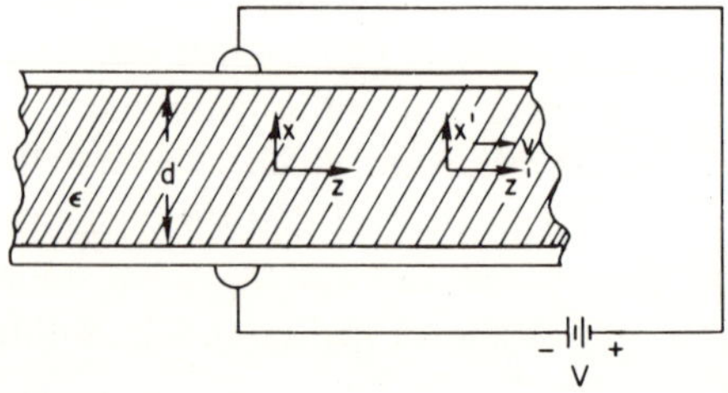

Fig. 3.3-2. A model of the device of Fig. 3.3-1 in which only uniform motion takes place. The dielectric material moves in the positive z-direction at velocity v. A primed frame of reference moves with the dielectric. The dielectric has permittivity ε and thickness d and is infinite in the y- and z-directions. A potential difference V is applied between upper and lower conducting plates. The plates can be held at rest in the unprimed frame or they can be moved with the dielectric.

in the laboratory frame

$$D_x = \gamma \varepsilon E_x', \qquad H_y = \gamma \beta c \varepsilon E_x', \qquad B_y = \gamma(\beta/c)E_x' \qquad (3.3.8)$$

It is this magnetic flux density B_y that is measured in the experiment of Fig. 3.3-1. For low velocities B_y is given by

$$B_y = \mu_0 \varepsilon_0 v E_x' = \mu_0 \varepsilon_0 v (V/d) \qquad (3.3.9)$$

which shows that the value of B_y and the corresponding current is independent of the value of the permittivity of the material when both the material and the metal plates move at the same velocity. Note that (B_y/μ_0) is not equal to H_y for a dielectric with permeability μ_0 when the dielectric is in motion. B_y is the quantity that is measurable experimentally, not H_y, as is evident from the fact that **B** appears in the Lorentz force equation rather than **H**, and it is the effect of this force that is observed in the flux meter in the experiment of Fig. 3.3-1.

Let us examine now the case, also described in Fig. 3.3-1, in which the metal plates stay at rest and only the dielectric moves. In this case, as in the previous case,

$$D_x' = \varepsilon E_x' \qquad (3.3.10)$$

but now in the rest frame of the moving observer we see a current due to the motion of the surface charges on the metal plates relative to the primed frame. The value of the surface charge on each plate is σ per unit area in the laboratory frame and for low velocities it is also σ in the rest frame of a moving observer, in which it is equal to the electric flux density, $\mathbf{D}' = \varepsilon \mathbf{E}'$, so $\sigma = \varepsilon V/d$. The current per unit width, J_l', is σv so

$$H_y' = J_l' = -\varepsilon v V/d, \qquad B_y' = \mu_0 H_y' \qquad (3.3.11)$$

and in this case B_y' is equal to $\mu_0 H_y'$ because the dielectric is at rest in this frame. If we now use Eqs. (3.1.11) and (3.3.5) to find the fields in the laboratory frame, we obtain

$$H_y = 0, \qquad B_y = (\mu_0 \gamma v V/d)\varepsilon_0[1 - (\varepsilon/\varepsilon_0)] \qquad (3.3.12)$$

In this case the measured flux density is a function of the relative permittivity $(\varepsilon/\varepsilon_0)$. It is also interesting to note that the system behaves exactly like a permanent magnet with $H = 0$ but $B \neq 0$.

A related series of experiments was performed by Wilson [7] in 1904 in which he moved a dielectric material through a dc magnetic field and

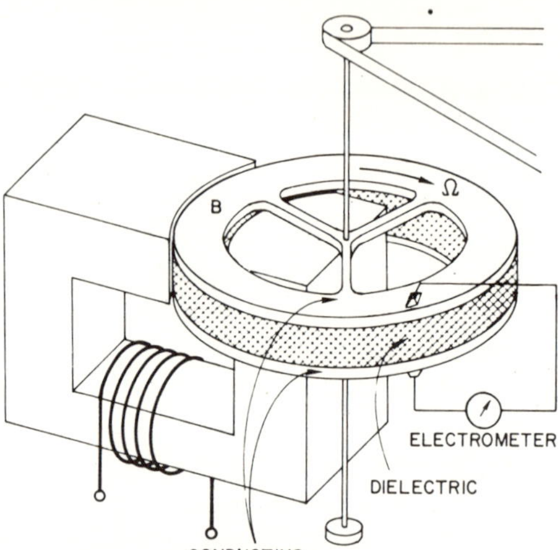

Fig. 3.3-3. An apparatus studied by Wilson in 1904. A dielectric slab is fastened between a pair of conducting plates and is rotated in the presence of a dc magnetic field normal to the direction of motion. The potential developed between the plates is measured as a function of the angular velocity and the magnetic field.

measured the potential difference using a sensitive electrometer, as shown in Figs. 3.3-3 and 3.3-4. Let us consider the case shown in Fig. 3.3-4 in which the dielectric moves between the metal plates at velocity v in the z-direction. Here there is a $\mathbf{v} \times \mathbf{B}$ force on the charges of the dipoles within the dielectric which is entirely equivalent to an applied electric field in the laboratory frame. The dipoles are aligned and produce a surface charge

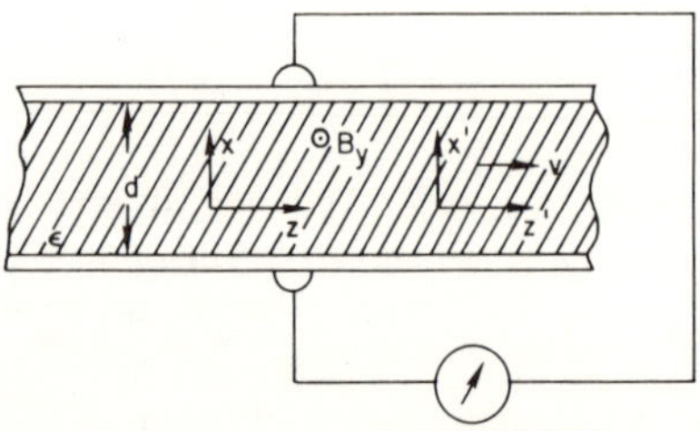

Fig. 3.3-4. A model of the device of Fig. 3.3-3 in which only uniform motion takes place. The dielectric material moves in the positive z-direction at velocity v. A primed frame of reference moves with the dielectric. The dielectric has permittivity ε and thickness d and is infinite in the y- and z-directions. A dc magnetic field is applied in the y-direction. A voltage V is measured on the electrometer connected between the upper and lower conducting plates which are at rest in the frame of reference of the dielectric.

on both sides of the dielectric material which results in a potential difference V between the metal plates which can be measured by an electrometer. To calculate the value of V in this case it is easiest to view the system from the laboratory frame and to use Gauss' law just as we would to calculate the capacitance of a condenser with the dielectric at rest. Here there is no external applied electric field, so there is no free charge within any volume which we might choose to integrate over and hence $D = 0$. We get the same value of D inside the dielectric as outside, using this argument. However, E is not equal to zero inside the dielectric. The situation is much like the previous case which resembled a permanent magnet. Here the system appears as an electret in the rest frame, i.e., $D = 0$ and $E \neq 0$ within the dielectric and there is no free charge. Electrets can be produced in media at rest by applying a dc electric field to a polarizable plastic in a liquid state and then allowing it to harden with its dipoles aligned by the field.

To calculate E and thus V we use Eq. (3.3.4) which for the configuration of Fig. 3.3-4 has the form

$$D_x = \varepsilon E_x + \gamma^2 \varepsilon \left(1 - \frac{1}{\mu \varepsilon c^2}\right)\left(-vB_y + \frac{v^2}{c^2} E_x\right) \tag{3.3.13}$$

Now since $D_x = 0$

$$E_x = \frac{c^2}{v} B_y \left(1 + \frac{c^2}{v^2} \frac{1}{\gamma^2[1 - (\varepsilon_0/\varepsilon)]}\right)^{-1} \tag{3.3.14}$$

which for $(v/c) \ll 1$ reduces to

$$E_x \cong vB_y \left(\frac{\varepsilon - \varepsilon_0}{\varepsilon}\right) \tag{3.3.15}$$

which when multiplied by the dielectric thickness d gives the potential $V = E_x d$. For $(\varepsilon/\varepsilon_0) \gg 1$ the field reduces to the value we obtain for motion of a conducting medium

$$E_x \cong vB_y \tag{3.3.16}$$

In a conducting medium moving through a magnetic field there are conduction electrons that can move under the influence of the $\mathbf{v} \times \mathbf{B}$ force and they move to one surface of the conductor, forming surface charges on both sides that yield a potential difference across the slab given by

$$V = E_x d = vB_y d \tag{3.3.17}$$

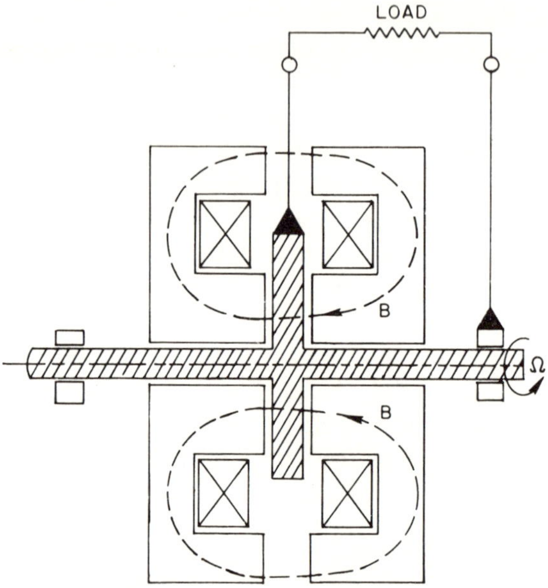

Fig. 3.3-5. A cross-section view of a homopolar generator. A highly conducting slab is rotated through a magnetic field. A high current at a low voltage can be drawn by an external load.

A further difference between the two cases is the ability of the conduction electrons to leave the slab and supply a dc current, something not possible for the polarization charges of the dielectric slab. The homopolar generator of Fig. 3.3-5 works on this principle and is of practical value where very low impedance loads are of interest. This type of generator was one of the first rotating machines to be built. In 1831 Faraday developed a demonstration machine that used this basic configuration [8].

PROBLEM 3.3-1 RÖNTGEN APPARATUS

A device of the type shown in Fig. 3.3-2 produces a certain value of B_y, as given by Eqs. (3.3.9) and (3.3.12), for the two cases of the plates moving with the dielectric and the plates at rest in the unprimed frame. Repeat these two calculations for the case of a dielectric of thickness d moving between a pair of plates spaced by a distance a, where $a > d$, so there is a finite air gap, of thickness $(a - d)/2$, on both sides of the dielectric, using $(v/c) \ll 1$.

PROBLEM 3.3-2 WILSON APPARATUS

A device of the type shown in Fig. 3.3-4 has a finite air gap. The dielectric has thickness d and the plates are spaced by a, where $a > d$, and the air gap is $(a - d)/2$ on each side. Calculate, for $(v/c) \ll 1$, the value of the measured potential.

3.4 Wave Transformations

Our discussion thus far has been limited to the transformation of static electric and magnetic fields. The Lorentz transformation was derived with reference to the behavior of a wave traveling at the velocity of light and so, of course, applies to the case of electric and magnetic fields that may vary in time and space as in a wave. The basic equation for the electric field of a wave of frequency ω' traveling in an arbitrary direction specified by the direction of the wave number $\mathbf{k}'$ is

$$\mathbf{E}' = \mathbf{E}_0' \exp[j(\omega' t' - \mathbf{k}' \cdot \mathbf{r}')] \tag{3.4.1}$$

where $\mathbf{r}'$ is the radius vector to the position from which the field is observed. We will obtain equations for the transformation of the frequency and for the wavenumber $\mathbf{k}'$ when the wave is observed from a moving frame of reference.

First let us consider the specific case of a TEM wave traveling in the positive z'-direction in free space, so that $\mathbf{k}' = \mathbf{1}_z' k'$. For such a wave $k' = \omega'/c$ and we can write $\mathbf{k}' \cdot \mathbf{r}'$ in terms of the angle θ' between the z'-axis which is the direction of the vector $\mathbf{k}'$, and the $\mathbf{r}'$-direction, as shown in Fig. 3.4-1, so that

$$\mathbf{k}' \cdot \mathbf{r}' = (\omega'/c) z' \cos \theta' \tag{3.4.2}$$

So we can express the electric field, as seen in the primed frame, as

$$\mathbf{E}' = \mathbf{E}_0' \exp\left[j\omega'\left(t' - \frac{z'}{c} \cos \theta'\right)\right] \tag{3.4.3}$$

where θ' specifies the angle between $\mathbf{k}'$ and $\mathbf{r}'$. If we look at this wave from the unprimed frame, taken as usual to be moving in the negative z'-direction with velocity v, we must see a wave with the same velocity c, but both the apparent frequency and the apparent angle between the wave normal and

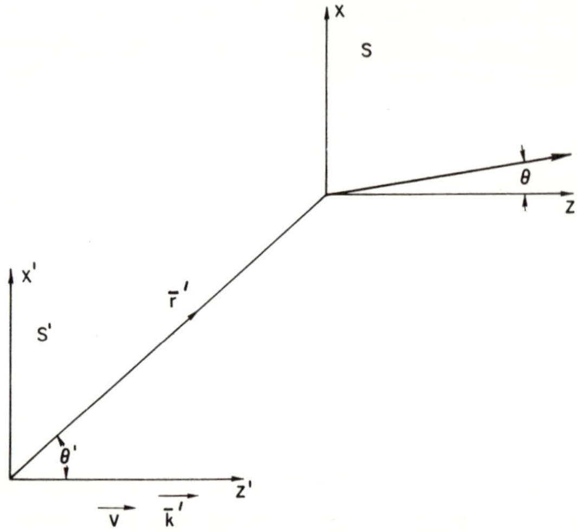

Fig. 3.4-1. A source of radiation at rest in the S' frame produces a plane wave traveling in the positive z'-direction that is observed from the origin of the S frame as having come from a direction at an angle θ to the z-axis. S' is moving at velocity v with respect to S in the z-direction. (A letter with an overbar in a figure is equivalent to a boldface letter in the text.)

the radius vector to the source position may be changed. Consider the wave field as seen in the unprimed frame.

$$\mathbf{E} = \mathbf{E}_0 \exp\left[j\omega\left(t - \frac{z}{c}\cos\theta\right)\right] \tag{3.4.4}$$

The Lorentz transformation for distance and time of Eq. (2.1.11) must transform the argument of the exponential of Eq. (3.4.3) into that of Eq. (3.4.4), if we are describing the same wave, i.e., if we are describing the same field as a function of time and space from both frames.

$$\begin{aligned}\omega[t - (z/c)\cos\theta] &= \omega'[t' - (z'/c)\cos\theta'] \\ &= \omega'\{\gamma[t - (\beta z/c)] - (\gamma/c)(z - \beta ct)\cos\theta'\}\end{aligned} \tag{3.4.5}$$

And, since this equation must hold for all z and t, we may equate coefficients of t and z separately to get

$$\omega/\omega' = \gamma(\cos\theta' + \beta)/\cos\theta = \gamma(1 + \beta\cos\theta') \tag{3.4.6}$$

Solving Eq. (3.4.6) for $\cos\theta$ we get the angle seen by the observer in the

S frame in terms of the angle in S' and $\beta = v/c$:

$$\cos\theta = (\cos\theta' + \beta)/(1 + \beta\cos\theta') \tag{3.4.7}$$

We see that as β becomes comparable to unity we get $\cos\theta > \cos\theta'$ for $0 < \theta' < (\pi/2)$ as in Fig. 3.4-1 and for $\beta \to 1$, $\cos\theta \to 1$ and $\theta' \to 0$. This effect is called aberration and is an important one in astronomical observation, as explained later in this section.

The frequency shift or Doppler effect is expressed in terms of the angle in S' and β from Eq. (3.4.6) as

$$\omega/\omega' = \gamma(1 + \beta\cos\theta') \tag{3.4.8}$$

For the source of radiation approaching the observer, as in Fig. 3.4-1, $\cos\theta'$ is positive and the observed frequency is greater than the transmitted frequency. For the source traveling away from the observer $\cos\theta'$ is negative and the frequency is apparently decreased. Note that there is an effect even if θ' is $\pi/2$ and the direction of propagation is at right angles to the direction of motion. This effect is not present when a Galilean transformation is used. For $\theta' = 0$ and an approaching observer, $\theta = 0$ and

$$\omega/\omega' = [(1 + \beta)/(1 - \beta)]^{1/2} = \gamma(1 + \beta) \tag{3.4.9}$$

For $\theta' = \pi$ and a receding observer, $\theta = \pi$ and

$$\omega/\omega' = [(1 - \beta)/(1 + \beta)]^{1/2} = \gamma(1 - \beta) \tag{3.4.10}$$

It is useful to consider the approximate form of these equations for small β. For $\theta' = 0$ and an approaching observer, $\theta = 0$ and

$$\omega/\omega' \cong 1 + \beta + \tfrac{1}{2}\beta^2 + \cdots \tag{3.4.11}$$

For $\theta' = \pi$ and a receding observer, $\theta = \pi$ and

$$\omega/\omega' \cong 1 - \beta + \tfrac{1}{2}\beta^2 - \cdots \tag{3.4.12}$$

An early direct experimental test of special relativity was performed by Ives and Stilwell [9] by measuring the Doppler shift in the frequency of light emitted by moving atoms, as shown in Fig. 3.4-2. They accelerated a hydrogen ion beam to an energy of about 10 kV and observed the emitted light from the drifting beam both directly from an observation post in the path of the beam ($\theta = 0$) and via a mirror that reflected the light emitted in the backward direction ($\theta = \pi$) to the same observation post for comparison of the wavelengths of the light from the two sources.

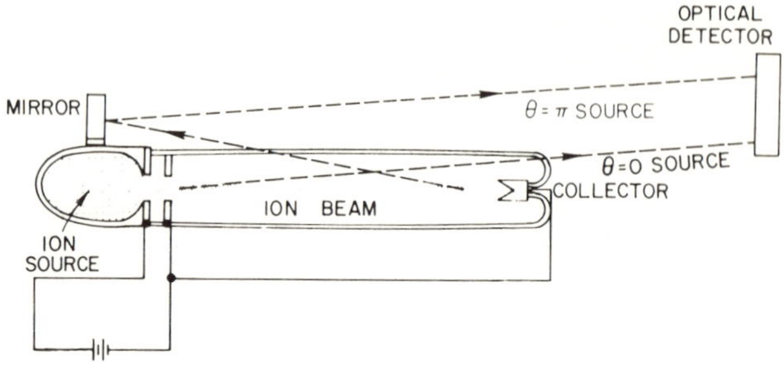

Fig. 3.4-2. The experiment conduced by Ives and Stilwell. A hydrogen ion beam is accelerated to 10 kV. Optical emission from the drifting beam is picked up at an optical detector lying essentially in the path of the beam and beyond the collector. The frequency of this light is compared with that of light emitted in the opposite direction and reflected to the same point by a mirror. The frequencies can be compared by comparing the wavelengths in the detector.

If the two observed frequencies are averaged, it is seen from Eqs. (3.4.11) and (3.4.12) that the first order shift cancels out leaving the second order shift which represents a net shift in the average frequency. The first order shift is what would be predicted by Newtonian mechanics, so a measurement of the second order shift is a direct confirmation of special relativity. In the Ives and Stilwell experiment the amount of observed second-order wavelength shift was about 0.05 Å. This is a very small shift, but even so, was measured to better than 10% accuracy, providing one of the most direct and graphic demonstrations of the correctness of special relativity.

Let us return now to Eq. (3.4.1) and generalize our derivation of the transformations of frequency and wave number. What we have derived for the transformation of frequency is perfectly general, but we have assumed the wave number k to be given by (ω/c), i.e., we have analyzed only the case of a wave traveling at the velocity of light. We will be interested in waves traveling at other velocities. Let v_p be the phase velocity of a wave, so that $k = (\omega/v_\mathrm{p})$ and denote the component of k parallel to the direction of the velocity $\mathbf{v}$ as $k_\parallel$ and the component of k perpendicular to $\mathbf{v}$ as $k_\perp$. Then, taking $\mathbf{k}'$ along the z'-axis as before

$$\mathbf{k}' \cdot \mathbf{r}' = k_\parallel' z' = k' \cos\theta' z' \tag{3.4.13}$$

Using the transformation of distance and time of Eq. (2.1.11) as before, but now regarding k' simply as an inverse length, and setting the arguments of the exponentials in the two frames equal to each other as before

$$(\omega' t' - k_\parallel' z') = (\omega t - k_\parallel z) \tag{3.4.14}$$

we obtain the transformation in the form

$$k_{\parallel}' = \gamma[k_{\parallel} - (\beta/c)\omega], \qquad k_{\perp}' = k_{\perp}, \qquad \omega' = \gamma(\omega - \beta c k_{\parallel}) \qquad (3.4.15)$$

which applies to waves of any velocity.

We are now in a position to study waves in a moving dielectric medium, a problem of fundamental importance in the history of relativity and electrodynamics [10–13]. We first examine this problem from the point of view that a wave has a velocity w which can be transformed just like the velocity of a particle or material medium.

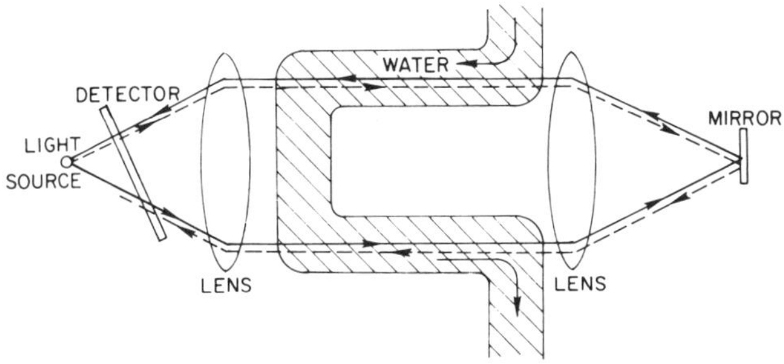

Fig. 3.4-3. The experiment conducted by Fizeau in 1853. The solid line represents a light path for a ray that travels in the same direction as the flow of water and the dashed line represents a light path in the opposite direction. Fringe patterns on the detector can be used to measure the difference in wavelength between waves traveling in opposite directions.

In 1853 Fizeau [14] conducted an experiment shown in Fig. 3.4-3 with a beam of light traveling through a moving stream of water and was able to observe a difference between the wavelength with the water moving and the water at rest. The first order theory of this experiment had been predicted by Fresnel [14] prior to 1853, and the coefficient by which the phase velocity is multiplied as a result of the motion is referred to as the Fresnel–Fizeau coefficient. It is much easier to derive their result from relativity theory than it is to go through a derivation based on the physical phenomena involved. However, it is of interest to at least make a plausible argument about the physical effect. We know that a plane wave is slowed down when it travels through a dielectric medium with a relative dielectric constant greater than unity, i.e., if ε_r is relative dielectric constant and ε_0 is the dielectric constant (permittivity) of free space, then we define ε_r by

$$\varepsilon = \varepsilon_r \varepsilon_0 \qquad (3.4.16)$$

where ε is the permittivity of the medium. This slowing down process can be thought of physically as one of setting the molecules of the medium into oscillation in such a way that the reradiation from these molecular dipoles combines with the incident wave to give a wave with a reduced phase velocity. The refractive index n of the medium is a measure of this effect, where n is defined by

$$n = c/w \tag{3.4.17}$$

where w is the wave velocity in the medium and c is the velocity of light in free space. Now if we cause the medium to move, we carry along the source of the reradiation in the direction of motion. Thus we might expect to speed up the wave velocity relative to its value in the absence of motion, if the motion were in the direction of the wave velocity and further slow it down, if motion were against the direction of wave velocity.

We may obtain the mathematical result directly from the transformation equation for velocity, Eq. (2.3.4), by interpreting u as the observed wave velocity in the laboratory frame, v as the velocity of motion of the dielectric medium, and u' as the wave velocity in the frame of the moving medium. Thus $u' = c/n$ is determined from Eq. (3.4.17) by the index of refraction and, assuming motion in the direction of wave propagation,

$$u = u_z = \frac{u_z' + \beta c}{1 + (\beta u_z'/c)} = \frac{(c/n)[1 + n(v/c)]}{1 + (1/n)(v/c)} \tag{3.4.18}$$

For low velocities of motion, $\beta \ll 1$,

$$u \cong (c/n) + v[1 - (1/n^2)] \tag{3.4.19}$$

which is the Fresnel–Fizeau result. If v is at an angle θ to u the relation can be shown to be

$$u \cong (c/n) + v \cos\theta[1 - (1/n^2)] \tag{3.4.20}$$

where v is a positive number and $\theta = 0$ for v in the same direction as u.

The Fizeau experiment and the Ives and Stilwell experiment provide a suitable background for a discussion of the Michelson–Morley experiment [15] which is probably the best known special relativity experiment. It has considerable historical significance and is a very simple experiment to understand. The Michelson interferometer that was used to conduct the experiment is shown in Fig. 3.4-4. Light from a single source is split into two rays that are caused to traverse different paths and the two rays are then recombined at a fringe detector that allows any difference between the lengths of the two paths, measured in wavelengths of light, to be dis-

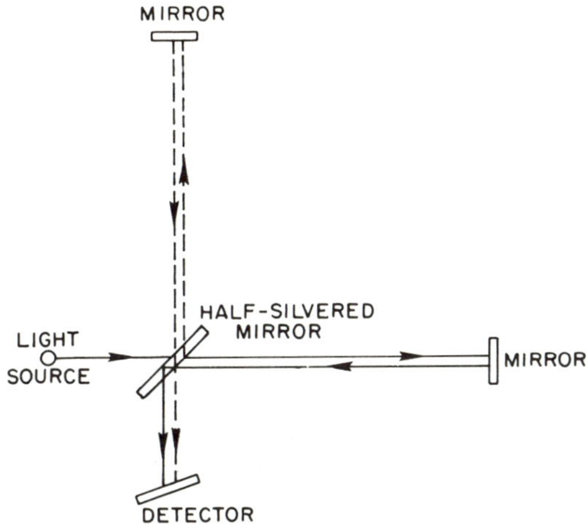

Fig. 3.4-4. The experiment conducted by Michelson and Morley in 1887. Light from the light source is split into two beams by the half-silvered mirror and the beams are combined at the detector after having traveled significant distances to the mirrors in two directions at right angles to each other.

tinguished. The experiment is thus identical in concept to the Fizeau experiment, except that, instead of sending the two rays through moving water, the rays are sent through free space along directions that are at right angles to each other. Michelson and Morley detected no difference between the wavelengths in any two directions, nor have any subsequent investigators, to a degree of accuracy of about 0.001 wavelength, and including tests at all times of year and all times of day. The velocity of the earth in its orbit is about 30 km sec^{-1} while the accuracy of the results of these experiments is about 3 km sec^{-1}.

This experiment does away with the possibility that there is a preferred or absolute frame of reference such as that of the fixed stars. The idea that there might be such a frame of reference is easy to understand from the point of view of a person familiar with the Doppler effect for sound waves as discussed in Example II. The name given to the postulated medium that supposedly carried light waves in a way analogous to the operation of air for sound waves was the ether. The ether frame which the Michelson–Morley experiment was intended to detect, if it were present, was the frame of reference from which the ether would appear stationary. Since there was no detectable difference between various orientations of the experiment, to an accuracy an order of magnitude greater than the velocity of the earth in its orbit it was clear that if there was an ether, it had in some way to be attached to the earth. This latter possibility was seriously proposed and is referred to as the idea of an ether drag. The idea was that massive bodies drag along with them a portion of the ether which then can be thought of as being somewhat like a gravitational field. This idea, however, is not consistent with the aberration of starlight as given by Eq. (3.4.7). Due to the earth's motion in its orbit a telescope must be tilted by an angle that depends on the velocity of the observer riding on the earth in order to see a star directly overhead. This angle can be computed from Eq. (3.4.7) and turns out to be about 10^{-4} radian using 30 km $\sec^{-1}$ for the earth's velocity, in agreement with experiment. If the ether were dragged along with the earth, no aberration should be observed.

The ether concept seemed important around the turn of the century and much of the relativity literature is written in terms of this concept. From our present vantage point, this concept is about on a par with the concept of phlogiston in the theory of chemical oxidation and other views of nature that have had to be left behind. The ether concept, however, served a very useful function is bridging the gap between Newtonian mechanics, relativistic mechanics, and Maxwell's equations by requiring the proponents of special relativity to be precise in their thinking and in their experiments.

Example III A Moving Resonator

A problem that is of interest in modern relativity research using lasers is the problem of the moving resonator [16]. Picture a resonator moving in the z-direction at velocity v, as shown in Fig. 3.4-5. The resonator operates in a TEM mode in free space, so the phase velocity of waves traveling back and forth in the resonator is the velocity of light in free space. The resonator is aligned with the z-direction, so the waves travel in the z-direction at the velocity of light. The resonator is excited by an oscillator

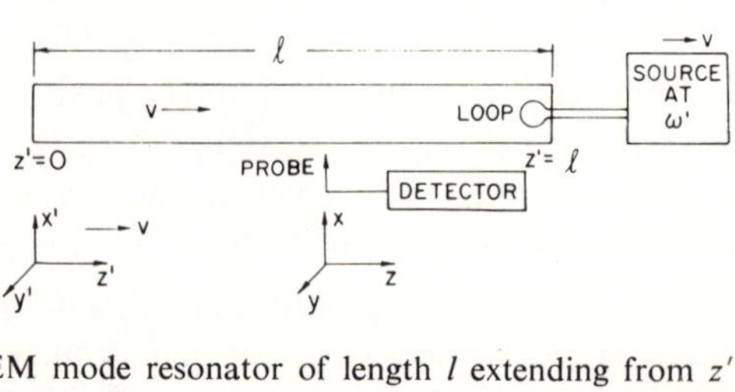

Fig. 3.4-5. A TEM mode resonator of length l extending from $z' = 0$ to $z' = l$ fed from a source having an angular frequency ω'. Both the source and the resonator are moving at velocity v with respect to the unprimed laboratory frame of reference.

that is lightly coupled to the resonator and that excites the resonator at a frequency ω' that is a resonant frequency at which the resonator is n wavelengths long. Physically this could be done at microwave frequencies by coupling a microwave oscillator such as a reflex klystron through a loop to a coaxial-line resonator. At optical frequencies this description could apply directly to a laser operating in a single mode of its optical resonator.

In the frame of reference of the resonator and the oscillator, the primed frame, there is a pure standing wave of electric field that can be described by the equation

$$E = E_0 \cos \omega' t' \sin(\omega' z'/c) \tag{III.1}$$

where ω' is the angular frequency, $\omega' = 2\pi f'$, and where the resonator is pictured as being short circuited at $z' = 0$ and $z' = l$ where $l = n\lambda'$ and λ' is the wavelength measured in the primed frame, $(2\pi/\lambda') = (\omega'/c)$. Let us now see how this standing wave looks to an observer in the unprimed laboratory frame. Experimentally we can imagine, in the microwave system, that we insert a small probe through the outer wall of the resonator and pick up a signal from the moving resonator that is passing by at a velocity v. In the optical system it would be more convenient to sample the two traveling waves emitted from either end of the resonator and to devise a detector in the rest frame to measure the frequencies or wavelengths of the light emitted from the moving system.

From the Lorentz transformation, Eq. (2.1.11) we can write down z' and t' and express E in the form

$$\begin{aligned} E &= E_0 \cos\{\omega'\gamma[t - (\beta z/c)]\} \sin[(\omega'/c)\gamma(z - \beta ct)] \\ &= \tfrac{1}{2}E_0\{\sin[\omega'\gamma(1-\beta)t + (\omega'/c)\gamma(1-\beta)z] \\ &\quad + \sin[\omega'\gamma(1+\beta)t - (\omega'/c)\gamma(1+\beta)z]\} \\ &= \tfrac{1}{2}E_0\{\sin[\omega_b t + (2\pi z/\lambda_b)] + \sin[\omega_f t - (2\pi z/\lambda_f)\} \end{aligned} \tag{III.2}$$

which is evidently a pair of traveling waves, still both traveling at the velocity of light, but having different frequencies $\omega_b = \omega'\gamma(1+\beta)$ and $\omega_f = \omega'\gamma(1-\beta)$, just as predicted by the Doppler effect, Eqs. (3.4.9) and (3.4.10). Evidently the wavelengths of the two waves are shifted by the same amounts as the frequencies, in order for $(\omega_f/c) = (2\pi_f/\lambda)$ and $(\omega_b/c) = (2\pi/\lambda_b)$.

The first question that arises at this point is, how can an integer number of wavelengths still fit into the resonator, when the wavelengths are different for forward and backward waves and both are different from the original primed frame wavelength? In other words, is the resonator still resonant when viewed from the unprimed frame? To answer this question, note that in the primed frame $E = 0$ at $z' = 0$ and $z' = l$ for all t, as required by the short-circuit boundary conditions. In the unprimed frame, the point $z' = 0$ which is $z = \beta ct$ yields zero field for all t in Eq. (III.2). The same holds for $z' = l$ which lies at $z = \beta ct + (l/\gamma)$ when the Lorentz contraction of resonator length is included. Thus the standing wave still "fits" the resonator.

The next question that we might ask is, what would the probe pick up as a function of time in the rest frame as the resonator moves past? Here we must set z equal to a constant and ask what is $E(t)$. Although we have pictured the physical experiment as a resonator lying between $z' = 0$ and $z' = l$ so that the field is zero elsewhere in space, our Eq. (III.2) gives us a field at all z', just as if we had an infinite set of resonators stacked end to end along the z- and z'-axis. Thus, if we set $z = 0$ or any other arbitrary value in Eq. (III.2) we will get a repetitive waveform corresponding to all these resonators passing by. A specific portion of the waveform then corresponds to what we would measure (with a very low Q measuring system). If we do then set $z = 0$ in Eq. (III.2) we get

$$\begin{aligned} E(0, t) &= \tfrac{1}{2}E_0[\sin \omega_b t + \sin \omega_f t] \\ &= E_0 \cos \omega'\gamma t \sin \omega'\gamma\beta t \end{aligned} \tag{III.3}$$

In the latter form we see that we should pick up a carrier at frequency $\omega'\gamma$ modulated at a low frequency $\omega'\gamma\beta$ and the low frequency envelope should have a frequency that is proportional to the velocity with which the resonator moves past us.

Problem 3.4-1 Wave Transformation

Compare the transformation equations for the wave number k in Eq. (3.4.15) with the transformation equations for length that lead to the Lorentz contraction.

3.5 TEM Waves in a Moving Lossless Dielectric Medium

In this section we treat the specific problem of plane waves in a moving dielectric medium more fully, making use of the transformations of the preceding sections for fields, frequency, and wave number. The problem

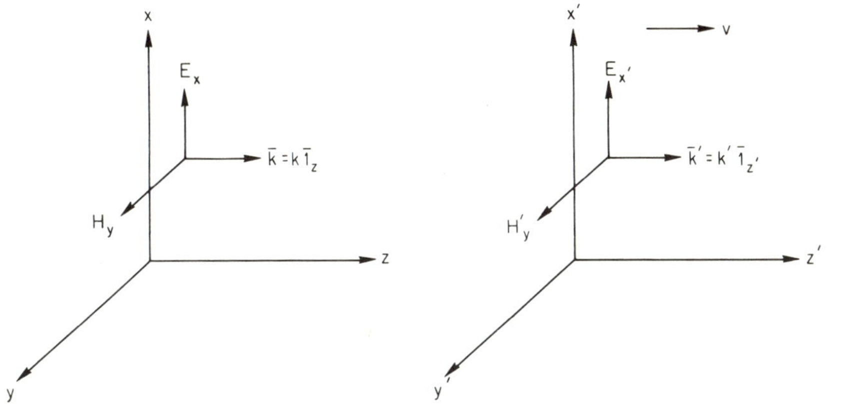

Fig. 3.5-1. Two coordinate systems. A TEM wave propagates in the positive z-direction. The primed system is moving to the right with constant velocity v relative to the unprimed system. A uniform medium characterized by μ and ε fills the space and is at rest with respect to the primed coordinate system [from D. A. Dunn, R. W. Wallace, and S. D. Choi, Waves in a moving medium with finite conductivity, *Proc. IEEE* **57**, 45–57 (1969)].

of interest here is described in Fig. 3.5-1 which shows two frames of reference with the primed frame moving at velocity v in the positive z-direction. A wave, TEM with respect to z, has a wave number $\mathbf{k}$ in the positive z-direction and is polarized with the electric field in the positive x-direction. A lossless dielectric medium filling the entire space is at rest in the primed frame, so that in this frame

$$\mathbf{D}' = \varepsilon\mathbf{E}', \qquad \mathbf{B}' = \mu\mathbf{H}' \tag{3.5.1}$$

where ε and μ are the permittivity and permeability of the material at rest. The wave fields viewed in this frame have the form

$$\begin{aligned} E_x' &= E_0' \exp[j(\omega't' - k'z')] \\ H_y' &= (E_0'/Z_0') \exp[j(\omega't' - k'z')] \\ D_x' &= \varepsilon E_0' \exp[j(\omega't' - k'z')] \\ B_y' &= \mu(E_0'/Z_0') \exp[j(\omega't' - k'z')] \end{aligned} \tag{3.5.2}$$

where $Z_0' = (\omega'\mu/k')$ is the wave impedance of the medium at rest and E_0' is an arbitrary constant [17]. The relation between E_x' and H_y' is obtained from Maxwell's curl equation. The dispersion relation, obtained from the wave equations, has the form

$$k'^2 = \omega'^2\mu\varepsilon = (\omega'^2/c^2)n^2 \tag{3.5.3}$$

Thus in the primed frame we have the familiar two solutions to the one-dimensional wave equation in a lossless medium representing waves traveling in the positive and negative z'-directions with velocity (c/n), corresponding to the two solutions to the dispersion relation Eq. (3.5.3). The wave impedance Z_0' is positive for the positively traveling wave and negative for the negatively traveling wave, implying different orientations of **E** with respect to **H** for excitation of the two waves, i.e.

$$Z_0' = \pm(\mu/\varepsilon)^{1/2} = \pm\eta \tag{3.5.4}$$

where the positive sign goes with the positive solution to Eq. (3.5.3) and the negative sign with the negative solution.

To obtain the dispersion relation in the laboratory frame, we use Eq. (3.4.15) in Eq. (3.5.3) to obtain

$$k^2(1 - \beta^2 n^2) - 2\omega k \frac{\beta}{c}(1 - n^2) + \frac{\omega^2}{c^2}(\beta^2 - n^2) = 0 \tag{3.5.5}$$

which can be solved for k to give

$$k = \frac{\omega}{c}\,\frac{\beta \pm n}{1 \pm \beta n} = \frac{\omega(\mu\varepsilon)^{1/2}}{(1 - \beta^2 n^2)}\left[\frac{\beta}{n}(1 - n^2) \pm \frac{1}{\gamma^2}\right] \tag{3.5.6}$$

which gives us again two waves, which we will designate k_1 and k_2, where k_1 corresponds to the positively traveling wave and k_2 to the negatively traveling wave in the primed frame.

$$k_1 = \frac{\omega}{c}\,\frac{\beta + n}{1 + \beta n}, \qquad k_2 = -\frac{\omega}{c}\,\frac{n - \beta}{1 - \beta n} \tag{3.5.7}$$

For $\beta n < 1$ these waves are also positively- and negatively traveling in the laboratory frame, respectively, but for $\beta n > 1$ both waves become positively traveling waves in the laboratory frame. The situation here is analogous to that with sound waves in which a wind is blowing with a velocity greater than the velocity of sound. In this case, even if a person shouts into the wind, expecting to excite a negatively traveling wave, he in fact excites a positively traveling wave that is carried along by the wind in the direction opposite to the direction in which he shouted. Figure 3.5-2 shows the dispersion diagram that goes with Eq. (3.5.7) which is for the specific case of $n^2 = 10$, so the negatively going wave reverses direction for $\beta > 0.316$.

To understand how these waves can be excited we must obtain the wave

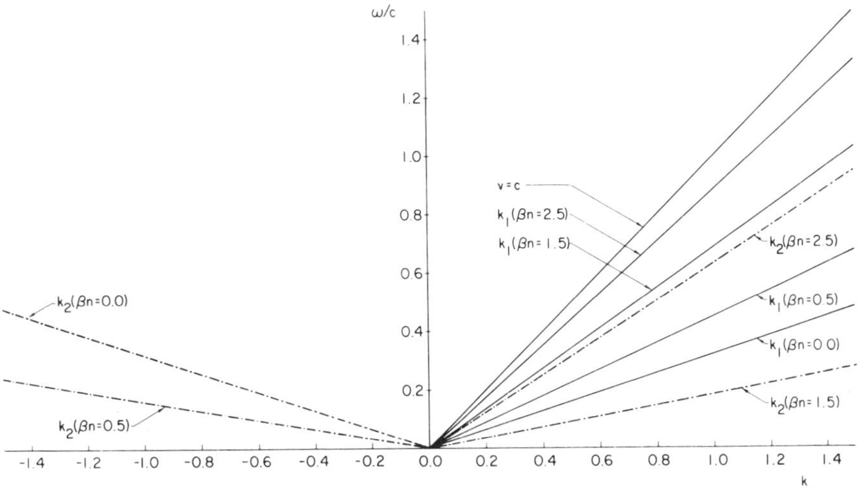

Fig. 3.5-2. Dispersion diagram showing ω/c vs. k for a uniform plane wave propagating in a moving lossless ($\sigma = 0$) dielectric medium with $n^2 = 10$, where n is the index of refraction in the medium at rest. The solid lines give $k(\omega)$ for the positive impedance wave with wave number k_1 which travels in the positive z-direction for all values of β. The dashed lines give $k(\omega)$ for the negative impedance wave with wave number k_2 which travels in the negative z-direction for $\beta n < 1$ and in the positive z-direction for $\beta n > 1$ [from D. A. Dunn, R. W. Wallace, and S. D. Choi, Waves in a moving medium with finite conductivity, *Proc. IEEE* **57**, 45–57 (1969)].

impedance Z_0 in the rest frame. This quantity can be found by Lorentz transforming the fields of Eq. (3.5.2) using Eqs. (3.1.11), (3.3.5) and (3.4.15).

$$\begin{aligned}
E_x &= E_0 \exp[j(\omega t - kz)] \\
H_y &= \frac{\gamma^2 E_0}{\omega\mu}\left[k(1 - \beta^2 n^2) - \frac{\omega\beta}{c}(1 - n^2)\right]\exp[j(\omega t - kz)] \\
D_x &= \gamma^2 \varepsilon E_0\left[\left(1 - \frac{\beta^2}{n^2}\right) - \frac{kv}{\omega}\left(1 - \frac{1}{n^2}\right)\right]\exp[j(\omega t - kz)] \\
B_y &= \frac{k}{\omega} E_0 \exp[j(\omega t - kz)]
\end{aligned} \tag{3.5.8}$$

where

$$E_0 = \frac{\gamma(\omega' + \beta c k')}{\omega'} E_0' = \frac{\omega}{\gamma(\omega - \beta c k)} E_0' \tag{3.5.9}$$

Written in this form it appears that the impedance Z_0 is a complicated function of n, ω, k, and β. However, if we use Eq. (3.5.5) for $k(\omega, \beta, n)$ we can reduce Eqs. (3.5.8) to the simple form

$$\begin{aligned}
E_x &= E_0 \exp[j(\omega t - kz)] \\
H_y &= \frac{E_0}{\pm\eta} \exp[j(\omega t - kz)] \\
D_x &= \frac{\varepsilon E_0}{n} \frac{(n \pm \beta)}{(1 \pm \beta n)} \exp[j(\omega t - kz)] \\
B_y &= \frac{E_0}{c} \frac{(\beta \pm n)}{(1 \pm \beta n)} \exp[j(\omega t - kz)]
\end{aligned} \tag{3.5.10}$$

where the plus signs all go together and the minus signs all go together, and correspond to the two wave solutions to Eq. (3.5.3), so that, surprisingly enough

$$Z_0 = Z_0' = \pm(\mu/\varepsilon)^{1/2} = \pm\eta \tag{3.5.11}$$

and the ratio of E_x to H_y in the laboratory frame is the same for each wave as in the primed frame. The only significant difference is the fact that in the laboratory frame the negative-impedance wave will travel in the positive z-direction if $\beta n > 1$.

In the next section we continue the discussion of this mode with a look at the power and energy in the two waves.

3.6 Poynting's Theorem and the Conservation of Energy

Energy and power involve products of two field quantities such as **E** and **H**. In this section we consider transformations of such product quantities. The conservation of electromagnetic field energy is expressed in Poynting's theorem. We will see that the usual form of Poynting's vector, which appears in Poynting's theorem and expresses the amount and direction of power flow, is invariant to a Lorentz transformation. So is the form of energy per unit volume in free space, but in a moving material medium the situation is more complex.

We begin by deriving Poynting's theorem in the usual way from Maxwell's equations in standard form as in Eq. (3.1.1). To obtain Poynting's theorem, take the scalar product of **H** with the third equation and the scalar product of **E** with the fourth equation and subtract the two equations:

$$\nabla \cdot (\mathbf{E} \times \mathbf{H}) = -\mathbf{H} \cdot \frac{\partial \mathbf{B}}{\partial t} - \mathbf{E} \cdot \frac{\partial \mathbf{D}}{\partial t} - \mathbf{J} \cdot \mathbf{E} \tag{3.6.1}$$

If we integrate this equation over a closed surface we obtain the usual result

$$\int_S \mathbf{E} \times \mathbf{H} \cdot d\mathbf{S} = -\int_V \left(\mathbf{H} \cdot \frac{\partial \mathbf{B}}{\partial t} + \mathbf{E} \cdot \frac{\partial \mathbf{D}}{\partial t} + \mathbf{J} \cdot \mathbf{E}\right) dV \tag{3.6.2}$$

It is from this equation that the interpretations are made that $\mathbf{E} \times \mathbf{H}$ represents the rate of energy flow out of the volume, that the first two terms on the rhs are time derivatives of field energy storage, and that the third term represents energy delivered into dissipation plus that delivered to coherent motion via terms of the form $\mathbf{f} \cdot \mathbf{v}$, where $\mathbf{f}$ is the force acting on the medium. If we are concerned with media at rest, the third term is purely a dissipation term. Note that Eqs. (3.6.1) and (3.6.2) must have exactly the same form in two frames moving at a constant velocity with respect to each other, because Maxwell's equations have the same form in such frames. Equation (3.6.1) is derived directly from Maxwell's equations and, if rederived in a primed frame, it would necessarily have the same form but would be expressed in terms of primed field quantities.

If ε and μ are constants in space and time and if $\mathbf{B} = \mu\mathbf{H}$ and $\mathbf{D} = \varepsilon\mathbf{E}$, Eq. (3.6.1) reduces to

$$\nabla \cdot (\mathbf{E} \times \mathbf{H}) + \frac{\partial}{\partial t}\left(\frac{\varepsilon E^2}{2} + \frac{\mu H^2}{2}\right) = -\mathbf{J} \cdot \mathbf{E} \tag{3.6.3}$$

which includes the field energy terms in their usual form for simple media at rest.

If we deal with problems in which there is a transformation from one frame of reference to another and if the medium is free space, the form of Eq. (3.6.3) is identical in the two frames of reference, because both Maxwell's equations and the constitutive equations have the same form in both frames. If the medium is other than free space, constitutive equations such as Eqs. (3.3.4) apply in the frame in which the material is in motion, so that Eq. (3.6.1) does not reduce to Eq. (3.6.3) in such a frame. Thus in moving media other than free space we must use Eq. (3.6.1) rather than Eq. (3.6.3) [18, 19].

It is often the case in practical problems that we are interested in time-average energy conservation only. For time-harmonic fields varying as $\exp(j\omega t)$ with real ω in simple media at rest, Eq. (3.6.3) can be further simplified. Denoting time-average by brackets $\langle\ \rangle$, the time-average of the energy storage term in Eq. (3.6.3) is

$$\left\langle \frac{\partial}{\partial t}\left(\frac{\varepsilon E^2}{2} + \frac{\mu H^2}{2}\right)\right\rangle = \frac{1}{T}\int_0^T \frac{\partial}{\partial t}\left(\frac{\varepsilon E^2}{2} + \frac{\mu H^2}{2}\right) dt \qquad (3.6.4)$$

Since these equations are written in terms of the instantaneous values of the fields, if we use subscript 1 to denote the complex multiplier of $\exp(j\omega t)$ and * to denote complex conjugate, we can write E as $\mathrm{Re}\{E_1 e^{j\omega t}\}$, etc., and so

$$\begin{aligned}\left\langle \frac{\partial}{\partial t}\left(\frac{\varepsilon E^2}{2} + \frac{\mu H^2}{2}\right)\right\rangle &= \frac{1}{T}\int_0^T \frac{\partial}{\partial t}\left[\frac{\varepsilon}{4}\,\mathrm{Re}(E_1E_1^* + E_1^2 e^{j2\omega t})\right. \\ &\qquad \left. + \frac{\mu}{4}\,\mathrm{Re}(H_1H_1^* + H_1^2 e^{j2\omega t})\right] dt \\ &= \frac{1}{T}\int_0^T \left[\frac{\varepsilon}{4}\,\mathrm{Re}(j2\omega E_1^2 e^{j2\omega t})\right. \\ &\qquad \left. + \frac{\mu}{4}\,\mathrm{Re}(j2\omega H_1^2 e^{j2\omega t})\right] dt \\ &= 0 \end{aligned} \qquad (3.6.5)$$

Thus on a time-average basis Eq. (3.6.3) reduces to

$$\langle \nabla \cdot (\mathbf{E} \times \mathbf{H})\rangle = -\langle \mathbf{J} \cdot \mathbf{E}\rangle \qquad (3.6.6)$$

Evidently, for free space the derivation of Eq. (3.6.6) is valid for any frame, whereas for material media, Eq. (3.6.6) has only been proven valid in the frame of the medium. We have to take the time-average of Eq. (3.6.1) rather than of Eq. (3.6.3) to deal with this case.

In general for material media in motion, **B** and **H** may not be in the same direction, even though μ and ε are scalars in the frame of the medium, as is evident from Eq. (3.3.4). Nevertheless, they may be in time phase, as in a plane wave. In this case, if

$$\mathbf{H} = \mathbf{H}_1 \cos \omega t \qquad (3.6.7)$$

where $\mathbf{H}_1$ is real, then

$$\mathbf{B} = \mathbf{B}_1 \cos \omega t \qquad (3.6.8)$$

where $\mathbf{B}_1$ is also real, even though **B** may not be in the same direction as **H** and $\mathbf{B} \neq \mu\mathbf{H}$. In this case, it is also true that the time-average of the energy storage terms is zero in general. For example, if Eqs. (3.6.7) and

(3.6.8) apply

$$\left\langle \mathbf{H} \cdot \frac{\partial \mathbf{B}}{\partial t} \right\rangle = \int_0^T \mathbf{H}_1 \cdot \mathbf{B}_1 \cos \omega t \left(\frac{\partial}{\partial t} \cos \omega t \right) dt$$

$$= - \int_0^T \mathbf{H}_1 \cdot \mathbf{B}_1 \cos \omega t \sin \omega t \, dt$$

$$= 0 \tag{3.6.9}$$

without regard to the directions and relative magnitudes of $\mathbf{H}_1$ and $\mathbf{B}_1$. Thus Eq. (3.6.6) may also be valid in the case of material media viewed from moving frames. The value of $\langle \mathbf{J} \cdot \mathbf{E} \rangle$ in this equation, however, may not be straightforward to calculate in general. If $\mathbf{J} = 0$, as in lossless media, the problem is indeed straightforward.

We now seek the transformation equations for power. For Poynting's theorem to have the same meaning in two frames, if $\mathbf{S}'$ is power in the primed frame and

$$\mathbf{S}' = \mathbf{E}' \times \mathbf{H}' \tag{3.6.10}$$

then, since Eq. (3.6.1) has the same form in terms of unprimed fields, power $\mathbf{S}$ in the laboratory frame must be

$$\mathbf{S} = \mathbf{E} \times \mathbf{H} \tag{3.6.11}$$

even in the case of moving material media having permeability and permittivity different from free space.

We now consider the example of the TEM mode in a moving lossless material medium discussed in the previous section. For this case we calculate the time-average value of Poynting's vector for each wave in the unprimed frame, using Eq. (3.6.10)

$$\mathbf{S} = \tfrac{1}{2} \operatorname{Re}\{\mathbf{E} \times \mathbf{H}^*\} \tag{3.6.12}$$

From Eq. (3.5.10) it is evident that

$$\mathbf{S}_1 = \tfrac{1}{2}(E_0{}^2/\eta)\mathbf{1}_z, \qquad \mathbf{S}_2 = -\tfrac{1}{2}(E_0{}^2/\eta)\mathbf{1}_z \tag{3.6.13}$$

where $\mathbf{S}_1$ and $\mathbf{S}_2$ are the time-average Poynting vectors for waves 1 and 2 with wave numbers defined by Eq. (3.5.7). An immediate question arises as to the interpretation of the situation for $\beta n > 1$ and k_2 positive, which as seen in the preceding section, implies a positive phase and group velocity. Since $\mathbf{S}_2$ is always negative, the meaningful interpretation of the situation is that the energy stored must become negative for $\beta n > 1$.

Let us try to check this interpretation. Although we are not sure what to define as stored energy for this case, we do know that the quantity $[\mathbf{H} \cdot (\partial \mathbf{B}/\partial t) + \mathbf{E} \cdot (\partial \mathbf{D}/\partial t)]$ is invariant and is equal to the time derivative of the stored energy for media at rest. Let W be the stored energy per unit volume. Then the relation

$$\partial W/\partial t = \mathbf{H} \cdot (\partial \mathbf{B}/\partial t) + \mathbf{E} \cdot (\partial \mathbf{D}/\partial t) \tag{3.6.14}$$

defines W to the extent that it is possible to do so from Poynting's theorem. If the medium is free space, it is easy to identify W from Eq. (3.6.3) as

$$W = (\varepsilon E^2/2) + (\mu H^2/2) \tag{3.6.15}$$

although even in this case, it is evident that all we really know from Poynting's theorem is $\partial W/\partial t$. For media other than free space we can also find a quantity that can be identified as W for fields that vary sinusoidally in time. For simplicity let all fields vary as $\cos \omega t$, as in Eqs. (3.6.7) and (3.6.8). Then

$$\partial W/\partial t = -\mathbf{H}_1 \cdot \mathbf{B}_1 \cos \omega t \sin \omega t - \mathbf{E}_1 \cdot \mathbf{D}_1 \cos \omega t \sin \omega t \tag{3.6.16}$$

Now we get this same form for $\partial W/\partial t$, if we define W as

$$W = \tfrac{1}{2}\mathbf{H} \cdot \mathbf{B} + \tfrac{1}{2}\mathbf{E} \cdot \mathbf{D} \tag{3.6.17}$$

The time-average value of W is then

$$\langle W \rangle = \tfrac{1}{4} \operatorname{Re}\{\mathbf{H} \cdot \mathbf{B}^* + \mathbf{E} \cdot \mathbf{D}^*\} \tag{3.6.18}$$

For the TEM mode of the preceding section we can then find $\langle W \rangle$ from Eqs. (3.5.10) as follows for wave 1:

$$\begin{aligned} \langle W \rangle_1 &= \frac{\varepsilon E_0^2}{4n} \frac{(n + \beta)}{(1 + \beta n)} + \frac{E_0^2}{4\eta c} \frac{(\beta + n)}{(1 + \beta n)} \\ &= \frac{E_0^2}{2\eta c} \frac{(\beta + n)}{(1 + \beta n)} \end{aligned} \tag{3.6.19}$$

For wave 2 the energy is

$$\begin{aligned} \langle W \rangle_2 &= \frac{\varepsilon E_0^2}{4n} \frac{(n - \beta)}{(1 - \beta n)} + \frac{E_0^2}{4\eta c} \frac{(n - \beta)}{(1 - \beta n)} \\ &= \frac{E_0^2}{2\eta c} \frac{(n - \beta)}{(1 - \beta n)} \end{aligned} \tag{3.6.20}$$

The energy for wave 2 thus becomes negative for $\beta n > 1$, which checks with the fact that the sign of the power is negative and the phase and group velocities are positive in this region. At $\beta n = 1$ the energy density becomes infinite. A wave launched at a particular value of z will be swept forward by the material that is moving at just the wave velocity in the medium at rest and at this critical value of β the wave will have zero phase and group velocity. The energy will thus be concentrated at the source position and will not propagate away from this position, thus corresponding to an infinite energy density at this position.

To summarize the results of this section, we have found that some fairly general statements about the form of the energy conservation equation can be made for moving media with arbitrary values of μ and ε. The form of Pointing's vector is invariant, as is the form of the time derivative of energy per unit volume. However, we were not able to make any very general statements about the form of the energy, except in the case of sinusoidal time variations. We were also able to find the time-average form of Poynting's theorem for sinusoidal time variations and arbitrary values of μ and ε.

PROBLEM 3.6-1 STORED ENERGY IN MOVING MATERIAL MEDIA

Calculate the time-average stored energy for the TEM mode of Section 3.5 in the primed frame. Is the energy negative for any value of β? Explain.

Example IV A Radiation-Pressure Motor

The problem of a moving mirror on which a light wave is incident presents a number of interesting features of the Doppler effect, and also illustrates the phenomenon of radiation pressure. The configuration to be discussed is illustrated in Fig. 3.6-1. A plane wave is traveling from left to right in the positive z-direction in free space. A mirror at $z' = 0$ is also moving from left to right at velocity v. The wave is normally incident on the mirror and all of the power is reflected from the mirror and so travels backward in the negative z-direction. We use subscript 1 to refer to the incident wave and subscript 2 to refer to the reflected wave. We can therefore write the equations for the fields in the primed frame, as follows, taking the

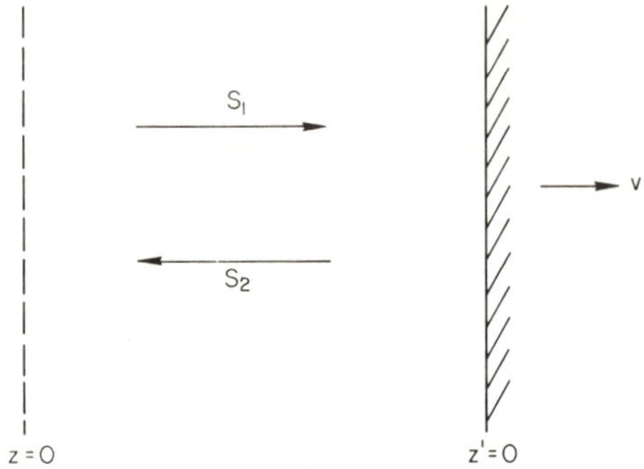

Fig. 3.6-1. A plane wave in free space carrying a power S_1 at frequency ω_1 is incident on a plane perfectly conducting wall at $z' = 0$ moving at velocity v in the positive z-direction. The reflected wave carries a power S_2 at frequency ω_2.

plane of polarization to be the x-plane:

$$\begin{aligned} E'_{x1} &= E_0' \exp[j(\omega' t' - k_1' z')] \\ H'_{y1} &= (E_0'/\eta) \exp[j(\omega' t' - k_1' z')] \\ E'_{x2} &= -E_0' \exp[j(\omega' t' - k_2' z')] \\ H'_{y2} &= (E_0'/\eta) \exp[j(\omega' t' - k_2' z')] \end{aligned} \tag{IV.1}$$

where η is the impedance of free space, $\eta = (\mu_0/\varepsilon_0)^{1/2}$, and E_0' is an arbitrary constant. We can impose the conditions in this frame, which is the frame of the mirror, that the frequencies of the incident and reflected waves are the same and equal to ω' and that the wave numbers are $k_1' = (\omega'/c)$ and $k_2' = -(\omega'/c)$. The electric field must be equal to zero at $z' = 0$. Thus all of the fields in Eq. (IV.1) have frequency ω' and $E'_{x2} = -E'_{x1}$ at $z' = 0$ which says that the reflection coefficient is -1 in this frame. Evidently also H_y' at $z' = 0$ is twice the value H'_{y1}. In the primed frame the problem is exactly that of a wave incident on a mirror at rest and there are no difficulties in understanding either the boundary conditions or the form of the phase factors for incident and reflected waves [20].

In order to obtain the fields in the laboratory frame, we simply perform a Lorentz transformation, using Eqs. (3.1.11). We first write the fields in terms of arbitrary constants as follows:

$$\begin{aligned} E_{x1} &= E_0 \exp[j(\omega_1 t - k_1 z)] \\ H_{y1} &= (E_0/\eta) \exp[j(\omega_1 t - k_1 z)] \\ E_{x2} &= rE_0 \exp[j(\omega_2 t - k_2 z)] \\ H_y^2 &= -r(E_0/\eta) \exp[j(\omega_2 t - k_2 z)] \end{aligned} \tag{IV.2}$$

where E_0 is an arbitrary constant that we can relate to E_0'; r is the reflection coefficient with a value to be determined; and ω_1, ω_2, k_1 and k_2 are the frequencies and wave numbers in the laboratory frame with values to be determined. We know that $|E_x/H_y| = \eta$ for both waves in this frame, just as in the primed frame, as was proved in Eq. (3.5.11).

Just as in the Doppler effect derivation above, we may set the arguments of the exponentials equal to their values in the primed frame for incident and reflected waves

$$\omega't' - k_1'z' = \omega_1 t - k_1 z, \qquad \omega't' - k_2'z' = \omega_2 t - k_2 z \tag{IV.3}$$

on the basis that the phase factor (the argument of the exponential) describes the same wave in both frames. Since we are dealing with waves in free space, the wave velocity for all waves is c, and so $k_1 = (\omega_1/c)$ and $k_2 = -(\omega_2/c)$. We can use the Lorentz transformation of ω and k of Eq. (3.4.15) to obtain the frequencies and wave numbers in the laboratory frame

$$\begin{aligned} \omega_1 &= \omega'\gamma(1+\beta), & \omega_2 &= \omega'\gamma(1-\beta) \\ k_1 &= (\omega'/c)\gamma(1+\beta), & k_2 &= -(\omega'/c)\gamma(1-\beta) \end{aligned} \tag{IV.4}$$

We see that the incident wave has a frequency higher than ω' in the laboratory frame and the reflected wave has a frequency lower than ω' in the laboratory frame.

To obtain the values of E_0 and r we use Eqs. (3.5.2) and (3.5.9) to obtain

$$E_0 = E_0' \frac{1}{\gamma(1-\beta)} \tag{IV.5}$$

$$r = -\frac{1-\beta}{1+\beta} \tag{IV.6}$$

Thus we can write the equations for the laboratory frame fields in terms of E_0' and ω' as follows:

$$\begin{aligned} E_{x1} &= \frac{E_0'}{\gamma(1-\beta)} \exp\left[j\gamma(1+\beta)\omega'\left(t - \frac{z}{c}\right)\right] \\ H_{y1} &= \frac{E_0'}{\eta\gamma(1-\beta)} \exp\left[j\gamma(1+\beta)\omega'\left(t - \frac{z}{c}\right)\right] \\ E_{x2} &= -\frac{E_0'}{\gamma(1+\beta)} \exp\left[j\gamma(1-\beta)\omega'\left(t + \frac{z}{c}\right)\right] \\ H_{y2} &= \frac{E_0'}{\eta\gamma(1+\beta)} \exp\left[j\gamma(1-\beta)\omega'\left(t + \frac{z}{c}\right)\right] \end{aligned} \tag{IV.7}$$

We see that in this frame the electric field is not equal to zero at $z' = 0$ at the mirror or at any value of t and z. If it is desired to establish a set of rules for boundary conditions at interfaces that are moving, one can do so, but it is much simpler to do the matching of boundary conditions in the frame of reference in which the interface between media is at rest, as we have done here. In this case no new rules are required beyond those used in solving problems for media at rest [20].

It is interesting to examine the energy balance in this problem, given a power input S_i at a reference plane in the rest frame, here taken to be at $z = 0$ as shown in Fig. 3.6-1. The time-average power per unit area carried by a plane wave polarized in the x-direction in a lossless medium is given by Eq. (3.6.4) and may be written in scalar form as

$$S = \tfrac{1}{2}\,\mathrm{Re}[E_x H_y{}^*] \tag{IV.8}$$

For our problem then we can write in the primed frame that

$$\begin{aligned} S_1{}' &= \frac{1}{2}\,\mathrm{Re}[E'_{x1} H'^*_{y1}] = \frac{E_0'^2}{2\eta} \\ S_2{}' &= \frac{1}{2}\,\mathrm{Re}[E'_{x2} H'^*_{y2}] = -\frac{E_0'^2}{2\eta} \end{aligned} \tag{IV.9}$$

where the sign of S indicates whether energy flow is in the positive or negative z'-direction. In this frame the power in each wave has an equal magnitude, the mirror is at rest and there is no mechanical work being done. In the laboratory frame on the other hand

$$\begin{aligned} S_1 &= \frac{1}{2}\,\mathrm{Re}[E_{x1} H^*_{y1}] = \frac{E_0'^2}{2\eta}\,\frac{1+\beta}{1-\beta} \\ S_2 &= \frac{1}{2}\,\mathrm{Re}[E_{x2} H^*_{y2}] = -\frac{E_0'^2}{2\eta}\,\frac{1-\beta}{1+\beta} \end{aligned} \tag{IV.10}$$

and we see that the power in the incident wave is greater than the power in the reflected wave. Also we see that the powers in both incident and reflected waves are different from their values in the primed frame. As we move through a wave traveling at the velocity of light, we therefore alter the apparent power carried by the wave, although we do not alter the apparent velocity with which energy is transported which remains equal to c. The physical explanation for where the additional power goes in a problem of this type depends on the boundary conditions. In the present case, it is instructive to examine the energy balance in the laboratory frame and to trace this out. In terms of S_1', the other powers are

$$S_2' = -S_1', \qquad S_1 = \frac{(1+\beta)}{(1-\beta)} S_1', \qquad S_2 = -\frac{(1-\beta)}{(1+\beta)} S_1' \qquad \text{(IV.11)}$$

The energy balance in the laboratory frame includes, beside the powers in incident and reflected waves, two other parts: (1) the power required to fill the region between the reference plane at $z = 0$ and the mirror at $z' = 0$ with field energy; (2) the power delivered to an external mechanical load which must be present and which is exerting a force F_s per unit area on the mirror. This external mechanical load is a very cleverly designed load that always exerts just the right force to keep the mirror moving at a constant velocity. We can calculate the force exerted on the mirror from the current flowing in the mirror and the magnetic field at the position of current flow. This is the radiation pressure and it must be exactly equal to the external force, if the mirror is to move at a constant velocity. We can also calculate this force from the energy balance, noting that the rate of delivery of energy to the external load S_m must be in this case

$$S_m = F_s v \qquad \text{(IV.12)}$$

where F_s is the force seen by an observer in the rest frame. Note that in the moving frame the mechanical power would be proportional to F_s', but is zero in this case because the velocity is zero in the moving frame.

The power S_e required to fill the region between the reference plane at $z = 0$ and the mirror at $z' = 0$ with field energy is

$$S_e = (S_1 - S_2)\beta \qquad \text{(IV.13)}$$

and the net input power S_i is

$$S_i = S_1 + S_2 \qquad \text{(IV.14)}$$

so, since there is no dissipation with a lossless mirror

$$S_i = S_e + S_m \qquad \text{(IV.15)}$$

and

$$S_m = F_s v = (S_1 + S_2) - (S_1 - S_2)\beta = 2\beta S_1' \qquad \text{(IV.16)}$$

so that

$$F_s = \frac{2S_1'}{c} = \left(\frac{1-\beta^2}{2\beta}\right)\frac{S_i}{c} = \left(\frac{1-\beta}{1+\beta}\right)\frac{2S_1}{c} \qquad \text{(IV.17)}$$

which, when β approaches zero, is seen to properly approach the value of radiation pressure on a mirror at rest, i.e., $2S_1/c$. In terms of a finite input power S_i, the value of F_s approaches infinity, because there is no net power transfer to a mirror at rest, even though there is a radiation pressure on it. Note also that for β nonzero the pressure is less for a given input power than it is for a mirror at rest, because the mirror is moving away from the source of radiation.

The radiation pressure can also be expressed in terms of the momentum of the wave. Evidently a wave must have momentum if it is to produce a force on a wall. The force F_s that we have computed above can be written in units of momentum as follows.

$$F_s = \text{momentum} \cdot \text{unit time}^{-1} \cdot \text{unit area}^{-1} \tag{IV.18}$$

If p_V is the momentum per unit volume

$$p_V = \text{momentum} \cdot \text{unit length}^{-1} \cdot \text{unit area}^{-1} \tag{IV.19}$$

and p_V is related to F_s through the relation

$$p_V = F_s(1/c) \tag{IV.20}$$

which for the problem discussed above is, from Eq. (IV.17)

$$p_V = 2S_1'/c^2 \tag{IV.21}$$

In this problem the force was produced by completely reversing the wave momentum in the primed frame, so the total momentum delivered to the mirror is just twice the momentum of the incident wave. For a plane wave the momentum per unit volume is thus simply

$$\mathbf{p}_V = \mathbf{S}/c^2 \tag{IV.22}$$

where $\mathbf{S}$ is the Poynting vector.

It is of interest to note that, for a given power, the wave momentum is inversely proportional to the square of the wave velocity. This fact suggests that a slow wave would have greater momentum per unit power than a TEM wave in free space.

In the discussion to this point we have pictured the mirror moving away from the source of radiation, i.e., we have taken v positive. In this case we picture the source of radiation as supplying power to an external load. If the radiation were at microwave frequencies we would have an elementary microwave motor. A generator is also possible, if the mirror moves in the negative z-direction, because then the external force necessary to keep the velocity constant acts in the same direction as the velocity, thus delivering power to the system which appears in the form of radiation. In this case the reflected wave will turn out to be larger than the incident wave and the difference in the powers in the two waves will be supplied by the external mechanical power source. This type of generator is a form of parametric amplifier and has been discussed in these terms by Pierce [21] and others [22–24].

PROBLEM 3.6-2 RADIATION PRESSURE MOTOR IN A DIELECTRIC MEDIUM (TYPE 1)

A perfectly conducting slab covers the x'–y' plane at $z' = 0$ and moves in the z-direction at velocity v. A plane wave traveling in the z-direction and polarized in the x-direction is completely reflected from this slab. The medium in which the wave travels is a lossless, uniform, isotropic dielectric with permittivity ε and permeability μ. Both the dielectric medium and the conducting slab are at rest in the primed frame moving in the positive z-direction at velocity v. A physical embodiment of this problem is a plasma column rapidly expanding into a vacuum with an electromagnetic wave being used to measure the velocity by bouncing it off the end of the expanding column. The frequency of the incident wave is ω', as seen from the primed frame. An external mechanical system maintains the slab at velocity v and can deliver or receive mechanical power.

Calculate in both frames:

1. The frequency of incident and reflected waves.

2. The reflection coefficient of the slab as a function of v.

3. The time-average value of Poynting's vector $\mathbf{S}$ for incident and reflected waves.

4. The energy stored per unit volume.

5. The force on the conducting slab from the power delivered to the slab, using the fact that $P = \mathbf{f} \cdot \mathbf{v}$ is the power delivered to the external mechanical system.

6. The momentum of the electromagnetic wave from the force exerted on the slab.

PROBLEM 3.6-3 RADIATION PRESSURE MOTOR IN A DIELECTRIC MEDIUM (TYPE 2)

Repeat problem 3.6-2 for the case of the dielectric at rest in the unprimed frame and the conducting slab at rest in the primed frame. One physical embodiment of this problem is a shock wave moving through an ionized gas at rest. Another embodiment is a moving interface between two regions of a crystal created by an acoustic wave in the crystal. A laser beam is bounced off the moving interface.

PROBLEM 3.6-4 PARTICLE PRESSURE MOTOR

A particle beam is incident on a perfectly absorbing wall at rest. Calculate the particle pressure on the wall and compare with the radiation pressure that would be produced by a light beam of the same incident power per unit area, for a particle beam having a number density of n particles per unit volume, mass per particle m, and velocity u. Assume that the wall is also a perfect absorber for light at the frequency of the incident light beam used as a basis for comparison.

REFERENCES

1. J. V. Jelley, "Cerenkov Radiation." Pergamon Press, Oxford, 1958.
2. P. Penfield, Jr. and H. A. Haus, "The Electrodynamics of Moving Media." MIT Press, Cambridge, Massachusetts, 1967.
3. H. Minkowski, *Gott. Nachr.*, p. 53 (1908).
4. H. A. Rowland, *Ber. d. Berlin Acad.*, p. 211 (1876).
5. W. C. Röntgen, *Ann. Phys.* (*Leipzig*) **35**, 264 (1888).
6. A. Eichenwald, *Ann. Phys.* (*Leipzig*) **11**, 1, 241 (1903).
7. H. A. Wilson, *Phil. Trans. Roy. Soc. London* **204**, 121 (1904). [Also H. A. Wilson and M. Wilson, *Proc. Roy. Soc., Ser. A* **89**, 99 (1913).]
8. L. P. Williams, "Michael Faraday, A Biography." Basic Books, New York, 1965.
9. H. E. Ives and G. R. Stilwell, An experimental study of the rate of a moving atomic clock, *J. Opt. Soc. Am.* **28**, 215 (1938).
10. A. Sommerfeld, "Electrodynamics." Academic Press, New York, 1964.
11. J. B. Marion, "Classical Electromagnetic Radiation." Academic Press, New York, 1965.
12. R. S. Elliott, "Electromagnetics." McGraw-Hill, New York, 1966.
13. W. F. Hughes and F. J. Young, "The Electromagnetodynamics of Fluids." Wiley, New York, 1966.
14. W. K. H. Panofsky and M. Phillips, "Classical Electricity and Magnetism," 2nd ed. Addison-Wesley, Reading, Massachusetts, 1962.

15. A. A. Michelson and E. W. Morley, *Amer. J. Sci.* **34**, 333 (1887); [*Phil. Mag.* **24**, 449 (1887)].
16. C. W. Carnahan, A proposed first-order relativity test using lasers, *Proc. IRE* **50**, 1976 (1962). [Also J. Shamir and R. Fox, *Proc. IEEE* **53**, 1155–1156, 2141–2142 (1965).]
17. D. A. Dunn, R. W. Wallace, and S. D. Choi, Waves in a moving medium with finite conductivity, *Proc. IEEE* **57**, 45–57 (1969).
18. J. A. Stratton, "Electromagnetic Theory." McGraw-Hill, New York, 1941.
19. L. Page and N. I. Adams, "Electrodynamics." Van Nostrand-Reinhold, Princeton, New Jersey, 1940.
20. R. C. Costen and D. Adamson, Three dimensional derivation of the electromagnetic jump conditions and momentum-energy laws at a moving boundary, *Proc. IEEE* **53**, 1181–1196 (1965). [Also F. J. Young, Electrodynamic boundary conditions at a moving boundary, *ibid.* **54**, 399–401 (1966).]
21. J. R. Pierce, Use of the principles of conservation of energy and momentum in connection with the operation of wave-type parametric amplifiers, *J. Appl. Phys.* **30**, 1341–1346 (1959). [Also Momentum and energy of waves, *ibid.* **32**, 3580–3584 (1961).]
22. P. A. Sturrock, In what sense do slow waves carry negative energy. *J. Appl. Phys.* **31**, 2052–2056 (1960).
23. R. J. Briggs, Transformation of small-signal energy and momentum of waves, *J. Appl. Phys.* **35**, 3268–3272 (1964).
24. T. Musha, Unified consideration of wave amplification in moving media, *J. Appl. Phys.* **35**, 3273–3279 (1964).

CHAPTER 4

Forces and Wave Interaction with Uniformly Moving Circuits and Continua

Another concept of great importance is that of force. In origin the concept doubtless arises from the muscular sensations of resistance experienced from external bodies. This crude concept may at once be put on a quantitative basis by substituting a spring balance for our muscles, or instead of the spring balance we may use any elastic body, and measure the force exerted by it in terms of its deformation. The concept of force so defined is limited to static systems; it is the task of statics to find the relation between the forces in systems at rest. We next extend the force concept to systems not in equilibrium, in which there are accelerations, and we must conceive that at first all our experiments are made in an isolated laboratory far out in empty space, where there is no gravitational field. We here encounter a new concept, that of mass, which as it is originally met is entangled with the force concept, but may later be disentangled by a process of successive approximations.

P. W. Bridgman, "The Logic of Modern Physics." MacMillan, New York, 1927. Reprinted by permission.

4.0 Introduction

In this chapter we study the interaction of electric and magnetic fields with moving circuits and continua. We restrict our study in this chapter to perfectly elastic solid media or to assemblages of conductors or current elements that are interconnected by elastic solid interconnections. In this chapter we further restrict the problem by the rather drastic assumption of constant velocity motion. We imagine an external "force balancer" to be present which is able to produce a time-varying force that is always exactly equal and opposite to the electromagnetic force, so that no acceleration takes place, either locally or of the entire medium. Thus the special theory of relativity applies exactly to this class of problems and a rigorously correct solution can be obtained. The solution will include an energy balance that tells us whether the external force balancer is absorbing energy or delivering energy to the system and thus whether we have a motor, a generator, or a brake. Because the form of the motion is so simple in the problems discussed here, we can deal rigorously with other complexities that arise. In the following chapters we will examine some of the problems associated with accelerated motion in a less rigorous and self-consistent fashion.

Here we deal with the two basic types of force that are expressed by the Lorentz force equation, i.e., the $q\mathbf{E}$ force and the $\mathbf{J} \times \mathbf{B}$ force. Other types of electrical force can be reduced to these two types, but it is often convenient to express the force on a material medium in other forms. For example, if there is a dielectric medium present with a gradient of permittivity in a region where there is a nonzero electric field there will be an induced polarization charge created wherever there is a gradient of permittivity. The force on this polarization charge is of course a result of q times $\mathbf{E}$, but it is often written as a separate term in the force equation. The same remark applies to forces due to electro- or magnetostriction, where the value of permittivity or permeability is a function of the density of the material. The formulation of the equations for these forces and their transformation for moving systems can be a subtle and complicated business [1, 2].

We will specifically limit our consideration of force generation to media in which the permeability and permittivity are not only constants but have the values of free space, i.e., $\mu = \mu_0$ and $\varepsilon = \varepsilon_0$. Under these conditions there is no doubt of the correctness of using the Lorentz force equation. However, in material media with other values of μ and ε, even at rest, the correct form of the force equation is a matter of controversy. Penfield and Haus [1] have recently summarized and commented on many of the issues in this field, but the problem is by no means resolved. For the case of media with the permittivity and permeability of free space, the force per unit volume $\mathbf{f}$ is

$$\mathbf{f} = \varrho\mathbf{E} + \mathbf{J} \times \mathbf{B} \tag{4.0.1}$$

It is possible to derive an equation for the conservation of momentum of electromagnetic fields that is analogous to Poynting's theorem, which is the statement of the conservation of energy. We will not make extensive use of this relation, but certain features of it are of general interest. The derivation of the conservation of the momentum equation for free space parallels the derivation of Poynting's theorem. If we take the cross product of $\varepsilon_0\mathbf{E}$ with the third Maxwell equation and the cross product of $\mu_0\mathbf{H}$ with the fourth equation and add, we get

$$\varepsilon_0(\nabla \times \mathbf{E}) \times \mathbf{E} + \mu_0(\nabla \times \mathbf{H}) \times \mathbf{H} = \mathbf{J} \times \mathbf{B} + \mu_0\varepsilon_0 \frac{\partial}{\partial t}(\mathbf{E} \times \mathbf{H}) \quad (4.0.2)$$

The lhs of Eq. (4.0.2) can be written in the form of the divergence of a tensor plus $\varrho\mathbf{E}$, yielding

$$\nabla \cdot \mathsf{T} = \varrho\mathbf{E} + \mathbf{J} \times \mathbf{B} + \mu_0\varepsilon_0 \frac{\partial}{\partial t}(\mathbf{E} \times \mathbf{H}) \quad (4.0.3)$$

where T is a tensor called the Maxwell stress tensor [2]. It has elements that are products of two field components such as E_xE_y. In purely mechanical systems the conservation of momentum equation can be written in this form also.

$$\nabla \cdot \mathsf{T}_{\mathrm{m}} = \mathbf{f}_{\mathrm{m}} + (\partial\mathbf{g}_{\mathrm{m}}/\partial t) \quad (4.0.4)$$

where T_{m} is a stress tensor, $\mathbf{f}_{\mathrm{m}}$ is a mechanical force, and $\mathbf{g}_{\mathrm{m}}$ is a mechanical momentum. It is the analogy between Eqs. (4.0.3) and (4.0.4) that allows us to identify the third term on the rhs of Eq. (4.0.3) as the time derivative of a momentum. There is thus an inertia property of electromagnetic fields that is similar to that of ordinary matter. Momentum can be transferred across empty space by an electromagnetic wave. As we have seen in the example at the end of the preceding chapter, the force acting on a plane mirror due to a TEM wave incident on it can be explained in terms of this wave momentum. Note that this momentum property is not fully equivalent to an additional volume force like $q\mathbf{E}$ or $\mathbf{J} \times \mathbf{B}$, because the momentum force is only present when there is a change in the momentum. On a time-average basis in a lossless medium in which a wave with $\mathbf{E}$ and $\mathbf{H}$ in time phase is propagating the momentum force is zero.

The integral over a volume of the stress tensor leads to an evaluation of the stresses transmitted across the surface surrounding the volume. It was a natural extension of this fact to attach physical reality to the medium that transmitted these stresses and to think of an ether as being required to fill this role.

Minkowski [3] in 1908 noticed that, if one is dealing with a medium other than free space characterized by scalar constants μ and ε, the derivation of Eq. (4.0.3) is identical except for the fact that μ_0 and ε_0 are replaced by μ and ε in the third term on the rhs. This discovery has been followed by sixty years of controversy about the proper value of this term in material media and almost no experiments [1, 4, 5]. We shall proceed in this chapter, first with a series of problems in simple lossy moving media with free space values of μ and ε, using Eq. (4.0.1) for the force. We will deal only with time-average forces and powers and so will not encounter the interesting problems associated with the momentum term of Eq. (4.0.3). Shockley and James [6] and Feynman [7] have discussed some of these momentum problems in an interesting way.

Perhaps the most important concept introduced in this chapter is that of self-consistency. We may wish to calculate the force on a moving current element in a magnetic field. There are two approaches to this problem. If we know the current and the magnetic field, we can solve for the force directly. However, if the current depends on the magnetic field and the magnetic field depends on the current, we must seek self-consistent values of field and current. As long as motion is uniform, self-consistency can be obtained rather easily. When there is accelerated motion, the problem of self-consistency becomes acute, as we will see in later chapters.

4.1 Self-Consistent Forces and Emfs in Wave Interactions with Moving Conducting Media—the TEM Mode

In this section we again use the technique of calculating the fields in a frame of reference moving with the medium and then use the Lorentz transformation to compute the fields in another frame [8]. In this section we apply this technique to the problem of a TEM wave propagating through a conducting medium characterized by conductivity σ, permeability μ, and permittivity ε, as shown in Fig. 4.1-1. As usual, we take the primed frame to be moving at velocity v with the medium in the positive z-direction. A TEM wave polarized in the x'-direction propagates in the positive z'-direction and has a wave number $\mathbf{k}' = k'\mathbf{1}_z$ and frequency ω'. The values of the fields are E_x' and H_y' and there is a current J_x' flowing in this frame, given by

$$J_x' = \sigma E_x' \tag{4.1.1}$$

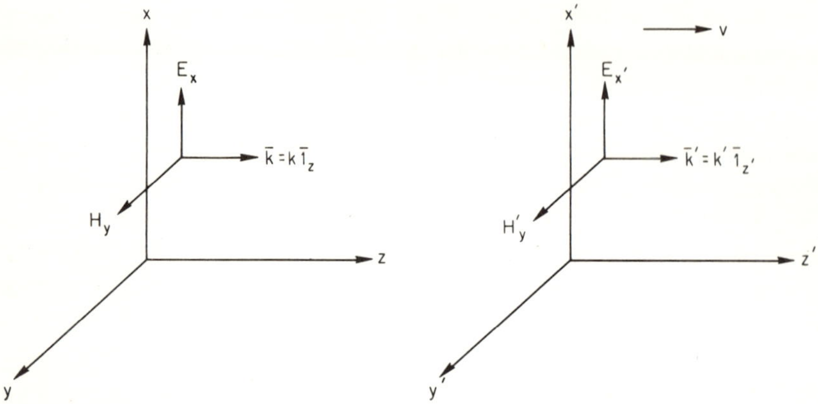

Fig. 4.1-1. Two coordinate systems. A TEM wave propagates in the positive z-direction. The primed system is moving to the right with constant velocity v relative to the unprimed system. A uniform medium characterized by μ, ε, and σ fills the space and is at rest with respect to the primed coordinate system. (A character with an overbar in a figure is equivalent to a boldface character in the text.) [From D. A. Dunn, R. W. Wallace, and S. D. Choi, Waves in a moving medium with finite conductivity, *Proc. IEEE* **57**, 45–57 (1969).]

The medium is assumed to be linear, homogeneous, isotropic, and charge free as viewed from this frame so the constitutive equations are

$$D_x' = \varepsilon E_x', \qquad B_y' = \mu H_y' \tag{4.1.2}$$

In the laboratory frame we can write the fields in the form

$$\begin{aligned} E_x &= E_0 \exp[j(\omega t - kz)], \qquad & D_x &= D_0 \exp[j(\omega t - kz)] \\ H_y &= H_0 \exp[j(\omega t - kz)], \qquad & B_y &= B_0 \exp[j(\omega t - kz)] \end{aligned} \tag{4.1.3}$$

thus representing a TEM wave with frequency ω and wave number $\mathbf{k} = k\mathbf{1}_z$. In the laboratory frame the constitutive equations are unknown and we will derive them by transforming the fields of Eq. (4.1.3) to the primed frame using Eqs. (3.1.11) and (3.3.5), obtaining thereby

$$\begin{aligned} E_x' &= \frac{\omega'}{\gamma(\omega' + k'v)} E_0 \exp[j(\omega' t' - k'z')], \qquad & D_x' &= \varepsilon E_x' \\ H_y' &= \frac{k' E_0}{\gamma\mu(\omega' + k'v)} \exp[j(\omega' t' - k'z')], \qquad & B_y' &= \mu H_y' \end{aligned} \tag{4.1.4}$$

in which we have made use of Eq. (4.1.2) to express all of the fields in terms of a single arbitrary constant, E_0. Now to evaluate the other constants in

Eq. (4.1.3) we can simply transform Eqs. (4.1.4) back to the laboratory frame, making use of the transformations of frequency and wave number in Eq. (3.5.15) which for our case can be written as

$$k' = \gamma[k - (\beta/c)\omega], \qquad \omega' = \gamma(\omega - \beta ck) \tag{4.1.5}$$

and substituting Eqs. (4.1.5) in (4.1.4) and again making use of Eqs. (3.1.11) and (3.3.5) we can express the fields in the laboratory frame in terms of E_0 alone.

$$\begin{aligned}
E_x &= E_0 \exp[j(\omega t - kz)] \\
D_x &= \gamma^2 \varepsilon E_0 \{[1 - (\beta^2/n^2)] - (kv/\omega)[1 - (1/n^2)]\} \exp[j(\omega t - kz)] \\
H_y &= (\gamma^2 E_0/\omega\mu)[k(1 - \beta^2 n^2) - (\omega\beta/c)(1 - n^2)] \exp[j(\omega t - kz)] \\
B_y &= (k/\omega) E_0 \exp[j(\omega t - kz)]
\end{aligned} \tag{4.1.6}$$

where $n^2 = c^2\mu\varepsilon$.

The next step in the analysis requires that we understand the relations between ω and k and ω' and k'. The dispersion relation in the primed frame for this problem is derived in most textbooks as a part of the standard treatment of waves in a conducting medium at rest. It is

$$k'^2 = \omega'^2 \mu\varepsilon - j\omega'\mu\sigma \tag{4.1.7}$$

The dispersion relation in the unprimed frame can be obtained directly from Eqs. (4.1.7) and (4.1.5).

$$\begin{aligned}
k^2(1 - \beta^2 n^2) = (\omega^2/c^2)(n^2 - \beta^2) &+ 2\omega k(\beta/c)(1 - n^2) \\
&+ j(\mu\sigma/\gamma)(kv - \omega)
\end{aligned} \tag{4.1.8}$$

Now we have to be a little more specific about what the problem is that we are solving. Let us return to Fig. 4.1-1 and imagine a source of power at $z = 0$ that has a constant amplitude, so that in the laboratory frame it is appropriate to regard ω as real. Without regard to whether the medium is in motion or not, the wave number in the laboratory frame will be complex, and there will be two values of k corresponding to the two waves that can exist in the medium. If the medium is at rest, these waves always consist of one forward and one backward wave, but as we saw for the case of a moving lossless medium this is not true for a medium moving faster than the velocity of light in the medium, i.e., for $\beta n > 1$. This is also the case here, where we have the added element that the waves decay away from the source. Here this results in a complex value of ω' if ω is real. This is easy to understand, if we picture a wave decaying in the z-direction and imagine how this wave looks to an observer moving in the same direction as the wave. Since the wave amplitude decreases in space in the laboratory frame, it must decay in time when observed from the primed frame, hence ω' must be complex.

Solving Eq. (4.1.8) for k with ω real results in

$$k = \frac{\omega(\mu\varepsilon)^{1/2}}{1-\beta^2 n^2}\left\{\frac{\beta}{n}(1-n^2) + \frac{j\beta n}{2\gamma}\frac{\sigma}{\omega\varepsilon} \pm \frac{1}{\gamma}\left[\frac{1}{\gamma^2} - \left(\frac{\beta n}{2}\frac{\sigma}{\omega\varepsilon}\right)^2 - \frac{j}{\gamma}\frac{\sigma}{\omega\varepsilon}\right]^{1/2}\right\} \tag{4.1.9}$$

for $\beta n \neq 1$ and

$$k = \frac{\omega(\mu\varepsilon)^{1/2}}{(\gamma\sigma/\omega\varepsilon)^2 + 4}\left\{2(1+\beta^2) + \left(\gamma\frac{\sigma}{\omega\varepsilon}\right)^2 - \frac{j}{\gamma}\frac{\sigma}{\omega\varepsilon}\right\} \tag{4.1.10}$$

for $\beta n = 1$. The two values of k given in Eq. (4.1.9) correspond to one positively- and one negatively traveling wave for $\beta n < 1$ and to two positively traveling waves for $\beta n > 1$. Before looking at the dispersion diagram that results from this equation it is convenient to introduce normalized frequency and wave number variables

$$\Omega = \omega\varepsilon/\sigma, \qquad \varkappa = (k/\sigma)(\varepsilon/\mu)^{1/2} \tag{4.1.11}$$

In terms of these variables Eq. (4.1.9) becomes

$$\varkappa = \frac{1}{(1-\beta^2 n^2)}\left\{\beta n\left[\left(\frac{1-n^2}{n^2}\right)\Omega + \frac{j}{2\gamma}\right] \pm \frac{1}{\gamma^2}\left[\Omega^2 - \left(\frac{\beta n}{2}\gamma\right)^2 - j\gamma\Omega\right]^{1/2}\right\} \tag{4.1.12}$$

Equation (4.1.12) has been plotted in Fig. 4.1-2 for $n^2 = 1$, which corresponds to a lossy medium with no wave slowing due to μ or ε. $\varkappa_1$ designates the solution with the plus sign in front of the radical and $\varkappa_2$ the minus sign. In this case, the real parts of $\varkappa_1$ and $\varkappa_2$ are symmetrical about the Ω axis and give the same magnitude for the phase velocities of the two waves. As β is increased, the positive z-traveling wave has a decreasing imaginary part and the negative z-traveling wave has an increasing imaginary part. It is also of interest to note the effect of increasing β on the shape of the real part $\varkappa_r$ of $\varkappa(\Omega)$. For $\beta = 0$ there is a region of very low wave velocity at small Ω. The slope of a radius vector to a point on one of these curves is equal to v_p/c, i.e., $\Omega/\varkappa_r = (\omega/k_r)/c$, and $v_p = (\omega/k_r)$ is the phase velocity. As β increases to 0.5 in Fig. 4.1-2, it is seen that the minimum slope of the $\varkappa_r(\Omega)$ curve is increased to 0.5. This is also true for smaller β, i.e., the minimum value of v_p/c is equal to β. It is even clearer as β increases to 0.9 in Fig. 4.1-2. Synchronism of wave and medium is thus not quite attainable in this system; it is impossible to move the medium faster than the wave.

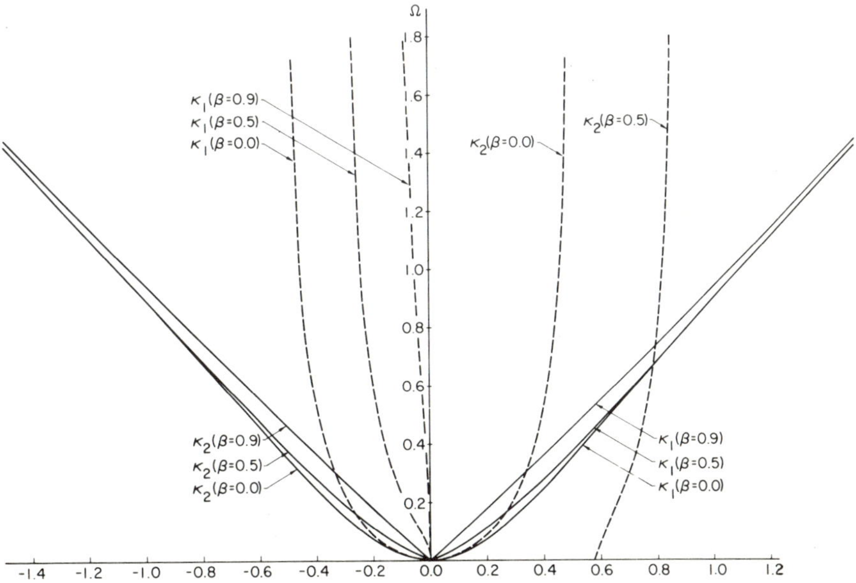

Fig. 4.1-2. Plot of Ω vs. $\varkappa$ for a uniform plane wave propagating in a moving conducting medium with $n^2 = 1$ for several values of β, where $\Omega = \omega\varepsilon/\sigma$ and $\varkappa = [k(\varepsilon/\mu)^{1/2}]/\sigma$. The real part of $\varkappa$ is given by the solid lines and the imaginary part is shown dashed. For $n^2 = 1$ the real parts are symmetrical but the imaginary parts are not [from D. A. Dunn, R. W. Wallace, and S. D. Choi, Waves in a moving medium with finite conductivity, *Proc. IEEE* **57**, 45–57 (1969)].

Figure 4.1-3 is a plot of Eq. (4.1.12) for $n^2 = 10$. In this case there are two waves with positive phase velocities when $\beta n > 1$. The imaginary parts are of opposite sign from the real parts for all β, however, so that both solutions imply waves that decay in the direction of propagation. As in the case of zero loss, the wave impedance remains negative for the wave that originally had negative $\varkappa$, even though the sign of $\varkappa$ becomes positive. To excite this wave it is thus necessary to maintain the phase of E_x and H_y in the relation appropriate to a negatively-traveling wave. Although in this case the material velocity can exceed the wave velocity of one of the waves, once $\beta n > 1$, in order to reach this condition in a physical system with typical values of μ and ε, it would be necessary to have a material velocity much greater than is achievable with ordinary techniques. Put a different

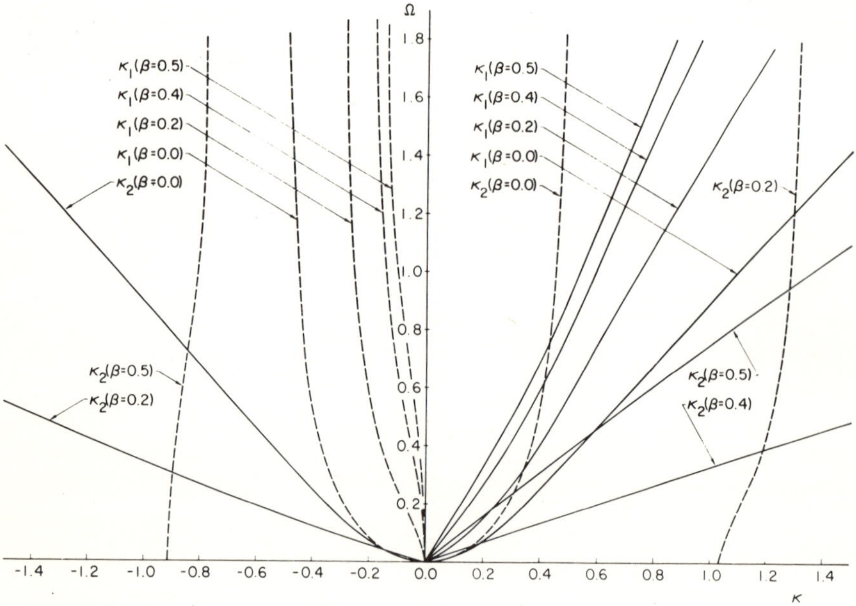

Fig. 4.1-3. Plot of Ω vs. $\varkappa$ for a uniform plane wave propagating in a moving conducting medium with $n^2 = 10$ for several values of β. The real part of $\varkappa$ is shown solid and the imaginary part is shown dashed. For $\beta n > 1$ the sign of $\varkappa_2$ becomes positive, corresponding to a positively-traveling wave with negative impedance [from D. A. Dunn, R. W. Wallace, and S. D. Choi, Waves in a moving medium with finite conductivity, *Proc. IEEE* **57**, 45–57 (1969)].

way, this is the hard way to build a slow-wave circuit, if a significant amount of wave slowing is desired. As will be seen, this condition, $\beta n > 1$, is a necessary condition for operation as a generator and can thus be significant.

In the above solutions β has only been allowed to take on values in the range $0 \leq \beta < 1$. It is convenient in later developments to extend the range of β to $-1 < \beta < 1$. If this is done, only $\varkappa_1$ solutions need to be considered, because $\varkappa_2(\beta = -|\beta|) = -\varkappa_1(\beta = |\beta|)$.

We next examine the currents and forces in the medium and the energy balance equations. Using the transformations of Eq. (3.1.16) and Eq.

(4.1.6) we can write for the charge and current density in the rest frame

$$\begin{aligned} J_x &= J_x' = \sigma E_x' \\ &= \sigma\gamma(E_x - \beta c B_y) \\ &= (\sigma\gamma E_0/\omega)(\omega - kv)\exp[j(\omega t - kz)] \\ \varrho &= \varrho' = 0 \end{aligned} \tag{4.1.13}$$

Thus, in the laboratory frame the force is due to J_x and B_y only, and on a time-average basis the force per unit volume $\mathbf{f}$ is given by

$$\begin{aligned} \mathbf{f} &= \tfrac{1}{2}\,\mathrm{Re}\{J_x B_y{}^* \mathbf{1}_z\} \\ &= (\gamma\sigma E_0{}^2/2\omega^2)(\omega k_\mathrm{r} - |\,k\,|^2 v)\exp(2k_\mathrm{i} z)\mathbf{1}_z \end{aligned} \tag{4.1.14}$$

where k_r and k_i are the real and imaginary parts of k, i.e.

$$k = k_\mathrm{r} + jk_\mathrm{i} \tag{4.1.15}$$

and where we have restricted the permeability and permittivity to their free space values, i.e., $\mu = \mu_0$, $\varepsilon = \varepsilon_0$, and $n = 1$, in order to use Eq. (4.0.1).

The energy balance in this type of system can be understood in terms of three quantities: P_f, the power delivered to an external mechanical load that maintains the medium at a constant velocity; P_l, the power dissipated in losses in the medium; and P_e, the total electrical power input, all on a per unit volume basis. The energy-balance equation is, from conservation of energy

$$P_\mathrm{e} = P_\mathrm{f} + P_l \tag{4.1.16}$$

We can define a motor, generator, or brake in terms of the signs of the three quantities in Eq. (4.1.16). When the equation is written in the form shown, a motor exists when the electrical power is the input ($P_\mathrm{e} > 0$) and the mechanical power is the output ($P_\mathrm{f} > 0$); a generator will exist when the electrical power is the output ($P_\mathrm{e} < 0$) and the mechanical power is the input ($P_\mathrm{f} < 0$); and a brake will exist when both the electrical power and the mechanical power are inputs ($P_\mathrm{e} > 0$, $P_\mathrm{f} < 0$). Of course, the power dissipated in losses must be nonnegative ($P_l \geq 0$) in all three cases. Note that we are not seeking a transition in the dispersion relation from a decaying wave to a growing wave, when we look for the transition to the generator condition. All that is necessary to obtain a generator is an appropriate change in the sign of the terms in the energy balance of Eq. (4.1.16). However, some care is required in the interpretation of what constitutes the electrical output and input, if any, in the generator case. It is helpful in this regard to note that in the generator case both waves have phase velocities in the direction of the material velocity and that the wave of interest which carries negative energy has a phase velocity less than the velocity of the material in order for the material to give energy to the wave. This is also true in the traveling-wave amplifier, as we will see in Chapter 7. One way of visualizing a generator of this type is shown in Fig. 4.1-4. P_1 and P_2

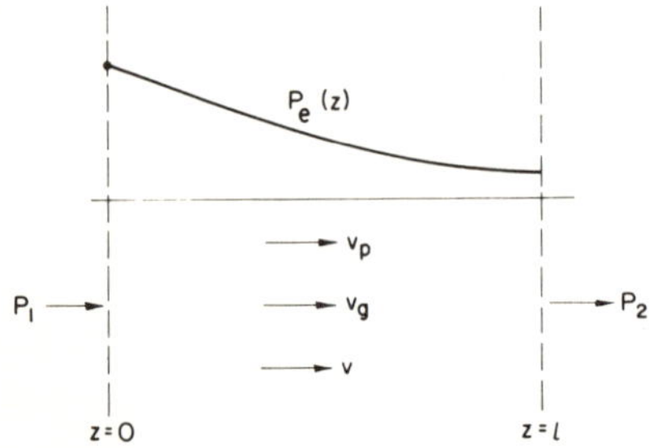

Fig. 4.1-4. Schematic drawing of a TEM mode system of length l propagating a positively-traveling wave, as viewed from the laboratory frame, for $\beta > 0$. P_1 is the electric power crossing the plane $z = 0$ and P_2 is the electric power crossing the plane $z = l$. The electric power per unit volume at any value of z is given by P_e and for this mode P_e decreases exponentially as z increases.

are the z-components of the Poynting vector, $\mathbf{P} = \mathbf{E} \times \mathbf{H}$, at planes $z = 0$ and $z = l$, and P_e is the power per unit volume, so it is related to P_1 and P_2 through the equation $P_2 - P_1 = \int_0^l P_e\, dz$. Now, if we regard the planes $z = 0$ and $z = l$ as accessible, we can regard P_1 and P_2 as electrical inputs or outputs. Of considerable interest is the analogy with the slow space-charge wave in a one-dimensional theory of electron beams which also is a negative energy wave, discussed in Chapter 7. In the system here there is the nice feature that one can see directly what field configuration is necessary to excite the wave of interest, because a complete field theory has been carried out. It is important to note here that, although positive power is supplied to the system at $z = l$ and positive power in a larger amount is removed at $z = 0$, the fields at $z = 0$ excite the wave that travels in the positive z-direction, i.e., causality and group velocity are from left to right in Fig. 4.1-4. Thus it is perhaps a matter of taste whether one refers to the wave in this case as a growing or a decaying wave. It decays in the direction of positive group velocity, but grows in the direction of positive power flow from input terminals to output terminals, because it carries negative energy.

The total electrical input per unit volume can be found from the rate at which the fields do work and on a time-average basis is, from Poynting's theorem,

$$\begin{aligned} P_e &= \tfrac{1}{2}\,\mathrm{Re}\{\mathbf{J} \cdot \mathbf{E}^*\} \\ &= (\gamma\sigma E_0^2/2\omega)(\omega - k_r v)\exp(2k_i z) \\ &= (\gamma/\Omega)(\Omega - \beta n \varkappa_r)(\sigma E_0^2/2)\exp(2k_i z) \end{aligned} \qquad (4.1.17)$$

The mechanical power delivered by the force **f** to an external load is

$$P_{\mathrm{f}} \stackrel{\Delta}{=} \mathbf{f} \cdot \mathbf{v}$$
$$= (\gamma\beta n/\Omega^2)[\Omega\varkappa_{\mathrm{r}} - \beta n \mid \varkappa \mid^2](\sigma E_0{}^2/2) \exp(2k_{\mathrm{i}}z) \qquad (4.1.18)$$

where we have written this equation in terms of n, even though the force **f** is only valid for $n = 1$. This procedure will be followed throughout this and the next two sections.

The power dissipated in losses must then be

$$P_l = P_{\mathrm{e}} - P_{\mathrm{f}}$$
$$= (\gamma/\Omega^2) \mid \Omega - \beta n\varkappa \mid^2 (\sigma E_0{}^2/2) \exp(2k_{\mathrm{i}}z) \qquad (4.1.19)$$

It is also interesting to derive the value of P_l in a more general form using the expression for the fields of Eqs. (3.1.11) and (3.3.5). We find that

$$P_l = \mathbf{J} \cdot \mathbf{E} - \mathbf{f} \cdot \mathbf{v} = (1/\gamma)(\mathbf{J}' \cdot \mathbf{E}') \qquad (4.1.20)$$

Hence, the losses are equal to the rate at which the fields do work in the moving frame. Because of the Lorentz contraction there is an apparent change in the unit volume when viewed from another frame, thus the appearance of the γ^{-1} factor in Eq. (4.1.20).

It is convenient to define a slip parameter by the relation

$$s = 1 - (v/v_{\mathrm{p}})$$
$$= (\Omega - \beta n\varkappa_{\mathrm{r}})/\Omega \qquad (4.1.21)$$

where v_{p} is the phase velocity of the wave ($v_{\mathrm{p}} = \omega/k_{\mathrm{r}}$). Note that v_{p} is a function of v and Ω determined through the dispersion relation. Arbitrary values of v and v_{p} and hence of slip s may or may not be allowable.

The sign of P_{e} is completely determined by the sign of s. Considering only $\varkappa_1$ solutions but allowing the range of β to be $-1 < \beta < 1$ the sign of P_{e} (or s) can be determined by the following argument: (a) For $0 < \beta < 1$, both v and v_{p} are positive with $v_{\mathrm{p}} > v$; hence $P_{\mathrm{e}} > 0$. (b) For $-(1/n) < \beta < 0$, $v < 0$ but $v_{\mathrm{p}} > 0$ approaching zero as β approaches $-(1/n)$; hence $P_{\mathrm{e}} > 0$. (c) For $-1 < \beta < -(1/n)$, both v and v_{p} are negative with $v_{\mathrm{p}} < v$; hence $P_{\mathrm{e}} < 0$.

A similar but more involved argument indicates that $P_{\mathrm{f}} > 0$ for $0 < \beta < 1$ and $P_{\mathrm{f}} < 0$ for $-1 < \beta < 0$. Therefore, the model does contain all three regions of operation associated with an ordinary induction machine. The divisions are: (a) motor, $0 < \beta < 1$; (b) brake, $-(1/n) < \beta < 0$; (c) generator, $-1 < \beta < -(1/n)$. However, since our force equation is limited to $n = 1$, it is evident that the generator region is not accessible.

It is natural to define an efficiency for these three regions as follows:

$$\text{motor:} \qquad \eta_{\mathrm{m}} \triangleq \frac{\text{mechanical power out}}{\text{electrical power in}} = \frac{P_{\mathrm{f}}}{P_{\mathrm{e}}} \tag{4.1.22}$$

$$\text{brake:} \qquad \eta_{\mathrm{b}} \triangleq \frac{\text{mechanical power in}}{\text{mechanical power in} + \text{electrical power in}} = \frac{|P_{\mathrm{f}}|}{|P_{\mathrm{f}}| + P_{\mathrm{e}}} \tag{4.1.23}$$

$$\text{generator:} \qquad \eta_{\mathrm{g}} \triangleq \frac{\text{electrical power out}}{\text{mechanical power in}} = \frac{P_{\mathrm{e}}}{P_{\mathrm{f}}} \tag{4.1.24}$$

The brake efficiency of Eq. (4.1.23) tells us how much electrical power input is required, in addition to the mechanical power input, to absorb the mechanical power at constant speed.

Figure 4.1-5 is a plot of the above defined efficiencies for $n^2 = 1$ and $\Omega = 0.1$ and 10. Note that reasonably high efficiencies are possible in all regions, but only for values of β that would be difficult to realize physically.

The analysis that we have been through here can be viewed as a form of radiation-pressure calculation. For the medium at rest, our results reduce

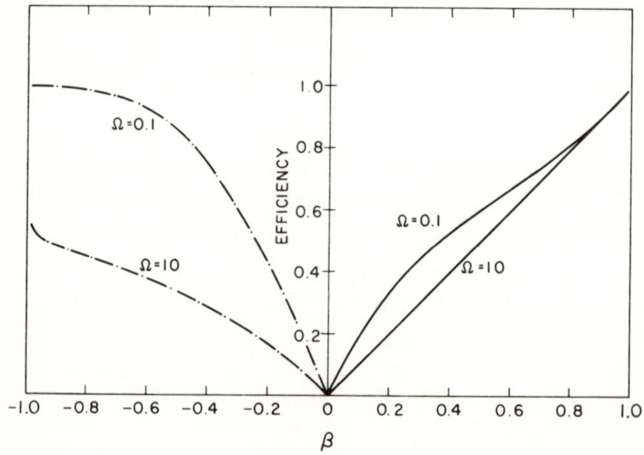

Fig. 4.1-5. Efficiency of the TEM mode vs. β for $n^2 = 1$, and for two values of Ω. The system behaves as a motor for $0 < \beta < 1$ and as a brake for $-1 < \beta < 0$ [from D. A. Dunn, R. W. Wallace, and S. D. Choi, Waves in a moving medium with finite conductivity, *Proc. IEEE* **57**, 45–57 (1969)].

to the usual result, if the power that enters our system at $z = 0$ is interpreted as the power transmitted into a medium that has a boundary at $z = 0$. Thus a wave incident from free space on to a conducting medium extending from $z = 0$ to $z = \infty$ would have some fraction of the incident power reflected and some fraction transmitted. The part that is transmitted would travel into the medium, produce a current, a magnetic field, and a corresponding force or radiation pressure that can be calculated using the equations derived here.

For $\beta = 0$, Eqs. (4.1.14) and (4.1.17) can be combined to give the equation relating force and electrical input power for a "locked rotor"

$$\mathbf{f} = (P_e/v_p)\mathbf{1}_z \tag{4.1.25}$$

where v_p is the phase velocity of the wave ($v_p = \omega/k_r$) inside the medium. It is well known that v_p can be much less than c for a conducting medium with small Ω, and this form of the equation makes it clear that in this region of low phase velocity, radiation pressure can be a large effect, in terms of the amount of force for a given power.

It has also been seen that, if we formulate the energy balance with $\beta \neq 0$, we find regions of operation for this plane-wave system that correspond to the usual regions of operation for an induction machine. In drawing this analogy we pictured the input fixed at $z = 0$ and allowed the wave to propagate into the moving medium extending from $z = 0$ to the right. The wave velocity in the moving medium was determined by the values of conductivity, frequency, and material velocity, and in this respect the situation was somewhat different from that in an ordinary induction machine in that the velocity was more restricted in the range of allowable values than is ordinarily the case. In particular, we were precluded from increasing the material velocity to a value higher than the wave velocity, except by raising the material velocity to a velocity greater than the velocity of light in the medium in the absence of loss.

In the next two sections we consider models of wave propagating systems that are closer to the usual machine situation and in which we will be able to more freely select wave and material velocities. We will see, however, that while the field patterns are different, the basic force mechanism and the form of the energy balance is not significantly altered from what is obtained here. Thus the relation between radiation pressure, ordinarily considered to be a very small effect, and force generation in an induction motor, known to be a very large effect, is very close. The significant difference lies in the ability to operate near synchronism in the ordinary machine at physically easily realizable material velocities. This is possible since the wave can be slowed down by the use of slow-wave circuits rather than by being restricted to the inherent wave slowing in a material medium with values of permittivity and permeability greater than those of free space, but typically not very much greater.

Problem 4.1-1 Radiation Pressure on a Slab at Rest

A plane wave is incident from free space on to a semiinfinite conducting region beginning at $z = 0$ and extending to $z = \infty$. The medium is characterized by σ, μ, ε and is uniform and isotropic. Calculate the time-average radiation pressure on this medium. Find the value of pressure for very high conductivity and compare with the result obtainable from an energy balance as in Problem 3.5-1 for a perfect conductor. Compare also with the value obtainable from the surface current and magnetic field at the surface of a perfect conductor.

Problem 4.1-2 Model of a TEM Mode Motor

Suggest a physically realizable device that would probably behave in the way that the one-dimensional model of this section behaves. In what ranges of frequency, velocity, etc. would your device be likely to behave differently from this model?

4.2 A Model of an Induction Motor—TE Mode in a Moving Conducting Medium

In an ordinary induction motor the field windings are distributed around the circumference of a circular stator. These field windings have connections brought out at a number of points around the circumference and all of these points can be fed with the same amplitude signal from a multiphase source of power. Thus, as we view the "rotating" field pattern at the radius of the rotor, we see a wave with real k traveling in the θ-direction. This problem is thus quite different from that of the model of the preceding section in which the wave was started out at a particular value of z and allowed to progress into a medium with whatever rate of decay and velocity was required by the properties of the medium.

In this section we wish to consider a problem in which we can arbitrarily impose any desired wave velocity that we wish, as in a physical induction motor with access to the windings at all positions. In a linear version of such a device we could also impose an arbitrary rate of wave growth or decay by varying the amplitude of the field excitation along the length of a linear machine, but we will not explore this possibility. In a circular machine we would expect the amplitude to be independent of θ, and in the following analysis we have considered only real k, as is appropriate to a model of a reentrant circular machine or a linear machine with constant amplitude excitation along its length. We treat the same basic configuration as in the preceding section, with a conducting medium filling all space and moving in the positive z-direction at velocity v as shown in Fig. 4.1-1. Here, however, we will not treat the mode shown in that figure, but will imagine that the y–z plane at $x = 0$ is covered with a slow-wave circuit representing the field windings of a linear induction machine. We will be able to specify the value of the current that flows in this plane. It is appropriate to think of this plane, representing the stator, as being fixed in the laboratory frame with the conducting media above and below this plane, representing the

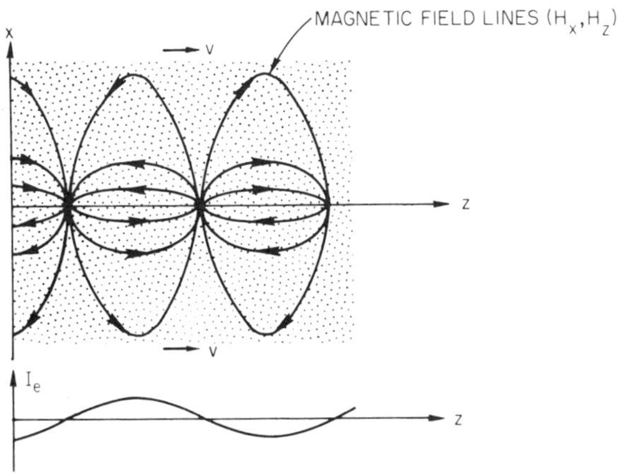

Fig. 4.2-1. Transverse-electric (TE with respect to z) field configuration.

The electric field is directed perpendicular to the plane of the paper, extending uniformly to infinity in the y-direction and decaying in the x-direction. The magnetic field is antisymmetric about the plane $x = 0$. A planar circuit covers the $x = 0$ plane. A uniform conducting medium extends to infinity above and below the $x = 0$ plane and moves to the right at velocity v with respect to the circuit. The excitation surface current on the circuit I_e is in the y-direction and takes the form of a wave traveling at velocity v_p in the z-direction to the right [from D. A. Dunn, R. W. Wallace, and S. D. Choi, Waves in a moving medium with finite conductivity, *Proc. IEEE* **57**, 45–57 (1969)].

rotor, moving past at the same constant velocity. We assume there are no variations in the y-direction, but allow variations in x and z. We therefore seek solutions of the form $\exp[j(\omega t - k_x x - k_z z)]$ where k_x and k_z are the wave numbers in the x and z directions, respectively.

We employ the same technique for solving this problem that we have used before of writing the fields in the primed frame where the material is at rest, and using the Lorentz transformation to transform to the laboratory frame [8]. We picture a current in the field windings that is in the y-direction that produces a TE mode with H_x and H_z in the form of loops and an associated electric field E_y, as shown in Fig. 4.2-1. Using Maxwell's equations we can immediately write down the fields in the primed frame, using $\mathbf{B}' = \mu\mathbf{H}'$, $\mathbf{D}' = \varepsilon\mathbf{E}'$, and $\mathbf{J}' = \sigma\mathbf{E}'$:

$$E_y' = E_{y0}', \qquad H_x' = -(k_z'/\omega'\mu)E_{y0}', \qquad H_z' = (k_x'/\omega'\mu)E_{y0}' \tag{4.2.1}$$

where $\exp[j(\omega' t' - k_x' x' - k_z' z')]$ is understood. The corresponding dispersion relation in the primed frame is also obtained directly from Maxwell's equations:

$$k_x'^2 + k_z'^2 = \omega'^2\mu\varepsilon - j\omega'\mu\sigma \tag{4.2.2}$$

From the Lorentz transformation it can be shown directly that

$$E_{y0}'/\omega' = E_{y0}/\omega \tag{4.2.3}$$

where E_{y0} is the value of the electric field in the rest frame. Consequently, the fields in the laboratory frame can be written as

$$\begin{aligned}
E_y &= E_{y0} \\
D_y &= \frac{\gamma^2\varepsilon}{\Omega}\left[\Omega\left(1 - \frac{\beta^2}{n^2}\right) - \frac{\beta}{n}\,\varkappa_z(n^2 - 1)\right]E_{y0} \\
H_x &= -\frac{\gamma^2}{\Omega\eta}\left[\varkappa_z(1 - \beta^2 n^2) - \frac{\beta}{n}\,\Omega(1 - n^2)\right]E_{y0} \\
B_x &= -\frac{n\varkappa_z}{\Omega c}\,E_{y0} \\
H_z &= \frac{\varkappa_x}{\Omega\eta}\,E_{y0} \\
B_z &= \frac{n\varkappa_x}{\Omega c}\,E_{y0}
\end{aligned} \tag{4.2.4}$$

in terms of normalized variables introduced in Eq. (4.1.11) and where $\exp[j(\omega t - k_x x - k_z z)]$ is understood.

Equations (4.2.4) and (4.2.1) are valid in both the region $(x > 0)$ above the current sheet and $(x < 0)$ below the current sheet. If we consider the case of $\sigma = 0$ and real k_z' in Eq. (4.2.2), we can immediately see that k_x' is pure imaginary, because $k_z'^2 > \omega'^2 \mu\varepsilon$ for a slow wave. Hence, the wave amplitude decays in the x-direction away from the $x = 0$ plane for both $x > 0$ and $x < 0$. Thus when we include finite conductivity, the imaginary part of k_x above the plane, $k_{x\mathrm{i}}^u$, must be negative and vice versa for $k_{x\mathrm{i}}^l$, the value of $k_{x\mathrm{i}}$ below the plane, i.e.,

$$k_{x\mathrm{i}}^u < 0, \qquad k_{x\mathrm{i}}^l > 0 \tag{4.2.5}$$

so that

$$\varkappa_x{}^l = -\varkappa_x{}^u \tag{4.2.6}$$

The current at $x = 0$ is taken to be

$$\mathbf{I}_\mathrm{e} = -2I_0 \exp[j(\omega t - k_\mathrm{e} z)]\mathbf{1}_y \tag{4.2.7}$$

where I_0 has the dimensions of current per unit length. Since the waves above and below the plane must be in synchronism with each other and the excitation current

$$k_z{}^u = k_z{}^l = k_\mathrm{e} \stackrel{\Delta}{=} k_z \tag{4.2.8}$$

If we now match the boundary conditions at the $x = 0$ plane we have, in general

$$\begin{aligned} \mathbf{1}_x \times (\mathbf{H}^u - \mathbf{H}^l) &= \mathbf{I}_\mathrm{e}, \qquad & \mathbf{1}_x \cdot [\mathbf{B}^u - \mathbf{B}^l] &= 0 \\ \mathbf{1}_x \times [\mathbf{E}^u - \mathbf{E}^l] &= 0, \qquad & \mathbf{1}_x \cdot [\mathbf{D}^u - \mathbf{D}^l] &= Q \end{aligned} \tag{4.2.9}$$

For the case considered here these conditions must be satisfied in the laboratory frame where the current sheet is at rest. The first two may be written using the fields from Eq. (4.2.4),

$$\begin{aligned} &\frac{\varkappa_x{}^u}{\Omega\eta} E_{y0}^u \exp[j(\omega t - k_z{}^u z)] - \frac{\varkappa_x{}^l}{\Omega\eta} E_{y0}^l \exp[j(\omega t - k_z{}^l z)] \\ &\quad = 2I_0 \exp[j(\omega t - k_\mathrm{e} z)] - \frac{n\varkappa_z{}^u}{\Omega c} E_{y0}^u \exp[j(\omega t - k_z{}^u z)] \\ &\quad = -\frac{n\varkappa_z{}^l}{\Omega c} E_{y0}^l \exp[j(\omega t - k_z{}^l z)] \end{aligned} \tag{4.2.10}$$

which immediately yield

$$E_{y0}^{u} = E_{y0}^{l}, \qquad E_{y0}^{u} = (\Omega\eta/\varkappa_x{}^u)I_0 \tag{4.2.11}$$

and the last two equations of Eq. (4.2.10) yield $Q = 0$.

We now define $\varkappa_x{}^u = \varkappa_x$ and write the fields only for $x > 0$ in terms of I_0

$$\begin{aligned}
E_y &= \frac{\Omega\eta}{\varkappa_x} I_0 \\
D_y &= \frac{\gamma^2 n}{\varkappa_x c}\left[\Omega\left(1 - \frac{\beta^2}{n^2}\right) - \frac{\beta}{n}\varkappa_z(n^2 - 1)\right]I_0 \\
H_x &= -\frac{\gamma^2}{\varkappa_x}\left[\varkappa_z(1 - \beta^2 n^2) - \frac{\beta}{n}\Omega(1 - n^2)\right]I_0 \\
B_x &= -\frac{\mu\varkappa_z}{\varkappa_x} I_0 \\
H_z &= I_0 \\
B_z &= \mu I_0
\end{aligned} \tag{4.2.12}$$

which is the final expression for the fields that are consistent with the current $\mathbf{I}_e$ given in Eq. (4.2.7).

Note that in deriving Eq. (4.2.12) we have not used the boundary conditions for a conductor at $x = 0$, even though in practice the current sheet would be established by running current through a set of conductors. While it is entirely appropriate to regard $\mathbf{I}_e$ as a source of excitation for this mode, it is important to note that we have not solved the general problem that would allow us to give the fields for an arbitrary form of $\mathbf{I}_e$. We have carefully selected a harmonic form of $\mathbf{I}_e$ representing a wave traveling in the z-direction and found the associated fields that match this form of $\mathbf{I}_e$ at the boundary and that satisfy Maxwell's equations. If we were interested, for example, in a square wave excitation, we could do a Fourier series analysis using a number of terms of this form.

It is also of interest that we were able to solve this two-dimensional problem by simply using wave functions of the form $\exp[j(\omega t - k_x x - k_z z)]$. This type of function we know satisfies the wave equation, and, provided we can match the boundary conditions with it, is a perfectly valid solution. It resembles a plane wave propagating at an angle to the z-direction, but, because k_x is nearly pure imaginary for real k_z, it is not a wave that propagates in x, but instead decays in x while propagating in z. This form of solution is characteristic of a slow wave propagating in the z-direction. It is a very important type of wave for many device applications, the induction motor being merely the simplest. The traveling-wave tube and the magnetron are examples of microwave tubes that make use of this type of wave. This type of wave is useful for two basic reasons. First, it is possible to synchronize the velocity of a moving medium with such a wave because it can be made to move at a low enough velocity to correspond to velocities of mechanical motion. Second, the fringing fields of such a wave extend out from the structure that carries it, allowing the moving medium to be put directly into these fields. At the same time, the fields do not propagate in the direction away from the structure, as they do in the case of an antenna, but decay, as in a cutoff waveguide, in this direction, and thus are not dissipated as we go along the length of the structure, except as a result of power delivered to the medium.

We turn now to the dispersion relation in the laboratory frame which can be derived directly from Eq. (4.2.2) using the Lorentz transformation for ω and k. It is

$$\varkappa_x^2 + \gamma^2(1 - \beta^2 n^2)\varkappa_z^2 = \gamma^2\Omega^2[1 - (\beta^2/n^2)] - 2\gamma^2\Omega\varkappa_z(n^2 - 1)(\beta/n) - j\gamma(\Omega - \varkappa_z\beta n) \tag{4.2.13}$$

The definition of phase velocity for this mode is expressed here as $v_p = (\omega/k_z)$ where here k_z is always real. It will be convenient to define the relative phase velocity V_p as

$$V_p = v_p/c \tag{4.2.14}$$

The slip parameter defined in Eq. (4.1.21) becomes

$$s = (\Omega - \beta n\varkappa_z)/\Omega \tag{4.2.15}$$

and β must satisfy the relation

$$\beta = V_p(1 - s). \tag{4.2.16}$$

The dispersion relation can then be written in the form

$$\varkappa_x^2 = (\gamma^2\Omega^2/n^2)\{s^2n^2 - [(1/V_p) - \beta]^2\} - j\gamma\Omega s \tag{4.2.17}$$

where $\varkappa_x$ always has a negative imaginary part. It is interesting to consider Eq. (4.2.17) for a situation where the excitation phase velocity v_p is much less than the velocity of light, i.e., $V_p \ll 1$, and where the frequency is high in the sense that $(\Omega/n^2V_p^2) \gg 1$. For these conditions the dispersion relation becomes to a very good approximation

$$\varkappa_x^2 \cong -\Omega^2/n^2 V_p^2 \tag{4.2.18}$$

hence

$$\varkappa_x \cong -j(\Omega/nV_p) \qquad \text{or} \qquad k_x \cong -j(\omega/v_p) \tag{4.2.19}$$

Therefore, to this approximation, k_x is purely imaginary and independent of the slip parameter s. To illustrate a case with V_p closer to 1, Fig. 4.2-2 is a plot of the dispersion relation given in Eq. (4.2.17) for $V_p = 0.5$ and $\Omega = 1.0$ and for $n^2 = 1$ and $n^2 = 10$. Because β is determined once V_p and s have been specified, through Eq. (4.2.16), both β and s have been used in labeling the abscissa of the plot. From this figure it is seen that even

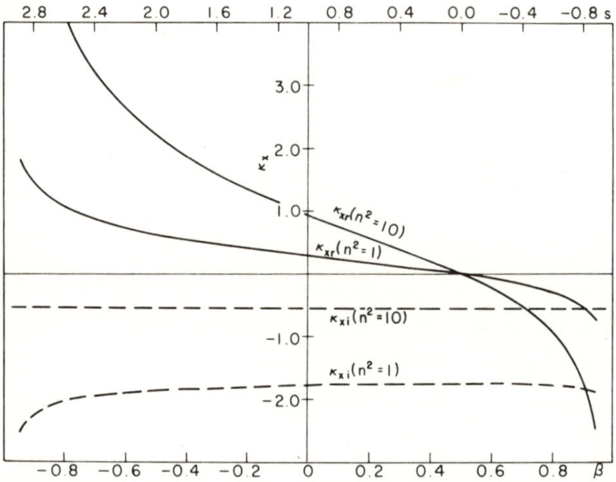

Fig. 4.2-2. Plot of $\varkappa_x$ vs. β (lower scale) or slip parameter s (upper scale) with $\Omega = 1$, $V_p = 0.5$, and for $n^2 = 1$ and $n^2 = 10$. $\varkappa_{xr}$ is the real part and $\varkappa_{xi}$ is the imaginary part of $\varkappa_x$ [from D. A. Dunn, R. W. Wallace, and S. D. Choi, Waves in a moving medium with finite conductivity, *Proc. IEEE* **57**, 45–57 (1969)].

over a rather extreme range of β the imaginary part of $\varkappa_x$ is still essentially constant. The real part shows, however, that the wave is traveling away from the excitation plane for $s > 0$ and toward the excitation plane for $s < 0$.

We turn now to the force and energy balance for this mode. Transformation of the conduction current, $J_y' = \sigma E_y'$, to the laboratory frame gives

$$J_y = \gamma\sigma\eta(\Omega - \beta n \varkappa_z) I_0/\varkappa_x \tag{4.2.20}$$

With excitation at real frequency ω, the time-average rate at which the fields do work is, from Eqs. (4.2.20), (4.2.12), and (4.1.17)

$$P_{\mathrm{e}} = [\gamma\sigma\Omega\eta^2 I_0^2(\Omega - \beta n\varkappa_z)\exp(2k_{x\mathrm{i}}x)]/2\,|\,\varkappa_x\,|^2 \tag{4.2.21}$$

The time-average force density is, again for μ_0, ε_0, and $n = 1$, from Eq. (4.0.1)

$$\mathbf{f} = \{[\gamma\sigma\mu\eta(\Omega - \beta n\varkappa_z)I_0^2\exp(2k_{x\mathrm{i}}x)]/2\,|\,\varkappa_x\,|\}^2[\varkappa_{x\mathrm{r}}\mathbf{1}_x + \varkappa_z\mathbf{1}_z] \tag{4.2.22}$$

Therefore, the time average power associated with this force density is

$$\begin{aligned} P_{\mathrm{f}} &\stackrel{\Delta}{=} \mathbf{f}\cdot v\mathbf{1}_z \\ &= \gamma\sigma\eta^2\beta n(\Omega - \beta n\varkappa_z)\varkappa_z I_0^2\exp(2k_{x\mathrm{i}}x)/2\,|\,\varkappa_x\,|^2 \end{aligned} \tag{4.2.23}$$

Introducing the slip parameter of Eq. (4.2.15) into Eqs. (4.2.21) and (4.2.23) we obtain

$$P_{\mathrm{e}} = [\gamma\sigma\Omega^2\eta^2 I_0^2 s\exp(2k_{x\mathrm{i}}x)]/2\,|\,\varkappa_x\,|^2 \tag{4.2.24}$$

$$P_{\mathrm{f}} = [\gamma\sigma\Omega^2\eta^2 I_0^2(1 - s)s\exp(2k_{x\mathrm{i}}x)]/2\,|\,\varkappa_x\,|^2 \tag{4.2.25}$$

These last two equations represent on a per unit volume basis the electrical power input and the mechanical power output. Using Eq. (4.2.16), the power dissipated in losses becomes

$$P_l = [\gamma\sigma\Omega^2\eta^2 I_0^2 s^2\exp(2k_{x\mathrm{i}}x)]/2\,|\,\varkappa_x\,|^2 \tag{4.2.26}$$

In the present case it would be meaningful to integrate all of these quantities over x in order to obtain total values for each. However, it should be noted that the last three expressions for P_{e}, P_{f}, and P_l all have the same spatial dependence in x; therefore, any integration or summation over a common volume will affect all three quantities in the same manner. Thus any type of ratio such as efficiency will give the same result regardless of whether densities or total quantities are used.

Using the same argument as above in determining what regions of operation predict a motor, brake, or generator analogous to an ordinary induction machine, Eqs. (4.2.24) and (4.2.25) indicate that the divisions are: (a) motor, $0 < s < 1$; (b) brake, $1 < s$; (c) generator, $s < 0$. Of course, this is identical to the situation in the standard induction machine.

Using the definitions of efficiency given in Eqs. (4.2.24)–(4.2.26),

$$\begin{aligned} \eta_{\mathrm{m}} &= 1 - s, & 0 < s < 1 \\ \eta_{\mathrm{b}} &= (s - 1)/s, & s > 1 \\ \eta_{\mathrm{g}} &= 1/(1 - s), & s < 0 \end{aligned} \tag{4.2.27}$$

It is interesting to note that these efficiencies are completely independent of the parameters n and Ω. Also they are independent of both the material velocity ratio β and the wave velocity ratio V_p, except insofar as these parameters appear in s. The efficiency of an induction machine is thus the same for any given value of slip, whether this value of s occurs for $\beta = 10^{-7}$ or for $\beta = 0.1$.

Figure 4.2-3 is a plot of the expressions in Eq. (4.2.27). Of some interest is the high efficiency region near zero slip. Even slips up to 40% allow high (>50%) efficiency for either motors or generators. For some applications it would be attractive to consider operation near unity slip, i.e., for wave velocities much greater than the medium velocity, but it is seen from Fig.

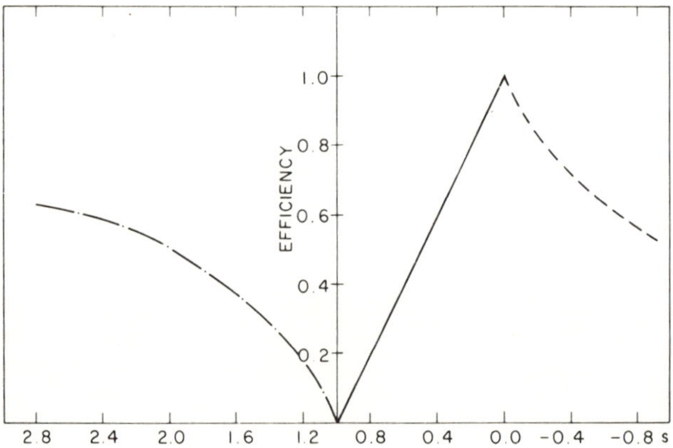

Fig. 4.2-3. Efficiency versus slip parameter s for the TE mode. The system is a motor for $0 < s < 1$, a brake for $s > 1$, and a generator for $s < 0$ [from D. A. Dunn, R. W. Wallace, and S. D. Choi, Waves in a moving medium with finite conductivity, *Proc. IEEE* **57**, 45–57 (1969)].

4.2-3 that the efficiency approaches zero in this region. Thus starting is inherently inefficient and running needs to be near synchronism for good efficiency.

Another feature of the present model that is of interest is the force balance that must exist between the force on the medium and the force on the circuit at $x = 0$ that we have postulated carrying the current I_e. These forces can be shown to be equal and opposite in general by integrating the force density in the medium over all x and comparing this result with the force per unit area exerted by the current I_e on the circuit at $x = 0$ in the presence of the magnetic field at $x = 0$.

PROBLEM 4.2-1 FORCE BALANCE IN AN INDUCTION MOTOR

Show that the force on the medium per unit area in the z-direction is equal and opposite to the force per unit area exerted on the circuit of the magnetic field at $x = 0$. The force per unit area on the medium can be found by integrating from $x = 0$ to $x = \infty$.

PROBLEM 4.2-2 HIGH-FREQUENCY MOTOR DESIGN

A linear induction motor of a certain type is adequately represented by the model in this section. Compare 60-Hz and 60-MHz designs for such a motor designed to operate with $\beta = 10^{-6}$ and a skin depth, a, where $k_{x\mathrm{i}}a = 1$, and $a = 1$ cm for both designs. Compare the values of σ, V_{p}, f, and η_{m} for values of slip near unity.

PROBLEM 4.2-3 LINEAR INDUCTION MOTOR

Laithwaite and Nasar [9] have suggested a law applying to linear motors that can be stated as follows. "As you make an object smaller you must ultimately melt it before you move it." Formulate this law more precisely and derive it.

4.3 The Surface-Charge Induction Motor—TM Mode in a Moving Conducting Medium

In this section we consider the dual of the mode in the preceding section and obtain a force in the rest frame that is due to $q\mathbf{E}$, in addition to the usual $\mathbf{J} \times \mathbf{B}$ force. This $q\mathbf{E}$ force results both from the surface charge on the interface between the medium and the air gap and also from the apparent volume charge density in the rest frame. This volume charge exists even though the charge density in the primed frame is zero, because there is an axial current density in this mode [8].

A. *TM Mode in a Moving Conducting Medium*

Figure 4.3-1 shows the field configuration in the TM mode. The electric field lines E_x, E_z, form loops around the magnetic field lines H_y which are into or out of the paper. The existence of an infinitesimal air gap is indicated between the circuit and the moving medium. As before, we start our analysis in the primed frame where the medium is at rest, assuming no variations

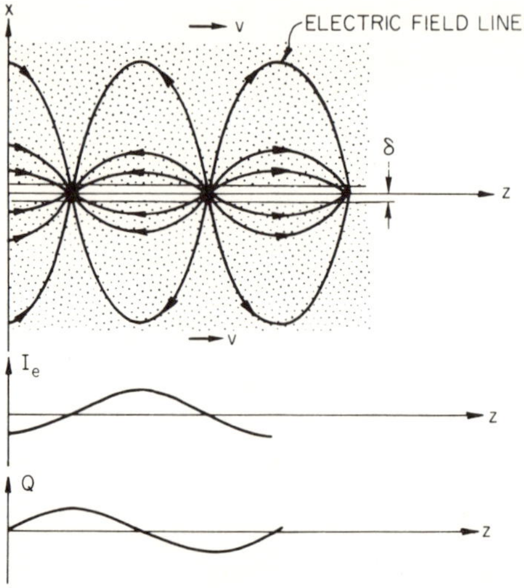

Fig. 4.3-1. Transverse-magnetic (TM with respect to z) field configuration. The magnetic field is directed perpendicular to the plane of the paper, extending uniformly to infinity in the y-direction and decaying in the x-direction. The electric field is symmetric about the plane $x = 0$. A planar circuit covers the $x = 0$ plane. A uniform conducting medium extends to infinity above and below the $x = 0$ plane and moves to the right at velocity v with respect to the circuit. An infinitesimal air gap of thickness δ separates the medium and the circuit. The excitation surface current on the circuit I_e is in the z-direction and takes the form of a wave traveling to the right at velocity v_p accompanied by a traveling-wave surface-charge distribution Q with a different phase [from D. A. Dunn, R. W. Wallace, and S. D. Choi, Waves in a moving medium with finite conductivity, *Proc. IEEE* **57**, 45–57 (1969)].

in y, and let the variations in x' and z' be $\exp[j(\omega' t' - k_x' x' - k_z' z)]$ as before. Maxwell's equations yield the same dispersion relation in the primed frame as we found for the TE mode, using $\mathbf{B}' = \mu \mathbf{H}'$, $\mathbf{D} = \varepsilon \mathbf{E}'$, $\mathbf{J}' = \sigma \mathbf{E}'$, as before

$$k_x'^2 + k_z'^2 = \omega'^2 \mu\varepsilon - j\omega' \mu\sigma \tag{4.3.1}$$

In the air gap of thickness δ there is a different dispersion relation

$$k_{x0}'^2 + k_z'^2 = \omega'^2 \mu_0 \varepsilon_0 \tag{4.3.2}$$

where k_{x0}' is the transverse wave number in the air gap. In the laboratory frame the dispersion relation in the medium is Eq. (4.2.13), just as for the TE mode. In the air gap the laboratory-frame relation can be written in either of the forms

$$k_{x0}^2 + k_z^2 = \omega^2 \mu_0 \varepsilon_0, \qquad \varkappa_{x0}^2 + \varkappa_z^2 = \Omega^2/n^2 \tag{4.3.3}$$

where both k'_{x0} and k_{x0} must be pure imaginary quantities because the medium is free space.

The fields and the current densities can be transformed to the laboratory frame as before, yielding in the laboratory frame, within the medium for $x \geq \delta$, in terms of the excitation current I_0

$$H_y = \frac{\gamma(\gamma + j\Omega)sI_0}{(1 + j\gamma s\Omega)}, \qquad B_y = \frac{\gamma\{\gamma n^2 + j\Omega[1 + \gamma^2 s(n^2 - 1)]\}\mu s I_0}{n^2(1 + j\gamma s\Omega)}$$

$$E_x = \frac{\gamma^2\{n^2\beta + j\gamma[\beta\Omega(n^2 - 1) + n\varkappa_z(1 - \beta^2 n^2)]\}\mu c s I_0}{n^2(1 + j\gamma s\Omega)}$$

$$D_x = \frac{\gamma(\gamma\beta + jn\varkappa_z)sI_0}{c(1 + j\gamma s\Omega)} \tag{4.3.4}$$

$$E_z = \frac{-j\gamma\varkappa_x\eta s I_0}{(1 + j\gamma s\Omega)}, \qquad D_z = \varepsilon E_z$$

$$J_x = \frac{j\gamma^2(n\varkappa_z - \beta\Omega)\sigma\eta s I_0}{n(1 + j\gamma s\Omega)}, \qquad J_z = \frac{-j\gamma^2\varkappa_x\sigma\eta s I_0}{(1 + j\gamma s\Omega)}$$

where the propagation constant $\exp[j(\omega t - k_x x - k_z z)]$ is understood and where the excitation current is taken to be

$$I_e = 2I_0 \exp[j(\omega t - k_e z)]\mathbf{1}_z \tag{4.3.5}$$

As before it may readily be shown that $k_e = k_z$.

In the air gap for $0 \leq x \leq \delta$, for the case of an infinitesimal δ

$$H_{y0} = I_0, \qquad E_{x0} = \frac{\mu_0 c n \varkappa_z I_0}{\Omega}, \qquad E_{z0} = \frac{-jn\varkappa_x\gamma s I_0}{c(1 + j\gamma s\Omega)\varepsilon_0}$$
$$D_{x0} = \varepsilon_0 E_{x0}, \qquad D_{z0} = \varepsilon_0 E_{z0}, \qquad B_{y0} = \mu_0 H_{y0} \tag{4.3.6}$$

where the propagation constant $\exp[j(\omega t - k_z z)]$ is understood. The fields for negative x are symmetrical to the above, and may be found by the same method used for the TE mode.

The important feature of this mode that distinguishes it from the TE mode is the role played by surface charge, both on the circuit at the plane $x = 0$ and at the air-gap–medium interface at $x = \delta$. There is also a volume-charge density in the medium resulting from the Lorentz transformation of the axial-current density which did not exist in the TE mode.

The surface charge at the circuit plane at $x = 0$ we refer to as Q and is given from the last of Eqs. (4.2.9) as

$$Q = \frac{2n\varkappa_z I_0}{c\Omega} \exp[j(\omega t - k_z z)] \qquad (4.3.7)$$

and is thus in phase with I_0. The surface charge at the medium–air-gap interface at $x = \delta$ is ϱ_s and

$$\varrho_s = -\frac{\gamma^2(n\varkappa_z - \beta\Omega)I_0}{\Omega c(1 + j\gamma s\Omega)} \exp[j(\omega t - k_z z)] \qquad (4.3.8)$$

The volume charge density within the medium for $x \geq \delta$ is ϱ and

$$\varrho = -j\frac{\gamma^2\varkappa_x\beta\sigma\eta s I_0}{c(1 + j\gamma s\Omega)} \exp[j(\omega t - k_x x - k_z z)] \qquad (4.3.9)$$

The current densities within the medium were given in Eq. (4.3.4). There is also a surface current per unit width J_{lz} at $x = \delta$ due to the convection of the surface charge at $x = \delta$ by the motion of the medium. It is

$$J_{lz} = \varrho_s v \qquad (4.3.10)$$

B. *Energy Balance*

In the TM mode there are three significant time average force terms: F_1 due to the surface charge ϱ_s, F_2 due to the volume charge in the medium ϱ, and F_3 due to the $\mathbf{J} \times \mathbf{B}$ volume force in the medium. In order to compare these and add them together to get the total z-directed force, it is necessary to integrate the volume forces over x. Thus in the TM mode we deal with the total force per unit area at any given value of z rather than the volume force.

The force due to surface charge is then, for the upper half of the system only, again using Eq. (4.0.1) and considering $\mu = \mu_0$ and $\varepsilon = \varepsilon_0$

$$F_1 = \frac{1}{2} \operatorname{Re}\{\varrho_s E_z^* |_{x=\delta}\}$$

$$= -\frac{\gamma^3(n\varkappa_z - \beta\Omega)\eta\varkappa_{xi} s I_0^2}{2\Omega c(1 + \gamma^2 s^2 \Omega^2)} \qquad (4.3.11)$$

The force due to the apparent volume charge is, also for the upper half only

$$F_2 = \frac{1}{2} \operatorname{Re}\left\{\int_\delta^\infty \varrho E_z^* \, dx\right\}$$

$$= -\frac{\gamma^3 |\varkappa_x|^2 \beta n s^2 I_0^2}{4c\varkappa_{xi}(1 + \gamma^2 s^2 \Omega^2)} \qquad (4.3.12)$$

The force due to the $\mathbf{J} \times \mathbf{B}$ force is, also for the upper half only

$$F_3 = \frac{1}{2} \operatorname{Re}\left\{\int_\delta^\infty J_x B_y^* \, dx\right\}$$
$$= -\frac{\gamma^3 \eta s^2 I_0{}^2 \Omega(n\varkappa_z - \beta\Omega)[1 + \gamma^2 s(n^2 - 1)]}{4cn^2 \varkappa_{xi}(1 + \gamma^2 s^2 \Omega^2)} \tag{4.3.13}$$

The total force per unit area F, is the sum of F_1, F_2, and F_3 as follows:

$$F = -\frac{\gamma\eta I_0{}^2 s n \varkappa_z (2\varkappa_{xi}^2 + \gamma^2 s^2 \Omega^2)}{2c\Omega\varkappa_{xi}(1 + \gamma^2 s^2 \Omega^2)} \tag{4.3.14}$$

The relative magnitudes of F_1, F_2, and F_3 can readily be compared for the case of low velocity $V_p \ll 1$ and for high frequencies $(\Omega/n^2 V_p{}^2) \gg 1$. For these conditions

$$k_{xi} \cong -k_z = -\omega/v_p \tag{4.3.15}$$

and when this is so

$$F_1/F_3 \cong c^2/v_p{}^2 \gg 1 \tag{4.3.16}$$

and also

$$F_1/F_2 \cong c^2/v v_p \gg 1 \tag{4.3.17}$$

which shows us that the force due to surface charge is the dominant mechanism. In addition it can readily be shown that F, the total force is always equal and opposite to the force on the circuit.

As before we calculate P_e and P_f, but now it is necessary to integrate over x to obtain the proper form of these quantities to use with Eq. (4.3.14). We use subscript s to denote the power per unit area, here calculating the total for both halves of the system:

$$P_{es} = \operatorname{Re}\left\{\int_\delta^\infty \mathbf{J} \cdot \mathbf{E}^* \, dx\right\}$$
$$= \operatorname{Re}\left\{\int_\delta^\infty J_x E_x^* \, dx + \int_\delta^\infty J_z E_z^* \, dx + J_{lz} E_z^* \big|_{x=\delta}\right\}$$
$$= -\frac{\gamma\eta I_0{}^2 s(2\varkappa_{xi}^2 + \gamma^2 s^2 \Omega^2)}{2\varkappa_{xi}(1 + \gamma^2 s^2 \Omega^2)} \tag{4.3.18}$$

The mechanical power per unit area is from Eq. (4.3.14)

$$P_{fs} = \mathbf{F} \cdot \mathbf{v}$$
$$= -\frac{\gamma\eta I_0{}^2 s(1 - s)(2\varkappa_{xi}^2 + \gamma^2 s^2 \Omega^2)}{2\varkappa_{xi}(1 + \gamma^2 s^2 \Omega^2)} \tag{4.3.19}$$

We can also obtain the lost power P_{ls} from Eqs. (4.3.18) and (4.3.19) using Eq. (4.1.16). The signs of P_{es}, P_{fs}, and P_{ls} have the same significance as for the TE case and the regions of motion, brake, and generator operation occur for the same ranges of slip, i.e., (a) motor, $0 < s < 1$, (b) brake, $s > 1$, and (c) generator, $s < 0$. Curiously enough the efficiencies defined in Eqs. (4.1.24)–(4.1.26) turn out to be identical in form to those for the TE case given in Eq. (4.2.27), as may be seen by inserting Eqs. (4.3.18) and (4.3.19) into Eqs. (4.1.24)–(4.1.26). Thus Fig. 4.2-3 applies to both the TM and the TE modes, although the physical force mechanisms are quite different.

The force given in Eq. (4.3.14) can be expressed very simply for the case of low velocity $V_p \ll 1$ and $(\Omega/n^2V_p^2) \gg 1$, using Eq. (4.3.4) to express I_0 in terms of E_z.

$$F \cong \varepsilon_0 \mid E_z \mid^2 n V_p / s\Omega \tag{4.3.20}$$

Thus the force, which is proportional to $\mid E_z \mid^2$ and V_p, simply decreases inversely with both s and Ω, implying that infinite force is obtained for zero frequency and zero slip. However, note that our approximation assumes high frequency, $(\Omega/n^2V_p^2) \gg 1$, so we cannot use Eq. (4.3.20) at frequencies that are low in this sense. A maximum force will occur at some nonzero frequency and there is also a nonzero slip that yields maximum force. Qualitatively, the optimum frequency is that for which the relaxation time of the dielectric is some constant times the period of the operating frequency. A complete solution to this problem has been given [11] which also includes the effects of a finite air gap.

The designs of machines of these two types will tend to be quite different, because the physical mechanisms are so different. The force mechanism of the TM mode is a surface-charge mechanism which requires a certain value of relaxation time in order to yield maximum force. Dielectric media with moderate values of loss will thus be typical rotor materials rather than metals. Also, because the TM mode force is proportional to the electric field, designs of TM mode devices will use high impedance rather than low impedance circuits to give a maximum electric field for a unit power rather than a maximum current and magnetic field as in the TE mode induction motor.

An experimental machine of this type is shown in Fig. 4.3-2 which used Plexiglas as the rotor material and operated at 60 Hz. The stator shown in Fig. 4.3-2(b) used 72 wires of size 18AWG connected in three groups to a three phase source through three circular rings on the ends to form a traveling-wave circuit. This machine was tested [11] and found to give

Fig. 4.3-2. (a) A surface-charge induction motor for 60-Hz operation. It is 14 in. wide, 14 in. high, and $19\frac{1}{4}$ in. long. (b) End view of the stator showing three-phase winding. (c) Exploded view of the motor showing Plexiglas rotor. [from S. D. Choi and D. A. Dunn, A surface charge induction motor, *Proc. IEEE* **59**, 737–748 (1971)].

approximately the theoretical value of torque given in this section. Similar effects were observed by Melcher [12] in a pump for slightly conducting liquids and by Grossetti [13]. The idea of a force of this type dates back to Arno [14] in 1892.

Problem 4.3-1 Surface-Charge Induction Motor

A linear motor of the type described in this section is to be designed for operation at 60 Hz with $\beta = 10^{-6}$ and $n = 1$. Compare this design with a TE motor operating with the same recirculating power at the same frequency, as in Problem 4.2-2.

4.4 Quasi-Static Formulation of Maxwell's Equations

In the preceding sections of this chapter we took the direct approach to force calculations on moving media, and we found that for some particularly simple configurations exact solutions were obtainable. In the case of the TE and TM modes, we saw that the results were independent of the absolute value of velocity. This result suggests that we might have arrived at the same point by an approximate method valid for low velocities only and indeed this is the case [15]. In this section we will take this approach and in the remainder of the chapter we will make use of this approach to develop a "circuit" model of an induction machine that connects with the conventional theory of such machines.

First we will look at the wave equation for the homogeneous case, i.e., for zero charge and current, which can be written in the form

$$\nabla^2 \mathbf{E} = \mu\varepsilon(\partial^2 \mathbf{E}/\partial t^2) \tag{4.4.1}$$

for the electric field. In a device such as a linear induction motor which has a direction of wave propagation which we may take to be the z-direction, we may write the wave equation using a transverse Laplacian ∇_t to refer to dimensions transverse to z such as x and y or r and θ, as follows:

$$\nabla^2 \mathbf{E} = \nabla_t^2 \mathbf{E} + (\partial^2 \mathbf{E}/\partial z^2) = \mu\varepsilon(\partial^2 \mathbf{E}/\partial t^2) \tag{4.4.2}$$

For a wave propagating in the z-direction we can write $\mathbf{E}$ proportional to $\exp[j(\omega t - kz)]$ and the wave equation can then be written in terms of ω and k as

$$\nabla_t^2 \mathbf{E} = (k^2 - \omega^2\mu\varepsilon)\mathbf{E} \tag{4.4.3}$$

If the wave of interest is much slower than the velocity of light in the medium

$$v_p/c_d \ll 1 \tag{4.4.4}$$

and since $c_d^2 = (\mu\varepsilon)^{-1}$ and

$$k = \omega/v_p \tag{4.4.5}$$

we have for this condition that

$$k^2 \gg \omega^2\mu\varepsilon \tag{4.4.6}$$

so that the wave equation can be written

$$\nabla_t^2 \mathbf{E} \simeq k^2 \mathbf{E} \tag{4.4.7}$$

and it is thus in this approximation entirely independent of the time derivative which gave the $\omega^2\mu\varepsilon$ term. The electric field thus is given by a solution to Laplace's equation, because each component of the vector Laplacian must separately be equal to zero if

$$\nabla^2 \mathbf{E} = 0 \tag{4.4.8}$$

Evidently, as is often stated, this is a zero curl field.

The shapes of the field lines in a quasi-static time-varying field are therefore essentially the same as the field shapes for a static field. This is a very important result, because very many electrical devices have wave velocities small in comparison with the velocity of light.

In actually applying this result, we can be concerned with either predominantly electric or magnetic devices. If we are picturing the typical magnetic device we start by assuming low velocities and thereby arrive at an equation for **H** that is identical to Eq. (4.4.8) for **E**. This tells us that we can neglect time derivatives in computing **H**, so we can use the form of Maxwell's equations that applies to static fields

$$\nabla \times \mathbf{H} = \mathbf{J} \tag{4.4.9}$$

to find **H** from **J** or we can use

$$\nabla^2 \mathbf{H} = 0 \tag{4.4.10}$$

Because it is a static field, there is no reason to have any electric field at all. However, it is not truly static, and normally there will be time variations in **H** and an associated value of electric field. If we wish to compute **E** we can use the value of **H** arrived at above in Faraday's law

$$\nabla \times \mathbf{E} = -\partial \mathbf{B}/\partial t \tag{4.4.11}$$

to get **E**. It would not be appropriate to seek strict self-consistency in such a problem by correcting **H** by adding in the $\partial \mathbf{D}/\partial t$ term in Eq. (4.4.9). But if this were done, it would presumably be a negligibly small correction.

In an electrostatic problem of this type we would start with the slow-wave assumption and arrive at Eq. (4.4.8) from which we could obtain **E**, or we could use

$$\nabla \times \mathbf{E} = 0, \qquad \mathbf{E} = -\nabla V \tag{4.4.12}$$

or find **E** from the charge distribution through Poisson's equation

$$\nabla \cdot \mathbf{D} = \varrho, \qquad \nabla^2 V = -\varrho/\varepsilon \tag{4.4.13}$$

As before one could, if desired, get a value of **H** from the quasi-static value of **E** determined, say, from Eq. (4.4.13). To get **H** it first would be necessary to find **J** from

$$\nabla \cdot \mathbf{J} = -\partial \varrho / \partial t \tag{4.4.14}$$

and then **H** could be found from

$$\nabla \times \mathbf{H} = \mathbf{J} + (\partial \mathbf{D}/\partial t) \tag{4.4.15}$$

As before, it would be inappropriate to use this value of **H** to correct the value of **E** from quasi-static theory.

In subsequent problems we will make use of exactly the procedures outlined above. In addition, we will have occasion to perform coordinate transformations using the Galilean transformation of Chapter 1 on electromagnetic waves. At first glance, such a procedure sounds wrong, but it is nothing more than another example of the applicability of quasi-static theory to slow waves. Certainly if we were to establish a dc magnetic field pattern that varied sinusoidally in space, we would have no objection to using a Galilean transformation to describe the view of such a wave pattern from a coordinate system that is moving at a velocity much less than c. Such a procedure would be valid, because the field obviously really is static in one frame. If it were quasi-static, i.e., if it were to include both **E** and **H** and to be time-varying, but to have a velocity much less than the velocity of light, from our discussion here it is evident that we could treat the fields exactly as if they were static. Hence, it is also correct in such a situation to use a Galilean transformation for low-velocity transformations, as we will see.

4.5 Faraday's Law for Moving Media

We have seen that Maxwell's equations are invariant under a Lorentz transformation so that Faraday's law has the same form in all uniformly moving frames of reference. We here wish to examine some applications of this result in the special case of motion at velocities much less than the velocity of light. The results are useful in the theory of devices such as induction motors.

First, we consider the case in which a loop is moving uniformly at velocity u through a magnetic field that is changing both as a function of time and position. In the frame in which the loop is at rest an observer measures a field $\mathbf{E}'$ and determines the emf V_e' by integrating around the loop. From

Eq. (3.1.2) the integral form of Faraday's law is, in the primed frame

$$V_e' = \oint \mathbf{E}' \cdot dl' = -\int (\partial \mathbf{B}'/\partial t') \cdot d\mathbf{S}' \tag{4.5.1}$$

We must of course measure $\mathbf{E}'$ and $\mathbf{B}'$ in the rest frame of the loop in performing this calculation.

We will be making calculations in a later section involving loops moving at velocities much less than the velocity of light, in which we may wish to calculate the value of the rhs of Eq. (4.5.1) in terms of quantities measured in the laboratory frame, although the emf is to be measured in the rest frame of the loop. Let us define $\mathbf{E}_m$ by the relation

$$\mathbf{E}_m = \mathbf{E} + \mathbf{u} \times \mathbf{B} \tag{4.5.2}$$

We therefore may write

$$\partial \mathbf{B}/\partial t = -\nabla \times \mathbf{E}_m + \nabla \times \mathbf{u} \times \mathbf{B} \tag{4.5.3}$$

Since $\nabla \cdot \mathbf{B} = 0$ we can also write

$$\partial \mathbf{B}/\partial t = -\nabla \times \mathbf{E}_m + \nabla \times \mathbf{u} \times \mathbf{B} - \mathbf{u}(\nabla \cdot \mathbf{B}) \tag{4.5.4}$$

which can be written in the form of the "convective operator" or "substantial derivative" D/Dt defined by

$$D\mathbf{B}/Dt = (\partial \mathbf{B}/\partial t) + \mathbf{u}(\nabla \cdot \mathbf{B}) - \nabla \times \mathbf{u} \times \mathbf{B} = -\nabla \times \mathbf{E}_m \tag{4.5.5}$$

The integral form of this equation is then

$$V_m = \int \mathbf{E}_m \cdot dl = -\int (D\mathbf{B}/Dt) \cdot d\mathbf{S} \tag{4.5.6}$$

It may readily be shown that the rhs of Eq. (4.5.6) is simply the total derivative of the flux through the loop. We now proceed to show that this is the case [16].

In Fig. 4.5-1 we illustrate a problem of this type in which the surface through which the flux is to be evaluated is S_1 at time $t + dt$. The surface

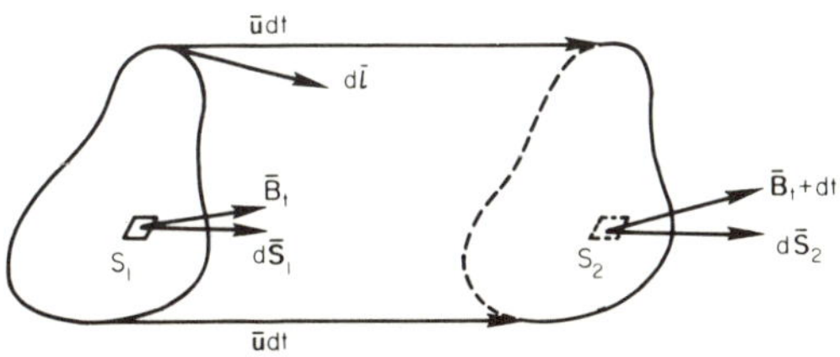

Fig. 4.5-1. A surface S_1 moves at velocity $\mathbf{u}$ to become S_2 at a time dt later. The magnetic field changes in time and space.

moves at velocity $\mathbf{u}$. During the same time interval the magnetic field changes from $\mathbf{B}_t$ to $\mathbf{B}_{t+dt}$ where we imply that $\mathbf{B}_t$ and $\mathbf{B}_{t+dt}$ are the fields defined as a function of position over both S_1 and S_2. We wish to evaluate the total derivative of the flux through the moving surface which we can express as

$$\frac{d}{dt}\int_S \mathbf{B}_t \cdot d\mathbf{S} = \frac{1}{dt}\int_S \mathbf{B}_{t+dt} \cdot d\mathbf{S}_2 - \frac{1}{dt}\int_S \mathbf{B}_t \cdot d\mathbf{S}_1 \tag{4.5.7}$$

which expresses the fact that the magnetic field at $t + dt$ must be used in combination with the surface element $d\mathbf{S}_2$ that is to be considered at that time, and similarly at t we must use $d\mathbf{S}_1$. We now wish to consider the closed surface defined by S_1, S_2, and the surface generated by the motion of the boundary of S from S_1 to S_2. We consider a single instant of time t and apply Gauss's theorem.

$$\int_V \nabla \cdot \mathbf{B}_t\, dV = \int \mathbf{B}_t \cdot d\mathbf{S}_2 - \int \mathbf{B}_t \cdot d\mathbf{S}_1 - \int \mathbf{B}_t \cdot (\mathbf{u}\, dt \times dl) \tag{4.5.8}$$

The flux change across the side surface is represented by the third integral on the rhs of Eq. (4.5.8). The signs in this equation are necessary to conform to our definitions of $d\mathbf{S}_1$, $d\mathbf{S}_2$, and $\mathbf{u}$ in Fig. 4.5-1. All three integrals are surface integrals, their sum being the total surface integral over the volume V. Now we know that

$$\mathbf{B}_{t+dt} = \mathbf{B}_t + (\partial \mathbf{B}_t/\partial t)\, dt \tag{4.5.9}$$

and we substitute Eqs. (4.5.8) and (4.5.9) in Eq. (4.5.7) and obtain

$$\begin{aligned}\frac{d}{dt}\int_S \mathbf{B}_t \cdot d\mathbf{S} &= \frac{1}{dt}\int \mathbf{B}_t \cdot d\mathbf{S}_2 + \int \frac{\partial \mathbf{B}_t}{\partial t} \cdot d\mathbf{S}_2 - \frac{1}{dt}\int \mathbf{B}_t \cdot d\mathbf{S}_1 \\ &= \int \frac{\partial \mathbf{B}_t}{\partial t} \cdot d\mathbf{S}_2 + \frac{1}{dt}\int_V \nabla \cdot \mathbf{B}_t\, dV + \oint \mathbf{B}_t \times \mathbf{u} \cdot dl \end{aligned} \tag{4.5.10}$$

where the last integral is a line integral around the perimeter of S_1. Using Stokes's theorem and

$$dV = \mathbf{u} \cdot d\mathbf{S}\, dt \tag{4.5.11}$$

as defined in Fig. 4.5-1 we find, using $\mathbf{B}$ for $\mathbf{B}_t$ and $d\mathbf{S}$ for $d\mathbf{S}_1$ or $d\mathbf{S}_2$,

$$\frac{d}{dt}\int \mathbf{B} \cdot d\mathbf{S} = \int \frac{\partial \mathbf{B}}{\partial t} \cdot d\mathbf{S} + \int (\nabla \cdot \mathbf{B})\mathbf{u} \cdot d\mathbf{S} + \int \nabla \times \mathbf{B} \times \mathbf{u} \cdot d\mathbf{S} \tag{4.5.12}$$

which is exactly the substantial derivative D/Dt of Eqs. (4.5.5) and (4.5.6).

These two equations define the relativistically correct form of Faraday's law for moving and deforming loops in terms of laboratory-frame fields. The quantity $\mathbf{E}_m$ is approximately $\mathbf{E}'$ for the case of low velocities so that $\gamma = 1$, as may be seen from Eqs. (3.1.18) and (3.1.20). Therefore, Eq. (4.5.1) can be written in the approximate form, valid only for low velocities,

$$V_e' = \oint \mathbf{E}' \cdot dl = -\frac{d}{dt}\int_S \mathbf{B} \cdot d\mathbf{S} = -\frac{d\Phi}{dt} \tag{4.5.13}$$

where $\mathbf{B}$ is measured in the laboratory frame and $\mathbf{E}'$ in the rest frame of the loop, the total derivative includes the effect of motion of the loop through a spatially varying field as well as the effect of a time variation seen in the rest frame, and Φ is the total flux through the loop. This is the usual form given in elementary texts.

It is of interest to express Eq. (4.5.13) in a slightly different form. Using Stokes's theorem in Eq. (4.5.12) we can reexpress the total derivative as follows:

$$V_e' = -\int (\partial\mathbf{B}/\partial t) \cdot d\mathbf{S} - \oint \mathbf{B} \times \mathbf{u} \cdot dl \tag{4.5.14}$$

Consider now a case where $\partial\mathbf{B}/\partial t$ is zero. Then for motion through a field that varies only in space

$$V_e' = \oint \mathbf{u} \times \mathbf{B} \cdot dl \tag{4.5.15}$$

In the event that one is dealing with a straight wire in a field that doesn't vary in the $\boldsymbol{l}$-direction, the equation becomes

$$V_e' = \mathbf{u} \times \mathbf{B} \cdot \boldsymbol{l} = \mathbf{B} \cdot \boldsymbol{l} \times \mathbf{u} \tag{4.5.16}$$

where the line integral is closed at infinity. Note that $\mathbf{u} \times \mathbf{B}$ is in the direction a positive charge would move using the Lorentz force law $\mathbf{F} = q(\mathbf{u} \times \mathbf{B})$, so this is the direction of current flow and the sense of this emf is to produce current flow in the $\mathbf{u} \times \mathbf{B}$ direction. If $\mathbf{B}$, l, and $\mathbf{u}$ are mutually perpendicular we obtain the familiar

$$V_e' = Blu \tag{4.5.17}$$

which can also be derived directly from the Lorentz force equation.

It is evident that there are two ways to calculate the emf around a uniformly moving loop from Faraday's law. Often the simplest approach is to take the partial time derivative of the flux as measured in the rest frame of the loop, as in Eq. (4.5.1). The other approach is to use the laboratory frame field and take the total derivative as in Eqs. (4.5.13) and (4.5.14).

If one is concerned with the emf generated along a moving wire in a magnetic field, the *Blu* equation of Eqs. (4.5.16) and (4.5.17) is applicable, but is evidently incomplete for a loop moving in time-varying fields.

Example V Emfs in Slowly Moving Loops in a Magnetic Field

We will consider a few simple examples in which a square loop with sides of length l lying in the x–z plane is moving at a uniform velocity $\mathbf{u} = \mathbf{1}_z u_0$ in the z-direction through a magnetic field $\mathbf{B} = \mathbf{1}_y B$ directed in the y-direction. B may vary in both time and space, but u_0 is a constant, as illustrated in Fig. 4.5-2, as is v, the velocity of the primed frame.

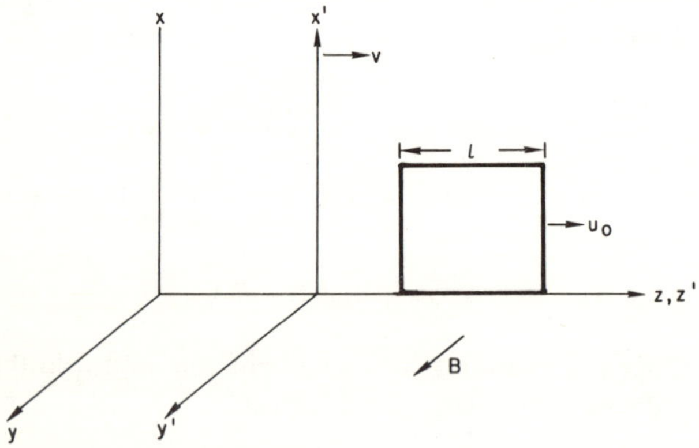

Fig. 4.5-2. A rigid square conducting loop with sides of length l lies in the x–z plane and moves in the positive z-direction at velocity u_0. A moving primed frame of reference also moves in the positive z-direction at velocity v. There is a y-directed magnetic field B present that can vary in time and in z, but not in x or y.

First, consider a field that varies only in time,

$$B = B_0 \sin \omega t \tag{V.1}$$

In the moving frame, using the transformation $t = t'$

$$B' = B_0 \sin \omega t' \tag{V.2}$$

and the emf observed in the primed frame is, from Eq. (4.5.1)

$$V_e' = -l^2(\partial B'/\partial t') = -l^2 B_0 \omega \cos \omega t' = -l^2 B_0 \omega \cos \omega t \tag{V.3}$$

In this case, in which B is independent of position, Eq. (4.5.14) gives the same answer immediately, because the second integral is zero.

Next, consider a field that varies only in space,

$$B = B_0 \sin kz \tag{V.4}$$

In the rest frame moving at velocity u_0 with the loop, we use the transformations $z = z' + u_0 t$ and $t = t'$ and obtain

$$B' = B_0 \sin(ku_0 t' + kz') \tag{V.5}$$

The emf in the rest frame of the loop is obtained from Eq. (4.5.1) by integrating over the loop surface. If one side of the loop is aligned with the x'-axis,

$$\begin{aligned} V_e' &= -l \int_0^l (\partial B'/\partial t')\, dz' \\ &= -lB_0 k u_0 \int_0^l \cos(ku_0 t' + kz')\, dz \\ &= -lB_0 u_0[\sin(ku_0 t' + kl) - \sin ku_0 t'] \end{aligned} \tag{V.6}$$

We can do the same problem in terms of laboratory frame quantities using Eq. (4.5.14). Since $\partial \mathbf{B}/\partial t$ is zero, only the second integral is nonzero, and considering the loop to lie between z and $z + l$, where z is increasing linearly with time,

$$\begin{aligned} V_e' &= -\oint \mathbf{B} \times \mathbf{u} \cdot dl \\ &= -B_0 u_0 l \sin kz + B_0 u_0 l \sin k(z + l) \\ &= -lB_0 u_0[\sin kz - \sin k(z + l)] \end{aligned} \tag{V.7}$$

Equation (V.7) can be transformed to the primed frame and Eq. (V.6) is, of course, obtained if we consider the loop to lie between $z' = 0$ and $z' = l$ in the primed frame as we did in deriving Eq. (V.6).

As a third example consider a traveling wave of magnetic field, traveling in the positive z-direction at velocity $v = \omega/k$. This is equivalent to a "rotating field," if we substitute θ for kz, and so is like the field of the stator of a synchronous machine or an induction motor. Let

$$B = B_0 \sin(\omega t - kz) \tag{V.8}$$

Here we may be interested in the view from two different moving frames, because there are two velocities of interest, v the wave velocity, and u_0 the loop velocity. Let us first transform to the rest frame of the loop as in the preceding examples. As before we use $z = z' + u_0 t$ and $t = t'$ and here we obtain

$$B' = B_0[\sin(\omega - ku_0)t' - kz'] \tag{V.9}$$

and we immediately can note the existence of a new frequency of interest, called the slip frequency, ω_s, given by

$$\omega_s = \omega - ku_0 = \omega[1 - (u_0/v)] \tag{V.10}$$

If u_0 is less than the wave velocity, the apparent frequency seen by the loop, ω_s, is evidently a lower frequency than ω, approaching zero as u_0 approaches v. Applying Faraday's law in the rest frame of the loop as before we find from Eq. (4.5.1) by the same type of integration used in Eq. (V.6)

$$V_e' = -lB_0(v - u_0)[\sin[(\omega - ku_0)t' - kl] - \sin(\omega - ku_0)t'] \tag{V.11}$$

Of interest is the fact that the emf is proportional to the slip speed $(u_0 - v)$ and that it is a sinusoidal voltage at the slip frequency ω_s. In an induction motor this emf produces a current that interacts with the field to produce a force which we will evaluate in another section of this chapter.

PROBLEM 4.5-1 EMF IN A MOVING LOOP

A square loop is moving through a magnetic field as in Fig. 4.5-2, and the magnetic field B is given by

$$B = B_0 \sin(\omega t - kz)$$

as in Example V. Compute the emf in a double-primed frame of reference moving at the phase velocity of the magnetic field wave, ω/k. Show that your result agrees with Eq. (V.11).

4.6 Forces on Prescribed Current Distributions in Prescribed Magnetic Fields—Conventional ac and dc Motors

In this section we assume that we know the magnetic field and the current distribution which is acted upon by whatever force is generated. Thus here we assume either that whatever fields are present as a result of the current distribution are small in comparison with the applied fields or that the field that is present includes the field due to the current distribution. We deal only with uniform motion of rigidly interconnected current elements.

We first examine the force on a single current element carrying a current $\mathbf{I}'$ at a point in space at which the magnetic field has the value $\mathbf{B}'$, as shown in Fig. 4.6-1. The element is considered part of an infinitely long wire, in

Fig. 4.6-1. A portion of an infinitely long wire carrying a current I' that moves at right angles to the direction of current flow at velocity $\mathbf{u}'$ in the presence of a magnetic field $\mathbf{B}'$. I', $\mathbf{u}'$, and $\mathbf{B}'$ are mutually perpendicular.

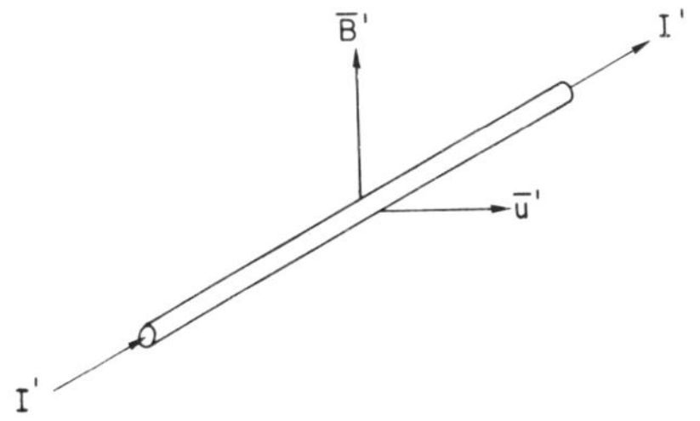

that we will neglect end effects. The wire is free to move only in the direction at right angles to the direction of current flow. The Lorentz force equation gives the force per unit length $\mathbf{F}_l'$ in terms of the charge per unit length q_l' and the velocity $\mathbf{u}'$ of the charge in the current element:

$$\mathbf{F}_l' = q_l'\mathbf{u}' \times \mathbf{B}' \tag{4.6.1}$$

The current $\mathbf{I}'$ is $q_l'\mathbf{u}'$ so

$$\mathbf{F}_l' = \mathbf{I}' \times \mathbf{B}' \tag{4.6.2}$$

Here it is evident that $\mathbf{B}'$ must be the value of the field at the position of the wire, viewed from the frame of reference of the wire or current element.

We now examine three specific cases of magnetic fields and currents corresponding to three standard types of motors, the dc commutator motor, the ac synchronous motor and the ac induction motor as shown in Figs. 4.6-2–4.6-4. We will describe the field of each type, first in the rest frame, and then as transformed to a suitable moving frame. We will then select a current variation that will yield a net force on a current element in the moving frame [17–19].

A. *dc Machine*

In the dc motor the stator field winding carries a dc current and so must provide a distribution of field that is fixed in space and time. The radial field seen at the radius of the conductors in the rotor in a real machine, as shown in Fig. 4.6-2, will be a square wave with rounded corners when plotted

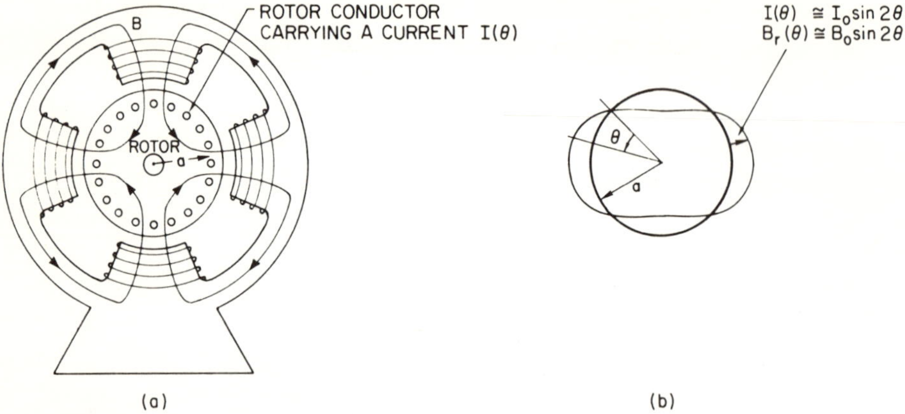

Fig. 4.6-2. (a) A dc commutator machine with four poles. A dc current is supplied to the stator field windings. (b) The radial magnetic field, $B_r(\theta)$, at the radius of the rotor conductors can be approximated by a sine wave. The current in the rotor conductors also varies approximately sinusoidally with angle and has the same phase as $B_r(\theta)$. Both wave patterns are at rest in the laboratory frame. The rotor moves through these patterns and the rotor current is switched from conductor to conductor by the commutator, maintaining the current constant in time at every value of θ, even though the current in any given conductor varies in time.

When used as a motor, input power is electrical and is supplied to the rotor while output power is mechanical and is delivered at the shaft. When used as a generator, the electrical output is removed from the rotor and mechanical power is delivered to the shaft.

as a function of angle. The field can be expressed as a Fourier series in space and, if we approximate it by the fundamental component only, we will be able to predict the main effects. If we do so, we can write the radial field in the laboratory frame as

$$B = B_0 \sin(p\theta/2) \tag{4.6.3}$$

where p is the number of poles in the machine. For example, a two pole machine has one full period of the field in 2π radians of the angle θ which is the physical angle. We now wish to transform this field to a moving frame moving at the velocity of the rotor which we will here assume to be moving at a constant angular velocity $\Omega = d\theta/dt$. We consider only low velocities, so we can use a Galilean transformation, as follows:

$$\theta = \theta' + \Omega t', \qquad t = t' \tag{4.6.4}$$

which says that if $\theta' = 0$ at the origin of the moving frame, the angle θ

increases linearly with time. We then have in the primed frame

$$B' = B_0 \sin[(p/2)(\theta' - \Omega t)] \tag{4.6.5}$$

which is a wave traveling in the negative θ'-direction at angular velocity Ω, as we might expect. Let us now imagine a single conductor at a fixed position, at say $\theta' = 0$ in the moving frame, which is carrying a current I'. Then from Eq. (4.6.2) the force per unit length on this conductor or current element will be

$$F_l' = I' B_0 \sin(p\Omega t/2) \tag{4.6.6}$$

making the assumption that B' and I' are mutually perpendicular, so F_l' is in the θ-direction. Evidently if we wish to make F_l' in Eq. (4.6.6) a unidirectional force, we must cause I' to vary in time with a period commensurate with the period of B'. This is readily accomplished with a commutator switch which produces a roughly square wave of current with time, starting with dc, and again we can approximate this wave with the fundamental only, so that if

$$I' = I_0 \sin[(p\Omega t/2) + \varphi] \tag{4.6.7}$$

we evidently get a unidirectional force

$$F_l' = I_0 B_0 \sin(p\Omega t/2) \sin[(p\Omega t/2) + \varphi] \tag{4.6.8}$$

unless φ is $n\pi/2$ and $n = \pm 1, \pm 3, \ldots$.

B. *Synchronous Machine*

Let us now consider the sort of field that is produced by mechanically rotating the dc field winding which is the rotor of a synchronous machine, as shown in Fig. 4.6-3, in the positive θ-direction. In the laboratory frame such a field appears to be a wave rotating in the positive θ-direction

$$B = B_0 \sin[(p/2)(\Omega t - \theta)] \tag{4.6.9}$$

where again p is the number of poles. If we transform this field to a primed frame moving at the wave angular velocity $(d\theta/dt) = \Omega$, the transformation equation is

$$\theta = \theta' + \Omega t' \tag{4.6.10}$$

and we have for the field in the primed frame

$$B' = B_0 \sin(p\theta'/2) \tag{4.6.11}$$

which as we expect is a non-time-varying field that varies with angle. Clearly a current element in a frame moving at the wave velocity will experience a unidirectional force, if it carries a dc current. The direction of force for a unidirectional current will reverse when θ' is varied through a

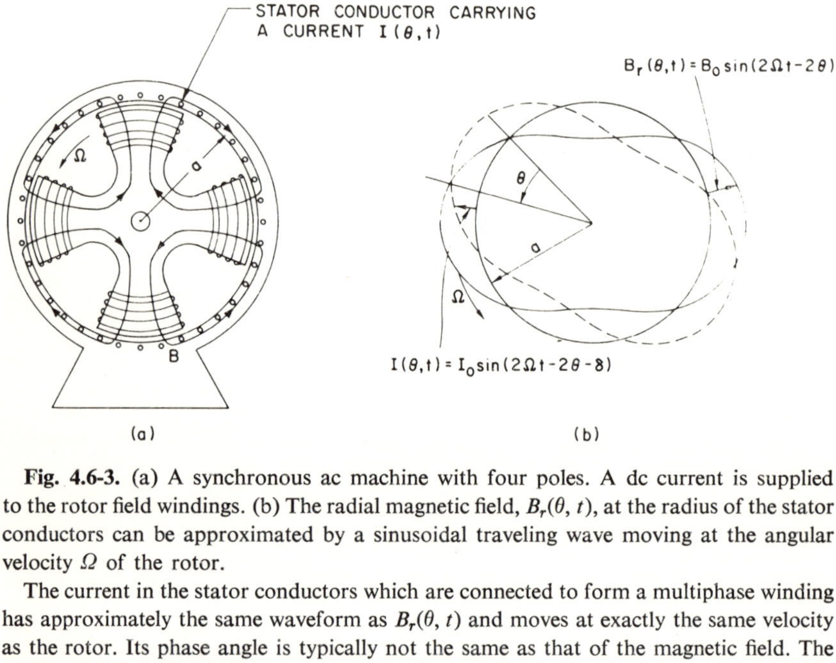

Fig. 4.6-3. (a) A synchronous ac machine with four poles. A dc current is supplied to the rotor field windings. (b) The radial magnetic field, $B_r(\theta, t)$, at the radius of the stator conductors can be approximated by a sinusoidal traveling wave moving at the angular velocity Ω of the rotor.

The current in the stator conductors which are connected to form a multiphase winding has approximately the same waveform as $B_r(\theta, t)$ and moves at exactly the same velocity as the rotor. Its phase angle is typically not the same as that of the magnetic field. The phase angle, δ, of the current depends on the impedance of the external load. When used as a motor, multiphase ac is supplied to the stator and mechanical power is delivered at the shaft; when used as a generator mechanical power is delivered to the shaft and multiphase ac is removed from the stator.

half period, which indicates that if current elements are to be placed regularly around the circumference of the stator at many values of a, some provision must be made to reverse the direction of current flow with the same periodicity as that of B'. Such an arrangement exists in a synchronous machine, since both the rotor which carries dc current and the stator which carries ac in the form of a rotating wave appear as non-time-varying patterns when viewed from a moving frame at the wave velocity. The patterns are synchronous, hence the name of the machine.

C. *Induction Machine*

In the induction motor there is also a rotating wave pattern in the stator that appears as a stationary pattern in a synchronously moving frame (Fig. 4.6-4). But the rotor in this case is nonsynchronous, moving at a velocity less than the wave velocity and so inducing an emf and an associated current pattern in itself. This current pattern, however, does not move at the speed of the rotor but slips around it and moves at synchronous

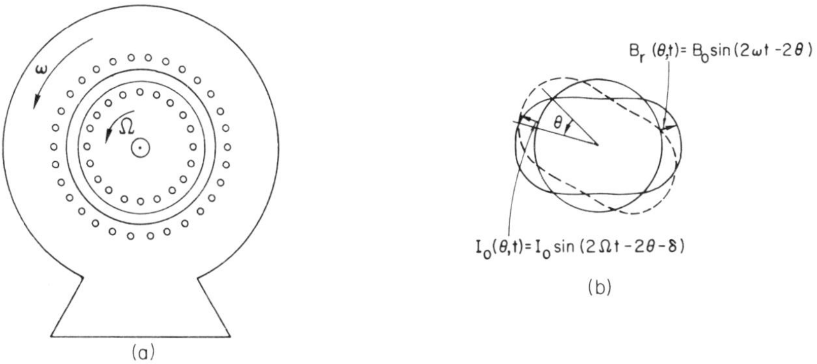

Fig. 4.6-4. (a) An ac induction motor with four poles. An ac multiphase current is supplied to the stator field windings and currents are induced in the rotor conductors by transformer action. (b) The radial magnetic field at the radius of the rotor conductors can be approximated by a sinusoidal traveling wave moving at the angular velocity, ω, where $(p\omega/2)$ is the applied frequency and p is the number of poles. In this figure $p = 4$.
The current in the rotor conductors has the same waveform and moves at the same velocity as the stator field pattern, when viewed from the rest frame. As in the synchronous machine there can be a phase lag, δ, between the stator magnetic field and the rotor current that depends, in this case, on the rotor impedance. When the rotor current pattern is viewed from a frame moving with the rotor, the apparent (slip) frequency and velocity obviously depend on the rotor angular velocity Ω. This slip parameter is zero when $\Omega = \omega$.

speed, thus appearing as a stationary pattern from a moving frame at the wave velocity, as required to produce a net unidirectional force. Thus in the induction motor there is a synchronous rotation of the stator field wave pattern and an induced current pattern in the rotor, even though the rotor itself is not synchronous.

D. *Force Calculations for a Continuous Current Distribution*

In order to compute the net force due to a number of current elements that are rigidly interconnected, it is evident that it is merely necessary to sum the forces due to each element separately. A related problem is that of a continuous current distribution. Let us here picture a conducting sheet carrying a current per unit width $\mathbf{J}_l'$ and compute the force per unit width per unit length which we will call $\mathbf{F}_s'$ the force per unit area. This force is by analogy to Eq. (4.6.2):

$$\mathbf{F}_s' = \mathbf{J}_l' \times \mathbf{B}' \tag{4.6.12}$$

in a magnetic field $\mathbf{B}'$. As usual, if $\mathbf{B}'$ is perpendicular to $\mathbf{J}_l'$ the force will be in a direction perpendicular to both and will thus be a vector lying in the plane of the current sheet.

Let us repeat the calculation of force for a dc motor with a field given by Eqs. (4.6.3) and (4.6.5). The force on a current sheet is evidently

$$F_s' = J_l' B_0 \sin[(p/2)(\theta' + \Omega t)] \tag{4.6.13}$$

if J_l' and B' are mutually perpendicular. The simplest way to get a unidirectional force in this case is evidently to make

$$J_l' = J_0 \sin[(p/2)(\theta' + \Omega t)] \tag{4.6.14}$$

to match the magnetic field. In the laboratory frame the current distribution thus appears stationary.

$$J_l = J_0 \sin(p\theta/2) \tag{4.6.15}$$

This is exactly what a commutator must accomplish by appropriate switching, i.e., it must produce a distribution of current as a function of angle in the rest frame that does not vary with time, even though the individual conductors are moving rapidly around the circumference. A similar technique may be used for analyzing the two other types of machine. The essential requirement is always that, in order to produce a net force, the current distribution, when viewed from the frame of reference in which the magnetic field is non-time-varying, must also be non-time-varying. This current distribution of course produces a magnetic field (90° out of phase with the current), so the above requirement can be stated in terms of the interaction between the magnetic poles of the rotor and stator. Regardless of whether the field pattern is at rest, as in the dc machine, or rotating as in the other machines, the requirement for net torque is that the rotor and stator poles should be at rest with respect to each other. In addition, for maximum torque they should be 90° out of phase, which they are if the current distribution and magnetic field are in phase.

4.7 Self-Consistent Forces and Emfs in Circuits Moving through Prescribed Magnetic Fields—the Induction Motor

In this section we assume, as in the previous section, the existence of prescribed magnetic fields that are not changed by the presence of currents that are present. Here, however, rather than specifying the current that is present, we specify the impedance of a circuit that can have a current induced in it. This form of analysis is much better suited to the problem of the induction machine than that of the preceding section. It introduces the problem of self-consistency in its simplest form.

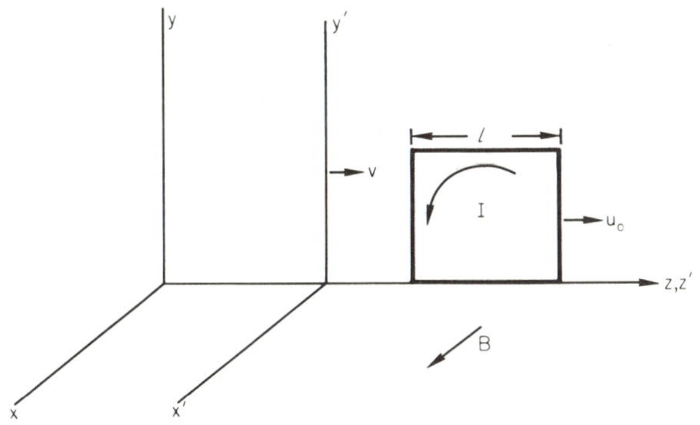

Fig. 4.7-1. A square conducting loop with sides of length l moves in the positive z-direction at velocity u_0 in a y-directed magnetic field B that may vary in time and z but not in x or y. It carries a current I and has an impedance Z at the frequency seen by an observer moving with the loop.

We begin with one of the simplest problems that illustrates the concepts we wish to introduce. In Fig. 4.7-1 a conducting loop lying in the x–z plane is shown moving in the positive z-direction at a uniform velocity u_0 in a y-directed magnetic field that may vary in time and space, but that has a known and prescribed value. We assign a value of impedance to the loop that depends on the frequency of current flowing in the loop. From the equations derived in Section 4.5 we can calculate the emf generated in such a loop for any given field. From the impedance we can calculate the current that flows in the loop. Next, we will calculate the force required to push the loop through the field at constant velocity from the current and the field. If we have a generator, the force will be in a sense to require the external force to deliver energy and, if the loop is acting as a motor, the force will be in a sense to absorb energy from the field. In either case we will obtain self-consistent values of current, emf, and force.

We will take a square loop with sides of length l and a static magnetic field in the laboratory frame of the form

$$B_y = B_0 \sin kz \tag{4.7.1}$$

The emf in the frame of the loop that is generated by moving through such a field at velocity u_0 is given by Eq. (V.6) as discussed in Example V:

$$V_e' = -lB_0 u_0[\sin(ku_0 t + kl) - \sin ku_0 t] \tag{4.7.2}$$

If the impedance of the loop at frequency $\omega' = ku_0$ is Z, the current in the loop is simply

$$I = V_e'/Z(\omega') \tag{4.7.3}$$

I has the same value at all points in the loop, since we here assume that the loop is small in comparison with a free-space wavelength at this frequency.

We now wish to find the external force on the loop required to maintain it at velocity u_0. From Eq. (4.6.2) the force on the sides of the loop is zero and the force on the front wire at $z' = l$ is in the z'-direction and has the value

$$F'(l) = IB'(l) = IB_0 \sin(ku_0 t + kl) \tag{4.7.4}$$

while the force on the rear wire at $z' = 0$ is

$$F'(0) = -IB'(0) = -IB_0 \sin ku_0 t \tag{4.7.5}$$

since the current is in the opposite direction in this wire. Assuming the loop is rigid, we simply add $F'(0)$ and $F'(l)$ to get the total force on the loop,

$$\begin{aligned} F' &= F'(0) + F'(l) \\ &= IB_0[\sin(ku_0 t + kl) - \sin ku_0 t] \end{aligned} \tag{4.7.6}$$

We can now substitute the value of I from Eqs. (4.7.2) and (4.7.3) obtaining

$$F' = -u_0 l B_0^2 [\sin(ku_0 t + kl) - \sin ku_0 t]^2 \tag{4.7.7}$$

which is the force required to maintain the loop at velocity u_0. It is interesting that this is a unidirectional force for all values of kl. For $kl = \pi$ the force is proportional to $\sin^2 ku_0 t$ and has a particularly simple form. This "rectification" of the force comes about because, whenever the magnetic field reverses sign, so does the sense of the induced current and so the force remains in the same direction.

It is interesting to consider the related problem of a moving conducting sheet which covers the x'–y' plane and is moving at the uniform velocity u_0 in the positive z-direction in the presence of a y-directed magnetic field which can vary in time and space as before. As in the preceding example let B be given by Eq. (4.7.1) and compute the induced current per unit width $\mathbf{J}_l'$. The induced emf per unit length due to this magnetic field is independent of x and can be written simply as the electric field E_x'

$$E_x' = u_0 B = u_0 B_0 \sin kz \tag{4.7.8}$$

If the impedance of the conducting sheet at the frequency observed in the moving frame is $Z_s(\omega')$ ohms per square, the current per unit width is

$$J_l' = E_x'/Z_s(\omega') \tag{4.7.9}$$

The force per unit area F_s' is in the z'-direction and has the value

$$F_s' = J_l' B' \tag{4.7.10}$$

The equation for the force is then, using Eqs. (4.7.8) and (4.7.9)

$$F_s' = (u_0 B_0^2/Z_s) \sin^2 kz = (u_0 B_0^2/Z_s) \sin^2(ku_0 t + kz') \tag{4.7.11}$$

In this case the force per period F_p' is of more interest than the force F_s'. If we integrate F_s' over a period of kz' we can obtain F_p', which is still a function of time, but not of z'.

To summarize, in the case of the loop or conducting sheet moving through a magnetic field, we have specified the magnetic field and have found that there is a force with which the loop or sheet must be pushed that is self-consistent with the emf that is produced by the motion.

REFERENCES

1. P. Penfield, Jr., and H. A. Haus, "The Electrodynamics of Moving Media." MIT Press, Cambridge, Massachusetts, 1967.
2. J. A. Stratton, "Electromagnetic Theory." McGraw-Hill, New York, 1941.
3. H. Minkowski, *Gott. Nachr.*, p. 53 (1908).
4. P. J. Cato, "The feasibility of testing the Minkowski electromagnetic momentum," M. S. Thesis. Electrical Engineering Department, MIT, 1967.
5. R. P. James, "Force on permeable matter in time-varying fields," Ph. D. Dissertation. Electrical Engineering Department, Stanford Univ., 1968.

6. W. Shockley and R. P. James, "Try simplest cases" discovery of hidden momentum" forces on "magnetic current," *Phys. Rev. Lett.* **18**, 876–879 (1967).
7. R. P. Feynman, R. B. Leighton, and M. Sands, "The Feynman Lectures on Physics," Vol. 2. Addison-Wesley, Reading, Massachusetts, 1964.
8. D. A. Dunn, R. W. Wallace and S. D. Choi, Waves in a moving medium with finite conductivity, *Proc. IEEE* **57**, 45–57 (1969).
9. E. R. Laithwaite and S. A. Nasar, Linear motion electrical machines, *Proc. IEEE* **58**, 531–542 (1970).
10. W. Shockley and R. P. James, Try simplest cases discovery of hidden momentum forces on magnetic currents, *Phys. Rev. Lett.* **18**, 876–879 (1967).
11. S. D. Choi and D. A. Dunn, A surface charge induction motor, *Proc. IEEE* **59**, 737–748 (1971).
12. J. R. Melcher, Traveling wave induced electroconvection, *Phys. Fluids* **9**, 1548–1555 (1966).
13. E. Grossetti, *Nuovo Cimento* **10**, 193 (1958).
14. R. Arno, *Rendi Atti Reale Accad. Lincei*, **1** (2), 284 (1892).
15. H. H. Woodson and J. R. Melcher, "Electromechanical Dynamics." Wiley, New York, 1968.
16. W. K. H. Panofsky and M. Phillips, "Classical Electricity and Magnetism," 2nd ed. Addison-Wesley, Reading, Massachusetts, 1962.
17. H. H. Skilling, "Electromechanics." Wiley, New York, 1962.
18. E. Levi and M. Panzer, "Electromechanical Power Conversion." McGraw-Hill, New York, 1966.
19. W. F. Hughes and F. J. Young, "The Electromagnetodynamics of Fluids." Wiley, New York, 1966.

CHAPTER 5

Nonuniform Motion of Charged Particles in Prescribed Electric and Magnetic Fields

Two great universal phenomena which must be understood in order to understand electronics are the motions of particles—that is, very small bodies—and the motion of waves . . . . The particles with which electronics deals are chiefly electrons. In order to understand the motion of electrons or ions through the emptiness of a vacuum tube, we must have some understanding of the laws of motion, laws which govern not only the motion of electrons and of ions in vacuum tubes but all motions in the world about us and the motions of heavenly bodies as well: the motion of baseballs and automobiles, the motions of satellites and planets in their orbits, and the motion of the stars in their courses.

To one familiar with the laws of motion, they come to seem simple and obvious. Certainly they are not obvious, for they eluded thoughtful and intelligent men for many centuries. Indeed, recently, during a trip by air, the man in the next seat next to mine marveled that when he dropped a pencil in the plane it did not fall behind on its way to the floor. I explained to him that in the seventeenth century Newton stated that a body continues at rest or in uniform motion in a straight line unless it is acted on by a force, and that this explained the behavior of the dropped pencil.

J. R. Pierce, "Electrons, Waves, and Messages." Doubleday, Garden City, New York, 1956. Reprinted by permission.

5.0 Introduction

In this chapter we picture a region of space within which the values of the electric and magnetic fields are given. We use the Lorentz force equation to calculate the motion of a single point charge within this region, given the initial conditions for the position and velocity of the charge. We also use the same approach to calculate the trajectories of a number of noninteracting charges with different initial conditions, and then superimpose these trajectories in order to obtain a first approximation to the behavior of a beam of interacting particles.

By restricting ourselves in this chapter to a treatment of a single particle or many noninteracting particles we sidestep one aspect of the central problem of electrodynamics, which is the problem of the self-consistent field due to all other charges and currents present. However, even so restricted, the problem of self-consistency remains in relation to the motion of a single particle when the possibility of radiation is included. We know from antenna theory that accelerated charges radiate. The classical first problem of antenna theory is that of a dipole charge distribution that is oscillating sinusoidally. Physically this problem corresponds to an electron and a positron moving back and forth through each other in accelerated motion. A similar but less symmetrical problem is that of an accelerated electron by itself. In either case the charge radiates a calculable amount of power, given the particle trajectory. From the loss of energy, one can calculate the first-order correction to the particle trajectory and presumably then compute the correction to the power radiated because of the change in the trajectory, and so on to a self-consistent solution. The corrected Lorentz force equation suggested by Rohrlich [1] would accomplish this result in a single integral equation.

In this chapter we will discuss some solutions to the Lorentz force equation without taking into account this "radiation reaction" force. Fortunately, in most physical problems the amount of acceleration experienced by a charged particle is so small that only very small corrections to a trajectory calculated neglecting radiation are required. The most extreme physical case is that of a high-energy electron being suddenly decelerated by impact or by undergoing a close collision with another particle or group of particles, as in x-ray generators, high-energy particle accelerators, and thermonuclear plasmas. It is possible that significant new results could be obtained in these areas, if a better force equation were available that would allow truly self-consistent trajectory calculations to be made for single particles including radiation. Feynman [2] has given an interesting discussion of his approach to this unsolved problem of classical physics. This contribution of Feynman and those of others in this field are discussed by Rohrlich [1].

In the case of most physical devices the problem stated in terms of single-particle radiation is not well-posed, however. In most devices, radiation in the usual sense of radiation into an infinite unbounded space does not occur, simply because most devices are not infinite and unbounded. In fact, at frequencies for which radiation occurs from accelerated charges with moderate amounts of acceleration, most devices form completely shielded enclosures. An example is a typical microwave electron tube, the klystron, discussed further in Chapter 6. In the region where the charges are accelerated and decelerated, the tube is specifically designed to act as a resonant cavity at a frequency near the frequencies at which the charges oscillate and might radiate. Consequently, if we were to calculate the radiation from the charges in such a tube, we would have to add the radiation fields due to many charges passing through the resonator at different times. These fields do not travel outward and disappear from the region of the charges, but instead are reflected back to the source and, because of the resonant effect, a field much greater than the radiation field of a single charge builds up in the region of the charges. The effect of this total field on the charge trajectories is much greater than the minor correction due to radiation from a single charge. The prescribed field that we deal with in calculating charge trajectories in this chapter may in some cases be thought of as a field that has been induced on the circuit by the passage of many charges through the circuit.

We therefore will proceed with our calculations of particle trajectories according to the Lorentz force equation, but with the realization that we are ignoring the effects of radiation. We proceed with some confidence, however, because the effect we omit is small in most of the devices considered here.

Before we proceed, however, one comment on the concept of a charged particle is in order. We have grown accustomed to the idea of fundamental particles, because they exist and have been studied in great detail. We are apt to forget what an unusual thing a charged particle really is. Something holds it together that, in the case of the electron, is, at this writing, not at all understood. If we think for a moment of the fantastic forces that exist between electrons that prevent us from assembling any number of them together without an equal number of positive charges, we see what an interesting thing a single electron is. If we were to attempt a study of "continuum electrodynamics" in which no "particles" existed, we would have great difficulty in collecting any finite amount of charge of the same sign in one place. We are, however, able to solve some problems by taking this point of view and we discuss this class of problems in Chapter 7. Here we examine only the motion of charged single particles.

Feynman [2] has given a nice example of the magnitude of the force between electrons. Picture standing at arms length from another person, and imagine that each of you has 1% of your body weight in the form of unneutralized charge. The force between you will be just equal to the "weight" of the earth. This example rather clearly points out the large difference in magnitude between the force of electricity and the force of gravity. It also suggests that if there are two signs of charge, as there are, that it will only be in special circumstances that we will find one sign of charge by itself as we assume to be the case in the problems of this chapter, because of the enormous forces of attraction between charges of the opposite sign. It is therefore only in regions where electric fields exist that push on one sign of charge and pull on the other that any significant charge separation can occur. Sheaths in plasmas, depletion layers in semiconductors, and interelectrode spaces in high vacuum tubes are examples of regions in which significant charge separation can occur in electrical devices and in which it is meaningful to consider motion of charges of a single sign. It is also clear why most of the universe is nearly neutral, either in the form of bound charges at low temperatures or in the form of plasmas at high temperatures.

5.1 Motion of a Point Charge in a Static Magnetic Field Uniform in Space

A particle of charge q and mass m is traveling at velocity $\mathbf{u}$ in a magnetic field $\mathbf{B}$. The velocity may be written in terms of the unit vector $\mathbf{1}_u$ in the $\mathbf{u}$ direction as

$$\mathbf{u} = \mathbf{1}_u u \tag{5.1.1}$$

The Lorentz force equation requires that in a magnetic field

$$\begin{aligned} \mathbf{F} = d(m\mathbf{u})/dt &= qu\mathbf{1}_u \times \mathbf{B} \\ &= \mathbf{1}_u[d(mu)/dt] + mu\, d\mathbf{1}_u/dt \end{aligned} \tag{5.1.2}$$

The part of the force in the $\mathbf{1}_u$ direction must be zero, because the force is in the direction $\mathbf{1}_u \times \mathbf{B}$. This feature of the magnetic force is basic and has the very important consequence that there is no energy gain or loss in a static magnetic field, because $\mathbf{F} \cdot \mathbf{u}$ must be zero. Energy can be transformed from translational kinetic energy to rotational energy, but there can be no gain or loss of total particle energy without an electric field.

In the present case, the fact that the force in the $\mathbf{1}_u$ direction is zero also implies that mu must be constant. Therefore, we may treat mu as a constant in what remains of Eq. (5.1.2)

$$mu\, d\mathbf{1}_u/dt = qu\mathbf{1}_u \times \mathbf{B} \tag{5.1.3}$$

From this equation we see that $d\mathbf{1}_u$ is normal to both $\mathbf{1}_u$ and $\mathbf{B}$. It is therefore correct to use Eq. (2.5.5) which applies when the velocity changes only in direction. A cylindrical coordinate system is defined in Fig. 5.1-1 with

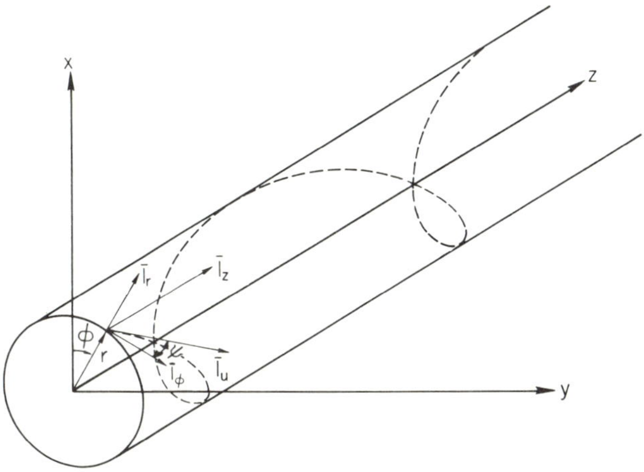

Fig. 5.1-1. A cylindrical coordinate system. A charge particle started in the $\mathbf{1}_u$ direction follows a helical trajectory in a z-directed magnetic field with a helix pitch angle ψ between $\mathbf{1}_u$ and $\mathbf{1}_\varphi$. (An overbar in a figure is equivalent to boldface in the text.)

coordinates (r, φ, z). For our problem, let $\mathbf{B} = \mathbf{1}_z B_z$ and let ψ be the angle between $\mathbf{1}_u$ and $\mathbf{1}_\varphi$, such that $\mathbf{1}_u \times \mathbf{1}_z = \mathbf{1}_r \sin[(\pi/2) - \psi]$. Then Eq. (5.1.3) becomes

$$m_0\gamma_0\, d\mathbf{1}_u/dt = \mathbf{1}_r q B_z \cos\psi \tag{5.1.4}$$

The projections of the vector $\mathbf{1}_u$ in the plane $z =$ constant at two instants of time, t and $(t + dt)$, define the angle $d\varphi$, so the radius vector $\mathbf{1}_u \cos\psi$ times $d\varphi/dt$ is $d\mathbf{1}_u/dt$. Therefore Eq. (5.1.4) may be written as

$$m_0\gamma_0\, d\mathbf{1}_u/dt = m_0\gamma_0 \mathbf{1}_r \cos\psi\, d\varphi/dt = \mathbf{1}_r q B_z \cos\psi \tag{5.1.5}$$

and if r_c is the radius of curvature in the plane $z =$ constant

$$r_c \, d\varphi/dt = u \cos \psi \tag{5.1.6}$$

From Eqs. (5.1.5) and (5.1.6) we can solve for r_c

$$r_c = \frac{m_0 u \mid \cos \psi \mid}{qB_z[1 - (u^2/c^2)]^{1/2}} \tag{5.1.7}$$

Thus r_c is a constant and the charge moves on a helical path with radius r_c and pitch angle ψ. If $\mathbf{u}$ is normal to $\mathbf{B}$, $\psi = 0$, and the path is circular. The radius r_c is the cyclotron radius and tells us the radius of the orbit of a relativistic particle, in the absence of radiation. It is seen from Eq. (5.1.7) that r_c is inversely proportional to the charge to mass ratio of the particle and so will be thousands of times larger for ions than for electrons. It is proportional to velocity for small velocities and increases faster than linearly for high velocities. Note that u, the total velocity, appears in the term in the denominator rather than the velocity component perpendicular to B. The product $B_z r_c$ is called the magnetic rigidity and for $\psi = 0$ is a characteristic of the charge to mass ratio of the particle and its energy.

The cyclotron frequency ω_c is just $(d\varphi/dt)$ and is, from Eq. (5.1.5)

$$\omega_c = \frac{qB_z[1 - (u^2/c^2)]^{1/2}}{m_0} \tag{5.1.8}$$

At low velocities ω_c is independent of velocity and is simply a characteristic of the particle charge to mass ratio (q/m_0) and the magnetic field. At high velocities the frequency is decreased from this value as u approaches c.

The time-average current produced by the particle itself is

$$I = \omega_c q/2\pi \tag{5.1.9}$$

so the particle is equivalent to a magnet with a magnetic moment of value μ_m, where

$$\mu_m = \pi r_c^2 I = m_0^2 u^2 \gamma_0^2 (\cos^2 \psi)/2\gamma_0 m_0 B_z = p_\perp^2/2\gamma_0 m_0 B_z \tag{5.1.10}$$

directed antiparallel to $\mathbf{B}$, where $\mathbf{p}$ is $\gamma_0 m_0 \mathbf{u}$ and $p_\perp = \gamma_0 m_0 u \cos \psi$ is the momentum perpendicular to the magnetic field. The flux Φ through this circular path may be written

$$\Phi = \pi r_c^2 B_z = (2\pi\gamma_0 m_0/q^2)\mu = \pi p_\perp^2/q^2 B_z \tag{5.1.11}$$

For the nonrelativistic case

$$\mu_{\mathrm{m}} \cong m_0 u_\perp^2 / 2B_z \tag{5.1.12}$$

where $u_\perp$ is the velocity $u \cos \psi$ perpendicular to the magnetic field.

5.2 Motion of a Point Charge in a Slowly Varying Magnetic Field Uniform in Space

If the magnetic field is uniform in space but is varying slowly with time, in the sense that the change in field is small during the time it takes for the particle to traverse one turn of its helical orbit, we can make what is called an adiabatic approximation to its trajectory [3]. We will show that, to this approximation, the magnetic moment is an adiabatic invariant and can be used to estimate the change in the energy and momentum of the particle corresponding to a given change in magnetic field [4–6].

We first calculate the energy gained during one turn around the helix, assuming that the radius is constant. Obviously this is not so, if the particle does gain energy, but under the adiabatic approximation the effect of a change in radius is neglected. Under this assumption it is meaningful to calculate the effective electric field acting on the particle from the macroscopic form of Maxwell's equations in which a line integral around a closed loop is taken. Let V_{c} be the change in energy per turn. Then for a field uniform in space

$$\begin{aligned} V_{\mathrm{c}} &= -\oint \mathbf{E} \cdot dl \\ &= \int_S (\partial \mathbf{B}/\partial t) \cdot d\mathbf{S} \\ &= \pi r_{\mathrm{c}}^2 \, \partial B_z/\partial t \\ &= \pi r_{\mathrm{c}}^2 \, dB_z/dt \end{aligned} \tag{5.2.1}$$

Note that V_{c} is positive for an increasing magnetic field. We now divide this energy change by the cyclotron period to get the time-average rate of change of energy with time.

$$d\mathscr{E}/dt = \omega_{\mathrm{c}} V_{\mathrm{c}}/2\pi = (p_\perp^2/2\gamma_0 m_0 B_z)(dB_z/dt) \tag{5.2.2}$$

We can also evaluate $d\mathscr{E}/dt$ by differentiating mc^2

$$\begin{aligned}\frac{d\mathscr{E}}{dt} &= \frac{d}{dt}(\gamma_0 m_0 c^2) \\ &= \frac{d}{dt} m_0 c^2\left(1 - \frac{u^2}{c^2}\right)^{-1/2} \\ &= \frac{d}{dt} m_0 c^2\left(1 + \frac{p^2}{m_0{}^2 c^2}\right)^{1/2} \\ &= \frac{1}{2\gamma_0 m_0} \frac{d}{dt}(p^2)\end{aligned} \tag{5.2.3}$$

Evidently only $p_\perp$ is a function of time, if the magnetic field is uniform in space, so

$$\frac{d\mathscr{E}}{dt} = \frac{1}{2\gamma_0 m_0} \frac{d}{dt}(p_\perp{}^2) \tag{5.2.4}$$

And, equating Eqs. (5.2.2) and (5.2.4) we find that

$$-\frac{p_\perp{}^2}{B_z{}^2}\frac{dB_z}{dt} + \frac{1}{B_z}\frac{dp_\perp{}^2}{dt} = \frac{d}{dt}\left(\frac{p_\perp{}^2}{B_z}\right) = 0 \tag{5.2.5}$$

so that

$$p_\perp{}^2/B_z = \text{constant} \tag{5.2.6}$$

and from Eq. (5.1.11) we see that this equation implies that the flux enclosed is constant. For the nonrelativistic case Eq. (5.2.6) can be reduced to

$$m_0 u_\perp{}^2/2B_z = \text{constant} \tag{5.2.7}$$

which states that the magnetic moment from Eq. (5.1.12) is also a constant or adiabatic invariant. We have thus obtained the result that the particle always surrounds the same flux and has the same magnetic moment and therefore will oscillate with the cyclotron frequency corresponding to whatever field is present at the position of the particle at any instant of time. This is not a surprising result in view of our assumption that the particle makes closed loops which allowed us to take a closed loop integration path. A similar result can be obtained for adiabatic motion in a non-time-varying slowly spatially-varying magnetic field traversed by an electron spinning around a flux line that does not curve [4, 5]. Additional complexities arise if the flux line curves as in a magnetic mirror.

PROBLEM 5.2-1 BETATRON

A point charge is injected into a z-directed magnetic field uniform in space with a velocity u_0 at right angles to the direction of the magnetic field. The magnetic field is increasing linearly with time and has a value of B_0 at $t = 0$ when the particle is injected.

a. Assume that the changing magnetic field is equivalent to an electric field $E_\theta = -(r/2)(dB/dt)$ tangential to the particle orbit. Show the relation of this assumption to Faraday's law. Compare the radius of the electron initially r_0 with its value when the flux has reached the value $2B_0$ using adiabatic theory.

b. Compare the results here with the problem of motion in a static field starting at the same radius r_0 in a field B_0 that is nearly uniform and traveling in a helical orbit with sufficient axial energy to penetrate to a region of higher field of value $2B_0$ which is also nearly uniform. The field changes very slowly from B_0 to $2B_0$. Find the value of final electron radius and compare with the result of part a, also using adiabatic theory.

5.3 Motion of a Point Charge in a Static, Nonuniform, Axially Symmetric Magnetic Field

In the preceding section we saw how an adiabatic theory could be developed for the motion of particles in magnetic fields that vary only "slowly" along an electron trajectory. The trajectories calculated from such a theory must also be "slowly" varying in their form, in order for the theory to give correct results. It is a complex and difficult analytical problem to estimate the degree of error in an adiabatic theory. It is also easy to think of problems in which adiabatic theory is obviously inapplicable. It is possible to obtain equations that describe particle motion in a static spatially varying magnetic field, if a definite equation for the field as a function of distance can be specified. We will examine here the solution to the problem of a field that varies sinusoidally with axial distance and that has axial symmetry. Before proceeding with this specific problem, we will examine the non-relativistic equations of motion in a general axially symmetric field and look at a couple of general theorems that are sometimes of use in this class of problems.

It is convenient to use the magnetic vector potential defined by $\mathbf{B} = \nabla \times \mathbf{A}$ to described fields with axial symmetry, because only a single component of **A**, A_θ, is needed to describe a magnetic field with components B_r and B_z. We will not treat the case here in which B_θ is nonzero, although, of course, this type of field can also be axially symmetric. For zero B_θ, the components of **B** can be expressed in terms of A_θ as

$$B_r = -\partial A_\theta/\partial z, \qquad B_z = (1/r)\partial/\partial r(rA_\theta) \tag{5.3.1}$$

It may be verified that these fields do satisfy $\mathbf{B} = \nabla \times \mathbf{A}$ and hence also $\nabla \cdot \mathbf{B} = 0$.

The Lorentz force equation for such a field can be written in terms of the magnetic field components as follows, for cylindrical coordinates [7]:

$$\begin{aligned} \frac{d^2r}{dt^2} - r\left(\frac{d\theta}{dt}\right)^2 - \eta B_z r\frac{d\theta}{dt} &= 0 \\ \frac{1}{r}\frac{d}{dt}\left(r^2\frac{d\theta}{dt}\right) + \eta B_z\frac{dr}{dt} - \eta B_r\frac{dz}{dt} &= 0 \\ \frac{d^2z}{dt^2} + \eta r B_r\frac{d\theta}{dt} &= 0 \end{aligned} \tag{5.3.2}$$

where η is (q/m) the charge to mass ratio of the particle. For electrons η is a negative quantity. Alternatively, in terms of A_θ

$$\begin{aligned} \frac{d^2r}{dt^2} - r\left(\frac{d\theta}{dt}\right)^2 - \frac{d\theta}{dt}\left(\eta A_\theta + \eta r\frac{\partial A_\theta}{\partial r}\right) &= 0 \\ \frac{d}{dt}\left(r^2\frac{d\theta}{dt} + \eta r A_\theta\right) &= 0 \\ \frac{d^2z}{dt^2} - \eta r\frac{d\theta}{dt}\left(\frac{\partial A_\theta}{\partial z}\right) &= 0 \end{aligned} \tag{5.3.3}$$

The second of Eqs. (5.3.3) can be integrated to give

$$p_\theta = r^2\left(\frac{d\theta}{dt}\right) + \eta r A_\theta \tag{5.3.4}$$

where p_θ is a constant determined by the initial conditions. Busch's theorem follows directly from Eq. (5.3.4) as follows. If we picture two planes denoted by subscripts 1 and 2 perpendicular to the z-axis at which the values of A_θ, r, and $d\theta/dt$ are known, the flux through a circle of radius r_1 may be denoted φ_1 and expressed in terms of $A_{\theta 1}$ the value of A_θ at plane 1

$$\begin{aligned} \varphi_1 &= \int_0^{r_1} 2\pi r B_z\, dr = \int_0^{r_1} 2\pi r\frac{1}{r}\frac{\partial}{\partial r}(rA_0)\, dr \\ &= 2\pi r_1 A_{\theta 1} \end{aligned} \tag{5.3.5}$$

and similarly $\varphi_2 = 2\pi r_2 A_{\theta 2}$. Then we may express Eq. (5.3.4) in terms of the values of the quantities at the two planes as follows, in terms of φ_1 and φ_2 from Eq. (5.3.5) which is Busch's theorem:

$$(d\theta/dt)_2 - (r_1{}^2/r_2{}^2)(d\theta/dt)_1 - (\eta/2\pi r_2{}^2)(\varphi_2 - \varphi_1) = -(\eta/r_2{}^2)(r_2 A_2 - r_1 A_1) \tag{5.3.6}$$

In this form Busch's theorem tells us that the difference in the angular velocity of a particle between two axial positions is completely determined by the difference in the fluxes enclosed by the radii of the particle at the two positions and the values of the radii of the particle at the two positions. The utility of this very general theorem is limited by the fact that one is often hard pressed to find the radii of the particle to put into the theorem.

Another general relation that can be obtained from the force equation is the relation between total energy and the velocity in each direction. As we might expect, the sum of the kinetic energies in each direction is equal to the total energy, as follows:

$$2\,|\,\eta\,|\,H_0 = (dr/dt)^2 + r^2(d\theta/dt)^2 + (dz/dt)^2 \tag{5.3.7}$$

where H_0 is the total kinetic energy in electron volts. The usefulness of this energy equation is greatest in examining the forms of solutions of interest. For example, if we imagine a particle injected into an increasing magnetic field with all its energy in axial velocity at plane 1, then at plane 1

$$2\,|\,\eta\,|\,H_0 = [(dz/dt)_1]^2 \tag{5.3.8}$$

As the particle crosses flux lines it will convert some or all of its energy into rotational energy. If it converts all of its energy into rotational energy, there will be a plane 2 at which, if the particle is rotating about the central axis of the field

$$2\,|\,\eta\,|\,H_0 = r^2[(d\theta/dt)_2]^2 \tag{5.3.9}$$

and the particle will be reflected from the "magnetic mirror" at this plane where its axial and radial velocities are both zero.

Let us now examine in more detail the specific case of a magnetic field that varies sinusoidally, as follows [8]:

$$B_\theta = 0, \qquad B_z = k(A_1 - A_2 \cos kz), \qquad B_r = k^2 r(A_2/2) \sin kz \tag{5.3.10}$$

The equation for A_θ that goes with this magnetic field is

$$A_\theta = (kr/2)(A_1 - A_2 \cos kz) \tag{5.3.11}$$

These fields do not satisfy Laplace's equation, because we have taken a radial variation that is linear in kr corresponding to the first term in an expansion of the modified Bessel function $I_1(kr)$, rather than using the complete equation. Such a procedure is accurate to within 10% out to kr equal to unity. If we restrict our analysis to small kr we therefore will be correct in using this approximate form of the field.

Let us now define the following normalized variables:

$$\begin{aligned}
\varrho &= r/r_0, \qquad r_0 = \text{particle radius for } t = 0 \text{ and } z = 0 \\
\zeta &= kz, \qquad \tau = \omega_0 t \\
\omega_0 &= (|\eta|/2)B_z(0) = (|\eta|\, k/2)(A_1 - A_2) \\
B_z(0) &= \text{axial field for } z = 0 \\
R &= \text{mirror ratio} = \frac{A_1 + A_2}{A_1 - A_2} = \frac{\text{maximum field}}{\text{minimum field}}
\end{aligned} \tag{5.3.12}$$

In terms of these variables the equations of motion, Eqs. (5.3.2) can be written in the form

$$\begin{aligned}
\frac{d^2\varrho}{d\tau^2} &= \frac{1}{\varrho^3} - \varrho\left(\frac{R+1}{2} - \frac{R-1}{2}\cos\zeta\right)^2 \\
\frac{d^2\zeta}{d\tau^2} &= \frac{(kr_0)^2}{2}(R-1)\sin\zeta\left[1 - \varrho^2\left(\frac{R+1}{2} - \frac{R-1}{2}\cos\zeta\right)\right] \\
\frac{d\theta}{d\tau} &= -\frac{1}{\varrho^2} + \frac{R+1}{2} - \frac{R-1}{2}\cos\zeta
\end{aligned} \tag{5.3.13}$$

where the signs are correct for electrons, i.e., η has been taken to be negative. These three simultaneous equations can be solved on a computer to obtain the particle trajectories in the form $\varrho(\zeta)$ and $\theta(\zeta)$ or $\varrho(\tau)$ and $\theta(\tau)$ for various initial conditions. A plot of a trajectory in two views giving $\varrho(\zeta)$ and $\theta(\zeta)$ as obtained on an analog computer is shown in Fig. 5.3-1 for typical initial conditions that lead to nonadiabatic trajectories. The trajectory is nonadiabatic in the sense that the latter part of the trajectory has a drastically different form from the initial part. Initially this particle is traveling axially, while at the end of the trajectory it is spinning about a flux line. The results are quite different from what is predicted from an adiabatic theory which gives good results only for particles that have a substantial fraction of their energy in rotational form initially.

This type of analysis can be applied to a magnetic mirror used for plasma containment, as shown in Fig. 5.3-2. The magnetic mirror is simply a region of magnetic field of the sort we are considering, formed by two coils and

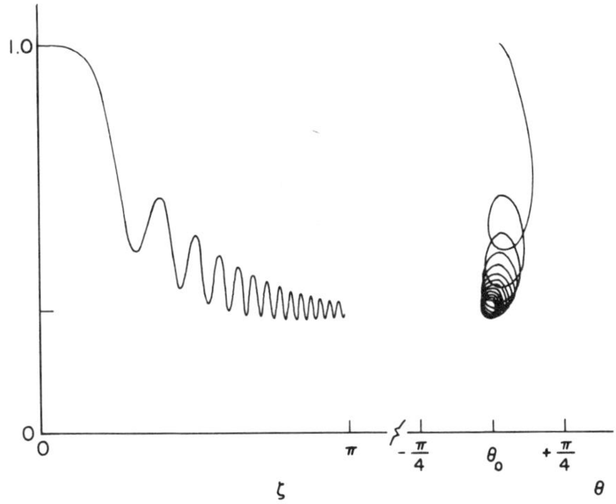

Fig. 5.3-1. Two views of a computed trajectory of a charged particle injected along a flux line in an axially symmetric magnetic field that increases in the direction of the initial velocity. The radius ϱ is shown as a function of both axial position ζ and angle θ [from D. A. Dunn and R. E. Holaday, Electron trajectories in a nonuniform axially symmetric magnetic field, *J. Appl. Phys.* **32**, 1612–1620 (1961)].

designed to make it difficult for particles that are created in the central region to escape out the ends. The adiabatic theory in its simplest form predicts that all particles will be contained that have velocities with a sufficiently large transverse component in comparison with their axial

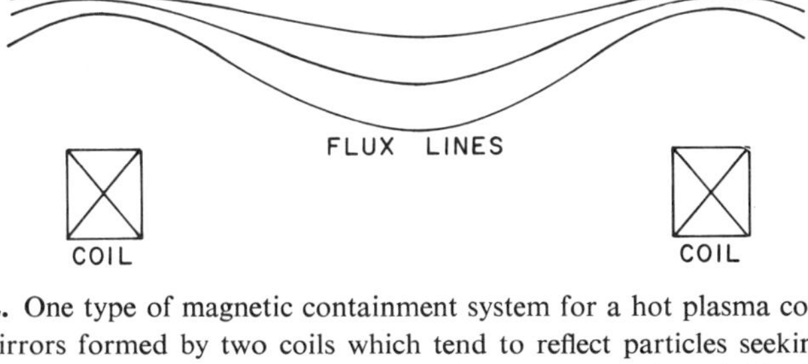

Fig. 5.3-2. One type of magnetic containment system for a hot plasma consists of the magnetic mirrors formed by two coils which tend to reflect particles seeking to escape along the axis of the system. Those particles with velocities lying within the adiabatic loss cone are not reflected by the mirrors and hence escape, according to adiabatic theory.

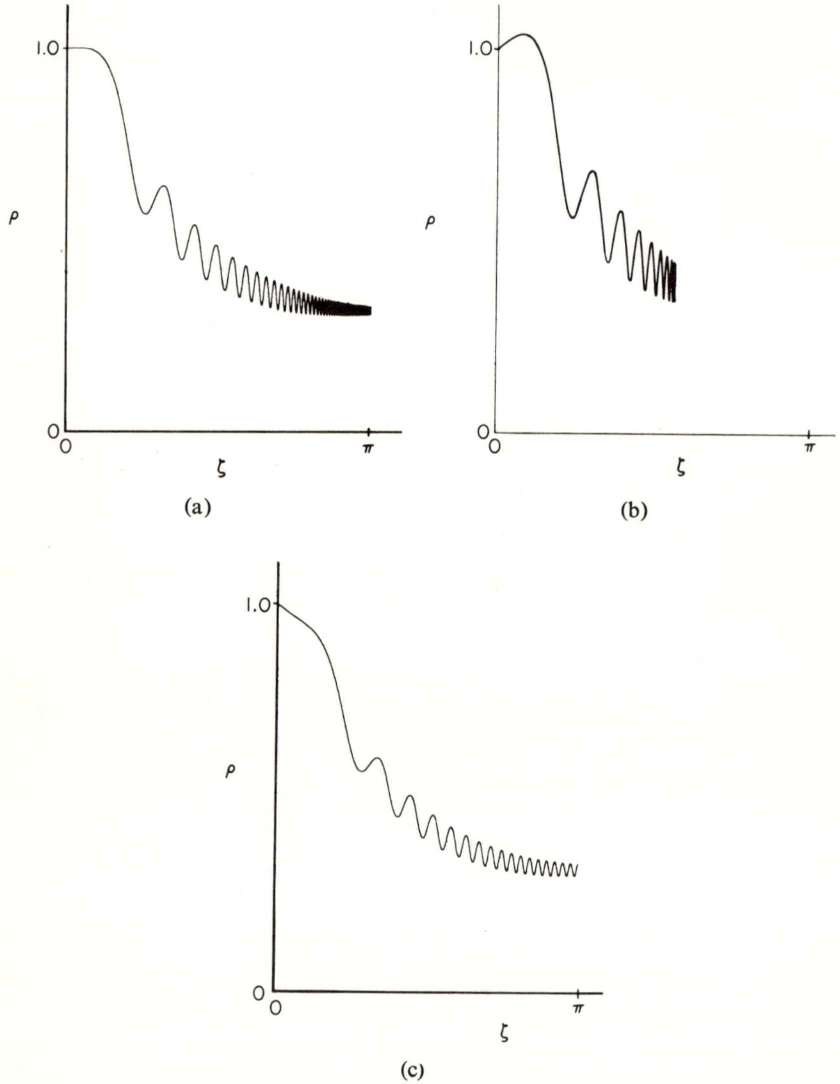

Fig. 5.3-3. Trajectories in the ϱ–ζ plane for a single particle in a magnetic mirror. In all three cases the magnetic flux lines are the same, with one line about which the particle spirals that starts parallel to the ζ-axis at $\zeta = 0$, $\varrho = 1.0$ where the particle is injected and ends up again parallel to the ζ-axis at $\zeta = \pi$. For this case the magnetic flux density is 10 times as great at $\zeta = \pi$ as it is at $\zeta = 0$.

(a) Here the particle was injected along the flux line and the magnitude of the initial velocity was adjusted to just produce mirroring at $\zeta = \pi$. Adiabatic theory predicts that all such particles should escape from the mirror.

component. Particles with too large a fraction of their energy in axial velocity will escape out the "loss cone" shown in Fig. 5.3-2, which according to the simplest adiabatic theory is independent of both the absolute velocity of the particle and whether or not it is spinning about the central axis or an off-axis flux line. The computer results for the nonadiabatic theory show a strong dependence on absolute velocity and the initial radius for particles injected along the axis. Many of the particles lying within the adiabatic loss cone are contained, if they are injected off-axis, because they convert much of their axial velocity to transverse velocity in crossing the inwardly-curved flux lines, as was shown in Fig. 5.3-1. There is also an inward tilt to the loss cone in the nonadiabatic theory, as shown in Fig. 5.3-3 which shows three trajectories for particles in the same field as Fig. 5.3-1 but with three differently directed initial velocities. The particle injected in the axial direction just converts all of its axial velocity to transverse velocity at the value of velocity chosen which was smaller than the value in Fig. 5.3-1. The particle directed inward penetrates the mirror, but the outward directed particle mirrors much closer to the injection plane than the axially directed particle.

We see that there are two qualitatively different effects in a magnetic mirror. One has to do with the fact that the flux density is increasing as the particle moves toward the mirroring point. This effect is closely analogous to the effect of increasing the magnetic flux as a function of time in a region of spatially uniform flux. The other effect has to do with the bending of the flux lines, as seen by an off-axis electron. The effect associated with this type of field could also be visualized in a region of uniform flux density with a changing direction. A particle in this type of field will convert axial energy to rotational energy, just as in the figures here. In fact, if we were to inject particles off-axis, but starting in the region of stronger field and directed toward the region of weaker field in a mirror configuration, we would expect to see trajectories not too different from those shown here. The principal difference would be that the spirals would increase in diameter rather than decrease, as the particle penetrated the weaker field region.

(b) Here the particle had the same initial speed as in (a), but its velocity was directed outward at an angle to the flux line. Mirroring occurred nearer to the injection plane.

(c) Here the particle had the same initial speed as in (a) and (b), but its velocity was directed inward. Mirroring did not occur. Evidently the exact loss cone is tilted inward and depends on parameters other than those of adiabatic theory [from D. A. Dunn and R. E. Holaday, Electron trajectories in a nonuniform axially symmetric magnetic field, *J. Appl. Phys.* **32**, 1612–1620 (1961)].

5.4 Motion of a Point Charge in a Static Electric Field Uniform in Space

We here consider the problem of motion in a non-time-varying electric field uniform in space. We consider particles with initial velocities along the direction of the field and at right angles to the field.

We will start with the case of motion in the direction of the field only, so we can use Eq. (2.5.4) for the longitudinal force. For an electric field **E**

$$\mathbf{F}_l = q\mathbf{E} = m_0\gamma_0{}^3\, d\mathbf{u}/dt \tag{5.4.1}$$

and since we have assumed motion in the direction of the field, **E** is in the **u** direction and we can write the scalar equation

$$qE = m_0\gamma_0{}^3\, du/dt = m_0[1 - (u^2/c^2)]^{-3/2}\, du/dt \tag{5.4.2}$$

which is an ordinary nonlinear first order differential equation in the velocity u which happens to be integrable directly. We can integrate Eq. (5.4.2) and obtain $u(t)$ in the form

$$(u/c)[1 - (u^2/c^2)]^{-1/2} = (qEt/m_0c) + \text{constant} \tag{5.4.3}$$

which for (u/c) much less than unity is the familiar nonrelativistic result.

The relativistic result can be rearranged and a direct solution obtained for $u(t)$ and $z(t)$ in this case. Let us take the case of u equal to zero at t equal to zero, so the constant in Eq. (5.4.3) is zero. Then

$$(u/c) = (qEt/m_0c)[1 + (q^2E^2t^2/m_0{}^2c^2)]^{-1/2} \tag{5.4.4}$$

and $z(t)$ is as follows, taking z equal to zero at t equals zero:

$$z = (m_0c^2/qE)[1 + (q^2E^2t^2/m_0{}^2c^2)]^{1/2} - (m_0c^2/qE) \tag{5.4.5}$$

It is seen that, for small t, the velocity increases linearly with t in Eq. (5.4.4), but for large t the velocity approaches c, as it should. Similarly, for small t, z increases as t^2 in Eq. (5.4.5), while for large t it increases linearly with t, because the velocity is a constant.

Let us now examine the case of motion with a component of velocity at right angles to the direction of the electric field so that **E** is no longer in the direction of **u**. We will take E to be in the x-direction and the velocity to have components in the x- and z-directions.

$$\mathbf{E} = \mathbf{1}_x E, \qquad \mathbf{u} = \mathbf{1}_x u_x + \mathbf{1}_z u_z \tag{5.4.6}$$

The electric field in this case cannot cause a change in the z-velocity or momentum directly, but because relativity limits the total velocity to c, there is a cross-coupling between these velocities that does not exist at low energies. We start with Eq. (2.5.2)

$$q\mathbf{E} = m_0\gamma_0^3\left[\left(1 - \frac{u^2}{c^2}\right)\frac{d\mathbf{u}}{dt} + \frac{u\mathbf{u}}{c^2}\frac{du}{dt}\right] \tag{5.4.7}$$

We can separate Eq. (5.4.7) into its x- and z-components as follows:

$$\begin{aligned} qE &= m_0\gamma_0^3\left[\left(1 - \frac{u^2}{c^2}\right)\frac{du_x}{dt} + \frac{uu_x}{c^2}\frac{du}{dt}\right] \\ 0 &= \left[\left(1 - \frac{u^2}{c^2}\right)\frac{du_z}{dt} + \frac{uu_z}{c^2}\frac{du}{dt}\right] \end{aligned} \tag{5.4.8}$$

The second of Eqs. (5.4.8) can be integrated in the form

$$\int_{u_0}^{u_z}\frac{du_z}{u_z} = -\int_{u_0}^{u}\frac{u\,du}{c^2 - u^2} \tag{5.4.9}$$

where u_0 is the initial velocity, here assumed to be in the z-direction. The initial velocity in the x-direction is assumed to be equal to zero. The result of the integration of Eq. (5.4.9) is

$$\frac{u_z}{u_0} = \left(\frac{c^2 - u^2}{c^2 - u_0^2}\right)^{1/2} \tag{5.4.10}$$

The definition of u is

$$u^2 = u_x^2 + u_z^2 \tag{5.4.11}$$

so that, from Eqs. (5.4.10) and (5.4.11) we can express u_x in terms of u as

$$u_x^2 = u^2 - \frac{u_0^2(c^2 - u^2)}{c^2 - u_0^2} = \frac{c^2(u^2 - u_0^2)}{c^2 - u_0^2} \tag{5.4.12}$$

and use Eq. (5.4.12) to substitute for u_x in the first of Eqs. (5.4.8), obtaining

$$\frac{qE}{m_0c^2(c^2 - u_0^2)^{1/2}} = \frac{u}{(u^2 - u_0^2)^{1/2}(c^2 - u^2)^{3/2}}\frac{du}{dt} \tag{5.4.13}$$

which may be integrated to obtain $u(t)$ in the form

$$u = \left(\frac{u_0^2 + c^2\Omega^2t^2}{1 + \Omega^2t^2}\right)^{1/2} \tag{5.4.14}$$

where

$$\Omega = \frac{qE(c^2 - u_0{}^2)^{1/2}}{m_0c^2} \tag{5.4.15}$$

Then u_x and u_z can be obtained from Eqs. (5.4.11) and (5.4.12) as follows:

$$u_x = \frac{dx}{dt} \frac{c\Omega t}{(1 + \Omega^2t^2)^{1/2}}, \qquad u_z = \frac{dz}{dt} = \frac{u_0}{(1 + \Omega^2t^2)^{1/2}} \tag{5.4.16}$$

Integrating Eqs. (5.4.16) yields $x(t)$ and $z(t)$, assuming $x(0) = z(0) = 0$,

$$x = (c/\Omega)[(1 + \Omega^2t^2)^{1/2} - 1], \qquad z = (u_0/\Omega)\sinh^{-1}\Omega t \tag{5.4.17}$$

from which t may be eliminated to give the trajectory in configuration space as follows:

$$\Omega x/c = \cosh(\Omega z/u_0) - 1 \tag{5.4.18}$$

For low velocities, $\Omega \cong qE/m_0c$, and, if we expand $\cosh(\Omega z/u_0)$ in a series valid for small values of the argument, we obtain

$$x \cong qEz^2/2m_0u_0{}^2 \tag{5.4.19}$$

which is the well known result for deflection of a particle in an electric field at low velocities, as in the region between the plates of a cathod ray tube.

Problem 5.4-1 A Heavy-Particle Accelerator

In a linear particle accelerator for ions or in the buncher section of an electron accelerator the velocity of the particle increases very substantially as a function of the distance traveled. The electric field of the traveling electromagnetic wave that accelerates the particles must be kept in synchronism with the particles throughout the entire length of the accelerator. In order to keep the electric field that the particle sees exactly constant over the entire length of a particle accelerator a great deal of effort would be required. In practice, it is sufficient to keep the particles in the same half-wavelength of the field so that they see a unidirectional field, although not a constant field. In this problem, assume that the field seen by the particles is constant over the length of an accelerator designed to accelerate protons with an initial energy of 100 keV to a final energy of 100 MeV using an electric field of 100 kV/cm in the frame of the protons. Calculate the velocity profile of the traveling wave as a function of distance needed to maintain synchronism with the protons.

5.5 Motion of a Point Charge in Crossed Static Electric and Magnetic Fields Uniform in Space

As we have seen, what appears as an electric field to an observer in one frame of reference will appear as a mixture of electric and magnetic fields when viewed by an observer moving with respect to the original frame. Similarly, what appears to us to be a mixed electric and magnetic field in the rest frame may appear as a pure electric field from a properly chosen moving frame. There is thus the possibility of simplifying some calculations of trajectories in mixed fields by selecting the proper frame of reference. We here examine some problems of this type which include the practically important case of motion in crossed electric and magnetic fields.

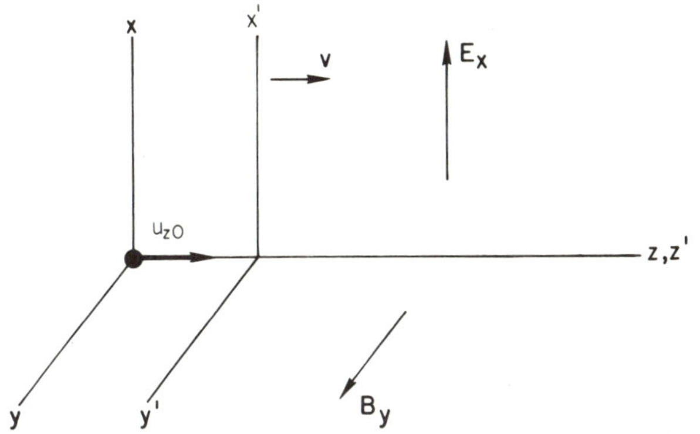

Fig. 5.5-1. Crossed electric and magnetic fields E_x and B_y. A particle has an initial velocity u_{z0}. A primed moving frame of reference has a constant velocity v in the positive z-direction.

Let us consider the problem shown in Fig. 5.5-1 in which there is a uniform electric field E_x and a uniform magnetic field B_y in the rest frame. Let us examine the equations for the field when viewed from a primed frame of reference moving with velocity v in the z-direction. From Eq. (3.1.11)

$$\begin{aligned} E_x' &= \gamma(E_x - \beta c B_y) \\ B_y' &= \gamma[B_y - (\beta/c)E_x] \\ E_y' &= E_z' = B_x' = B_z' = 0 \end{aligned} \tag{5.5.1}$$

If we choose β such that $E_x = \beta c B_y$, we have in the primed frame

$$E_x' = 0, \qquad B_y' = B_y/\gamma \tag{5.5.2}$$

and if we now solve for motion in the primed frame we need only solve the problem of motion in a uniform magnetic field and then transform back to the rest frame to get the trajectory in the mixed fields. A limitation on this approach is reached when E_x/cB_y becomes equal to unity, because then the electric field can no longer be transformed away with β less than unity. For $E_x > cB_y$, it is convenient to choose β such that $(\beta/c)E_x = B_y$ which yields a pure electric field in the moving frame:

$$E_x' = E_x/\gamma, \qquad B_y' = 0 \tag{5.5.3}$$

We have considered the necessary equations of motion for each of these cases in preceding sections.

Here we will consider further the case for which $E_x' = 0$, $B_y' = B_y/\gamma$, and $v = E_x/B_y$. We will consider only the case of an initial velocity in the z-direction u_{z0}, so that in the primed frame the initial velocity is u'_{z0} where from Eq. (2.3.4)

$$u'_{z0} = \frac{u_{z0} - \beta c}{[1 - (\beta u_{z0}/c)]} \tag{5.5.4}$$

In the primed frame we will then have circular motion with radius r_c' as given by Eq. (5.1.7)

$$r_c' = \frac{m_0 u'_{z0}}{qB_y'[1 - (u_{z0}^{2\prime}/c^2)]^{1/2}} \tag{5.5.5}$$

where here we allow negative values of r_c' corresponding to negative u'_{z0} for convenience. The radius is of course always the absolute value of r_c'. The frequency of rotation is ω_c' as given by Eq. (5.1.8)

$$\omega_c' = \frac{qB_y'[1 - (u_{z0}^{2\prime}/c^2)]^{1/2}}{m_0} \tag{5.5.6}$$

The value of u_{z0} can be either greater or less than v. Let us take initial conditions $z' = x' = 0$ at $t' = 0$. Then the z' vs. x' trajectories which are circles in the primed frame can be written in the form

$$z' = r_c' \sin \omega_c' t', \qquad x' = -r_c'(1 - \cos \omega_c' t') \tag{5.5.7}$$

where r_c' is positive for $u_{z0} > v$ and negative for $u_{z0} < v$. These equations can be readily transformed to the rest frame using Eqs. (2.1.12) and defining

$$\alpha = \frac{\omega_c' r_c'}{\beta c} = -\frac{[1 - (u_{z0}/\beta c)]}{[1 - (\beta u_{z0}/c)]}, \qquad \theta = \omega_c' \gamma [t - (\beta z/c)] = \omega_c' t' \tag{5.5.8}$$

so that we obtain for the rest frame trajectories

$$z = \gamma r_c' \sin\theta + \gamma(r_c'/\alpha)\theta, \qquad x = -r_c'(1 - \cos\theta) \tag{5.5.9}$$

where r_c' and ω_c' can be written in terms of rest frame quantities as follows

$$\begin{aligned} r_c' &= (\gamma^2\gamma_0 m_0/qB_y)(u_{z0} - \beta c) \\ \omega_c' &= (qB_y/\gamma^2\gamma_0 m_0)[1 - (\beta u_{z0}/c)]^{-1} \end{aligned} \tag{5.5.10}$$

The motion is a uniform translation in z as θ increases with a superimposed circular motion in x and z.

An interesting feature of these equations is the fact that the radius r_c' goes to zero for u_{z0} equal to v leaving a simple translation in z with no circular motion. On the other hand, if we have u_{z0} initially in the negative z-direction in Fig. 5.5-1, there is no value of $|u_{z0}|$ for which r_c' is zero, and also the term that is linearly proportional to θ in Eq. (5.5.9) still results in a positively increasing z. Equations (5.5.9) can be renormalized as follows:

$$Z = z\alpha/\gamma r_c' = \alpha \sin\theta + \theta, \qquad X = x\alpha/r_c' = \alpha(\cos\theta - 1) \tag{5.5.11}$$

where α has the same sign as r_c', i.e., α is negative for $u_{z0} < v$ and positive for $u_{z0} > v$. Figure 5.5-2 is a plot of Z vs. X for α between -1 and 1, corresponding to initial velocities u_{z0} between 0 and $2\beta c/(1 + \beta^2)$. For negative values of u_{z0}, α varies between -1 and $-\infty$. Figure 5.5-3 is a plot of Z vs. X for several negative values of u_{z0}. Here it is seen that the trajectories all start out with negative values of Z, but the particles are very rapidly turned around and carried off to positive Z. A crossed-field region of this type can thus act as a trap for particles, allowing particles traveling from left to right in Fig. 5.5-2 to get through, but turning back any particles that try to go from right to left [9]. The particles turned back are also carried off to the right even though they start to the left. Physically this is a result of the fact that, if we think of the electric and magnetic force terms separately, they just cancel for particles traveling to the right at $u_{z0} = v$. For any initial velocity to the left, the magnetic force is in the opposite sense and so adds to the electric force, resulting in a large deflection which turns into a circular motion plus a drift to the right.

This velocity selective feature of crossed electric and magnetic fields also makes them very useful for mass spectrographs and related devices. A very simple test of special relativity can be made using this type of system which is well suited to the determination of the charge to mass ratio of a particle. Since the charge is constant and the mass increases according to

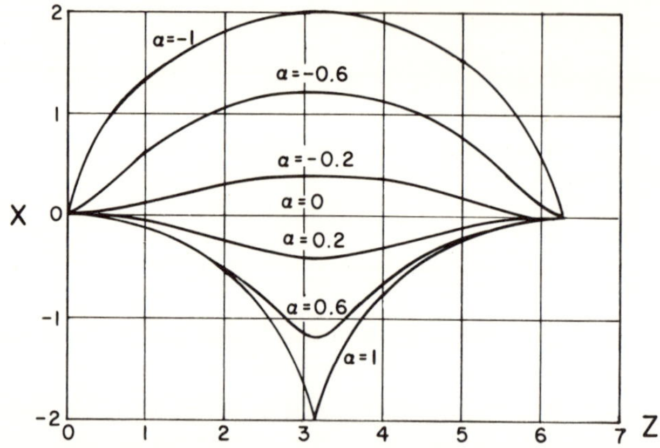

Fig. 5.5-2. Particle trajectories Z vs. X in crossed electric and magnetic fields for various initial velocities, all directed in the positive z-direction. For $\alpha = 0$ the trajectory is a straight line in the direction of the initial velocity [from D. A. Dunn, R. P. Borghi, and G. Wada, A crossed-field multisegment collector for beam-type tubes, *IRE Trans. Electron Devices* **ED-7,** 262-267 (1960)].

special relativity, this type of experiment allows a simple measure of the mass increase with velocity and has been so used with considerable accuracy.

Perhaps the most important present practical application of this field configuration is in the magnetron and other crossed-field microwave tubes in which electrons are given energy by the dc electric field and give up this

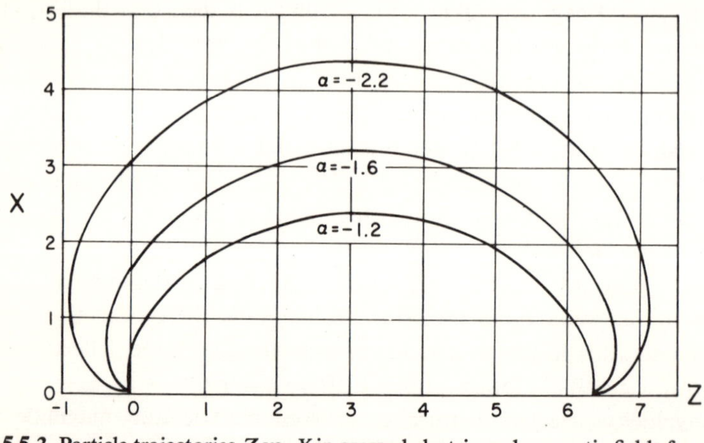

Fig. 5.5-3. Particle trajectories Z vs. X in crossed electric and magnetic fields for various initial velocities, all directed in the negative z-direction. In this case all particles end up going to the right after a short initial excursion to the left of $z = 0$ [from D. A. Dunn, R. P. Borghi, and G. Wada, A crossed-field multisegment collector for beam-type tubes, *IRE Trans. Electron Devices* **ED-7,** 262–267 (1960)].

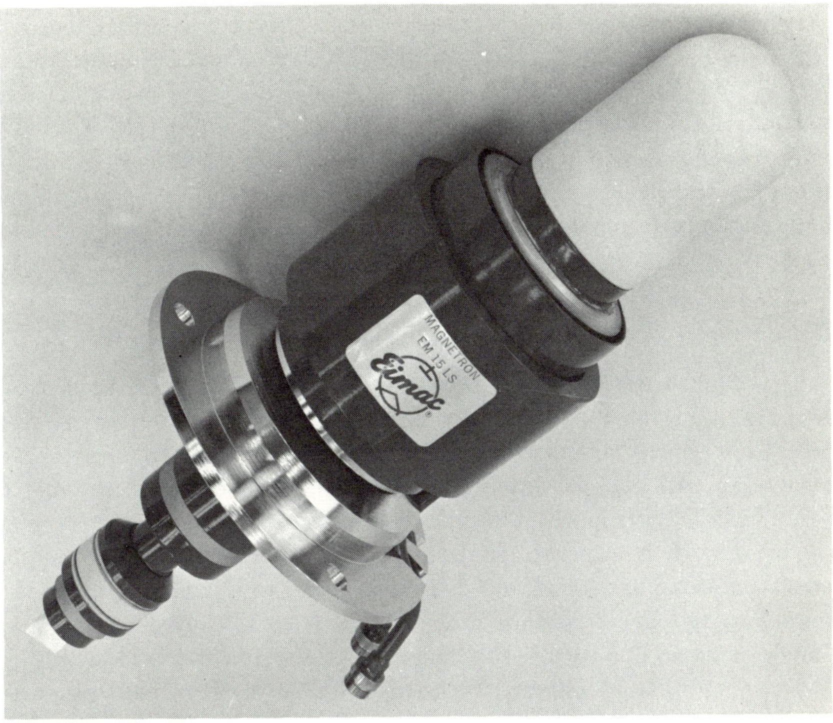

Fig. 5.5-4. A commercial magnetron used in industrial heating applications. It provides 25 kW output at 915 MHz at an efficiency of 85%. [Courtesy Varian Associates.]

dc energy by interacting with rf fields produced by a series of resonators or other rf circuits. Fig. 5.5-4 is a photograph of a magnetron oscillator used for industrial heating applications. Most devices of this class are reentrant with the rf circuit arranged in a circular configuration around a cylindrical cathode. The electron trajectories in such a device are strongly influenced by the rf fields and an understanding of these devices is therefore dependent on the solution of particle trajectory problems in time-varying fields. In the next section we examine some problems of this type.

5.6 Motion of Many Noninteracting Particles in a Static Electric Field Uniform in Space

Much of the early microwave tube work was done on the basis of an understanding obtained by considering the motion of many noninteracting particles. Such an approach is bound to lead to errors, both in understanding the behavior of the particles when they are squeezed together tightly and in numerical values for charge densities and currents. Nevertheless this type of analysis is very simple and does give a qualitatively correct view of many processes of a very fundamental nature [10]. The basic approach that is used is to obtain the general equation for a single particle trajectory and then to plot a number of these trajectories for various initial conditions, corresponding to different particles having different injection times.

A. *The Case of Zero dc Field—the Klystron Amplifier*

The simplest case is that of the klystron amplifier of Fig. 5.6-1. An electron beam traveling at a high velocity passes through a buncher cavity and receives a velocity modulation from the action of the electric field in the cavity gap. All particles leaving the buncher gap then travel through a field-free drift space in which the velocity modulation is converted to current modulation. This current modulated beam is passed through an output cavity, inducing an rf field and delivering power to a load coupled to the output cavity. The trajectory problem that is to be solved is extremely simple; it is that of a particle traveling through a region of zero electric field.

The analysis is as follows. We assume that a beam of electrons with energy V_0 passes through a very short buncher gap with an rf voltage

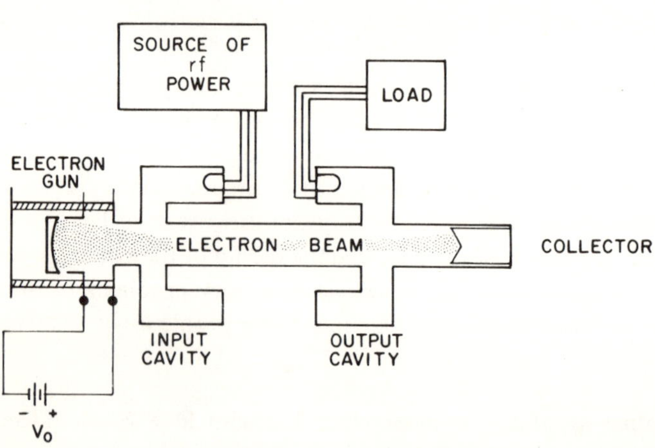

Fig. 5.6-1. A two-cavity klystron amplifier in which particles at a velocity corresponding to the potential V_0 traverse the input cavity gap across which there is a potential $V_1 \sin \omega t$, and then travel at a constant velocity through the drift space to the output cavity gap where the ac current resulting from velocity modulation drives the second cavity and delivers power to the load.

$V_1 \sin \omega t$ across it. The energy of electrons leaving the gap and entering the drift space at $z = 0$ is then

$$V = V_0 + V_1 \sin \omega t \tag{5.6.1}$$

and the corresponding velocity is $u = (2eV/m)^{1/2}$ if the gap is short enough, so

$$u = u_0[1 + (V_1/V_0) \sin \omega t]^{1/2} \tag{5.6.2}$$

where $u_0 = (2eV_0/m)^{1/2}$ is the dc beam velocity. For $(V_1/V_0) \ll 1$, we may write Eq. (5.6.2) approximately as

$$u = u_0[1 + (V_1/2V_0) \sin \omega t] \tag{5.6.3}$$

A sinusoidal velocity modulation of the form of Eq. (5.6.3) could, in principle, be produced by a nonsinusoidally varying gap voltage of arbitrary amplitude, so our ability to produce this simple form of velocity modulation is not restricted to small signals. Nor will our analysis that follows be a "small-signal" analysis in the usual sense. In fact the very reason for performing this type of analysis is to obtain some estimate of the large-signal behavior of the klystron. Properties such as efficiency cannot be obtained from a true small-signal theory in which, at all positions and times, the time-varying portion of all quantities is small in comparison with their dc values. Here the ac velocity is normally small in comparison with the dc velocity at the injection plane, but the ac current can and does become large at other positions downstream, as we will see.

We will analyze specifically now the model of the klystron amplifier shown in Fig. 5.6-2. Within the region from $z = 0$ to $z = d$, the dc electric field is zero, and we wish to calculate the position and velocity as a function of time for particles injected with velocities specified by Eq. (5.6.3) at $z = 0$. The Lorentz force law tells us that the acceleration is zero in a field-free space so the velocity and momentum of each particle remains constant after injection, exactly as given by Eq. (5.6.3). Evidently then the trajectories in distance–time space will be straight lines with sinusoidally varying slopes. If we wish to use Eq. (5.6.2) to specify the slopes corresponding to a sinusoidal voltage variation, it is very little additional trouble. However, as we will see, there are other things such as the space-charge field that have been left out of this analysis that are more important than this matter of the velocity variation, so we will stick with Eq. (5.6.3).

Figure 5.6-3 is a distance–time diagram for the case $(V_1/2V_0) = 0.2$. The distance d across the diode is taken to be unity. A particle injected at

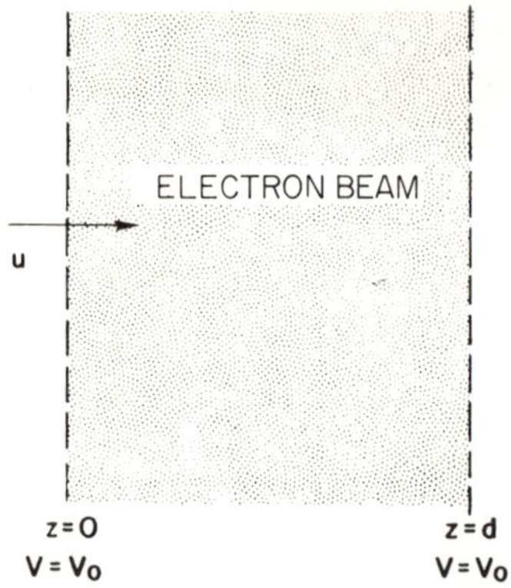

Fig. 5.6-2. A one-dimensional model of the drift space of a two-cavity klystron. The beam enters with velocity $u = u_0[1 + (V_1/2V_0) \sin \omega t]$. In this model the space charge of the beam is neglected, so the drift space is field free, both walls being held at the same dc potential V_0.

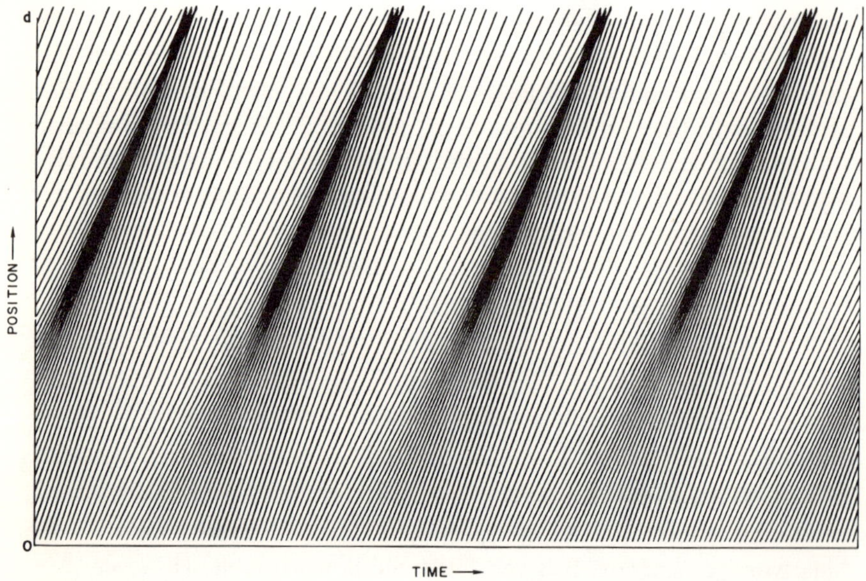

Fig. 5.6-3. A distance–time diagram for many noninteracting particles in a field-free space that are injected uniformly in time with a sinusoidally varying injection velocity $u = u_0(1 + 0.2 \sin \omega t)$. The total distance shown is the distance d, and the transit time across this distance of a particle traveling at u_0 is t_0. The modulation frequency is determined by the relation $\omega t_0 = 2\pi$. The number of particles per period $(2\pi/\omega)$ was 40.

the dc velocity will traverse the diode in a time $t_0 = d/u_0$, and the period of the rf modulation, $(2\pi/\omega)$, is normalized to this time, i.e., specified as a factor times t_0. For the case of Fig. 5.6-3, $\omega t_0 = 2\pi$. Note that $\omega t_0 = (\omega d/u_0)$ which is the total phase change experienced by a particle traversing the region at a velocity u_0 with a frequency ω. In this figure the number of particles per modulation period was taken to be 40.

The use of this sort of diagram is as follows. If we position ourselves at some point in z such as at $z = d$ and measure the number of particles crossing this plane per unit time (the current) we will find that it varies in time at the frequency of the signal applied to the buncher gap. The amplitude of the current does not vary sinusoidally with time, however, and it can have strong components at harmonics of the drive frequency. In a klystron amplifier, a resonant output cavity is driven by this current, and, being resonant at the fundamental frequency, it contains only the fundamental frequency in its output. The optimum place to put the output cavity for maximum current can be found by Fourier analyzing the current waveform as a function of z. In Fig. 5.6-3, one suspects that a maximum might occur between $z = 0.8d$ and $z = d$.

The predicted optimum position from Fig. 5.6-3 obviously is not exact, because of space-charge forces acting between the electrons which prevent the degree of bunching predicted by this theory from being reached and cause a shift in the position of the maximum current. It is surprising, at first glance, that this elementary theory comes out close enough to be interesting, when it is considered that the only field acting on the electrons in the drift space is their own space-charge field which we have neglected. If there were a strong external field acting to bunch the electrons and, at the same time, there were space-charge debunching fields present, we might feel more comfortable with this assumption. In the present case, however, there are strong forces that tend to drive the bunch together just as if there were an external field present, and they are simply the inertia forces of the electrons. The electrons have been injected into the drift space with initial velocities and momenta that will carry them together even with space-charge forces acting to keep them apart. The relative strength of the space-charge debunching forces and these bunching forces can be evaluated by the methods to be explained in the next chapter.

Here we also wish to examine another type of trajectory diagram that is widely used in studying many-particle problems. This is the phase diagram in which we plot the velocity of the particles versus the phase angle at which the particles arrive at a given point in space. We wish to choose a particular point in space and as time runs through a full period of the input signal waveform, we wish to record both the time at which each particle passes us and its velocity as it passes us and to record these in a phase versus velocity diagram. We can obtain this information graphically from Fig. 5.6-3 or we can, in this case, get an analytical solution. At any position z the time at which an electron arrives t_b is related to its starting time t_a through the equation

$$t_b - t_a = z/u_a \tag{5.6.4}$$

where u_a is the velocity at t_a. Then from Eq. (5.6.3)

$$\omega t_b = \omega t_a + \frac{\omega z}{u_0[1 + (V_1/2V_0) \sin \omega t_a]} \tag{5.6.5}$$

Evidently the velocity of a particle at position z at time t_b is the same as its velocity when it left at t_a. So if we wish to plot (u_a/u_0) as a function of ωt_b we need only Eqs. (5.6.3) and (5.6.5). Such a plot is given in Fig. 5.6-4 for the positions $z = 0.8d$ and $z = d$. This diagram tells us not only how the particle velocity is varying during a cycle, but also how the particles are bunched and how fast the bunch (if any) is moving with respect to the background particles. In many devices an examination of this type of diagram can lead to an understanding of where the energy is going and can often suggest ways of improving device efficiency.

B. *The Case of a Retarding dc Field—the Reflex Klystron*

We now wish to consider a further example of noninteracting particle analysis in which a dc retarding field is applied to a stream of electrons which has been initially velocity modulated, as in the previous example. If this dc retarding field is sufficient, the particles will all be returned to the plane, but they will not return uniformly spaced in time. Again there will be a bunching effect and an ac current will be produced. The physical device that uses this effect is the reflex klystron shown schematically in Fig. 5.6-5. An electron beam with energy V_0 is passed through a resonator where it is velocity modulated and then into a dc retarding field produced by a repeller electrode operated at a potential below the cathode potential of the beam. The beam is turned around and the velocity modulation is converted to current modulation during its transit from the resonator to the turn around point and return, where the current is used to drive the same resonator that modulated the beam initially. The system can oscillate, if the phasing of the current relative to the voltage across the gap is correct. The amplitude of the rf gap voltage is controlled by the load. If the loading is too heavy

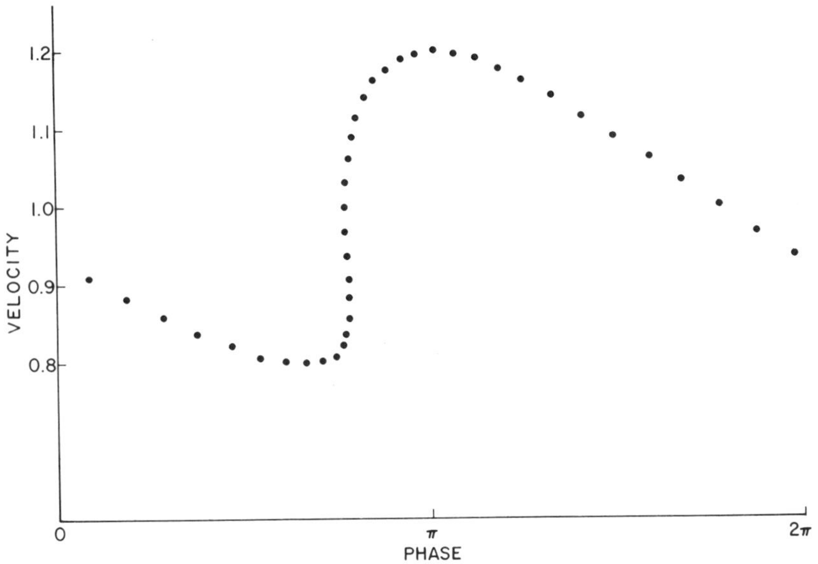

(a)

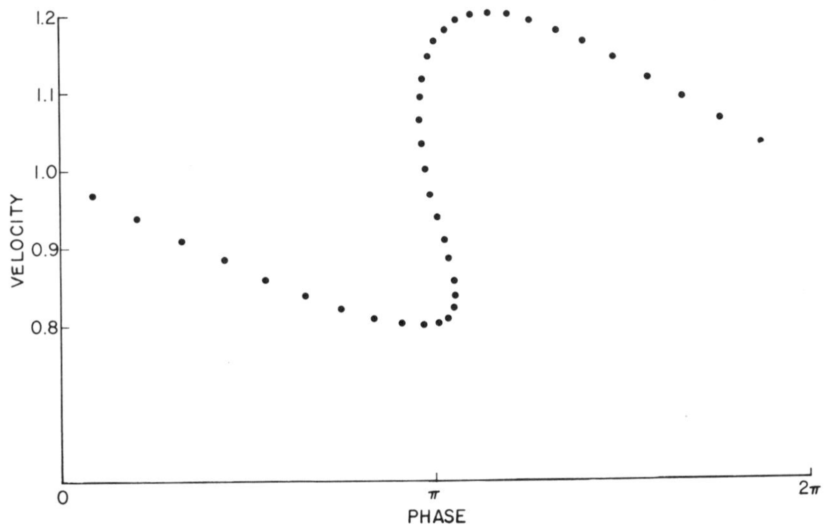

(b)

Fig. 5.6-4. (a) The velocity as a function of the arrival phase of a sequence of particles during one cycle of the modulation frequency, for the case of Fig. 5.6-3 at the position $(z/d) = 0.8$. The velocity is normalized to the dc velocity u_0 and phase is ωt_b where t_b is time of arrival at the position $(z/d) = 0.8$. Note that a considerable number of particles arrive at essentially the same time, corresponding to a sharp pulse of current at about 0.8π.

(b) The velocity as a function of the arrival phase of a sequence of particles during one cycle of the modulation frequency, for the case of Fig. 5.6-3 at the position $(z/d) = 1.0$. Here the current pulse has split into two smaller peaks than was the case at $(z/d) = 0.8$, corresponding to an "overbunched" condition.

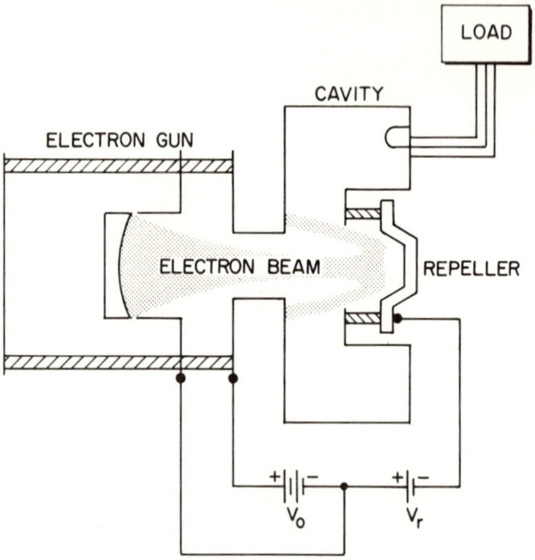

Fig. 5.6-5. A reflex klystron oscillator in which particles at a velocity corresponding to the potential V_0 traverse the cavity gap across which there is a potential $V_1 \sin \omega t$. The beam is stopped and returned back through the same gap by the negative dc field created by the repeller electrode at potential V_r. The ac current resulting from velocity modulation drives the cavity and delivers power to the load.

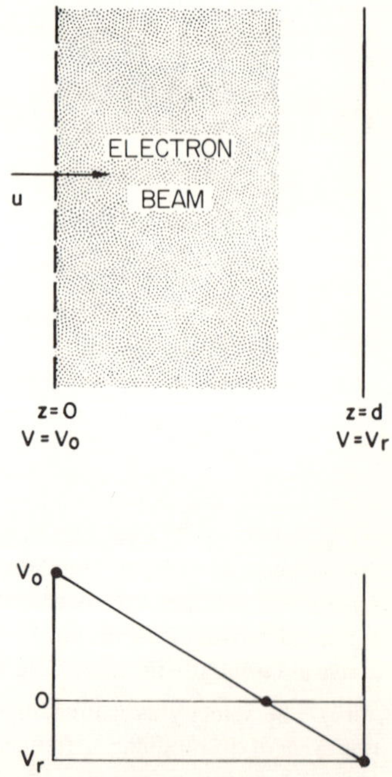

Fig. 5.6-6. A one-dimensional model of the repeller space of a reflex klystron and the corresponding dc potentials. The beam enters with velocity $u = u_0[1 + (V_1/2V_0) \sin \omega t]$. In this model the space charge of the beam is neglected, so the space has a uniform field with one wall at V_0 and the other at V_r.

the tube will not oscillate; in such a case it can be thought of as a single-port regenerative amplifier.

Figure 5.6-6 is a diagram of the model of this device that we analyze. As in the amplifier calculation above, we consider the case of sinusoidal velocity modulation with the injection velocity being that given by Eq. (5.6.3). We assume a uniform dc electric field in the repeller space E_0 equal to $(V_0 - V_r)/d$, where V_r is repeller potential relative to cathode and V_0 is resonator potential relative to cathode. V_r is negative so $(V_0 - V_r)$ is the total voltage across the repeller space, here taken of length d. The equation of motion is, for nonrelativistic velocities

$$du/dt = -(e/m)[(V_0 - V_r)/d] \tag{5.6.6}$$

which is the same as the equation for free fall in a gravitational field. The solution may be written

$$u = -(e/m)[(V_0 - V_r)/d]t + u_i, \qquad z = -(e/m)[(V_0 - V_r)/2d]t^2 + u_i t \tag{5.6.7}$$

where z is position measured from the resonator at the injection plane and u_i is initial velocity at this plane, as given by Eq. (5.6.3).

Figure 5.6-7 shows two distance time diagrams for cases of $(V_1/2V_0) = 0.1$ and 0.2. For both cases the electric field $(V_0 - V_r)/d$ was set equal to $2V_0/d$. For the case $(V_1/2V_0) = 0.1$ the number of particles injected per modulation period was 40. For the case $(V_1/2V_0) = 0.2$ there were 35 particles per period. In both cases $\omega t_0 = 2\pi$, i.e., the modulation period was equal to (d/u_0), the transit time across the distance d at the injection velocity with the diode field free. The trajectories in this case are all parabolas described by Eq. (5.6.7) rather than straight lines. Both the depth of penetration and the time at which particles return to the injection plane are functions of the initial velocity.

Figure 5.6-8 gives two phase diagrams corresponding to the distance–time diagrams of the preceding figure. Both phase diagrams are for the position $z = 0$ for the beam on its return transit past this point. Obviously the phase diagrams at this position for the entering beam are pure sine waves with amplitude 0.1 and 0.2, respectively. Note that the velocity amplitude for the two cases is still about what it was on injection, but the shape of the wave is nonsinusoidal with a large current pulse at the center of each diagram where the dots line up at the same time. The $(V_1/2V_0) = 0.1$ case is very near the peak current condition, while for $(V_1/2V_0) = 0.2$ the current is beyond its peak.

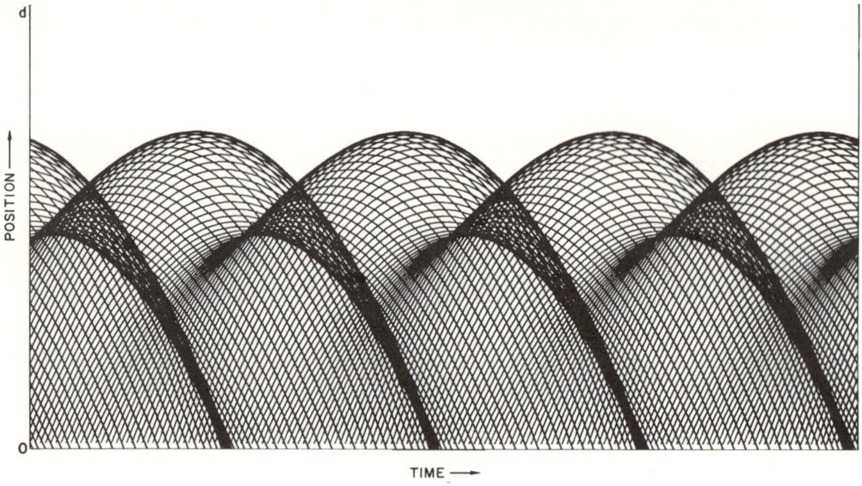

(a)

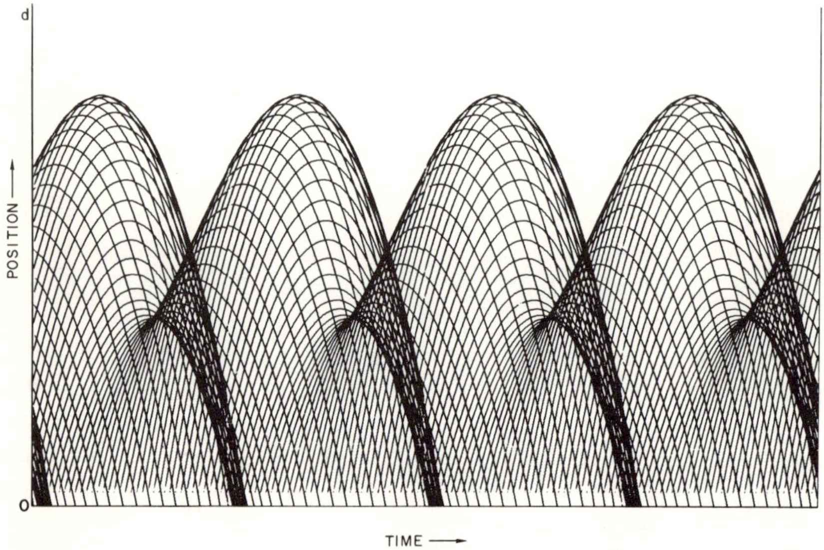

(b)

Fig. 5.6-7. (a) A distance–time diagram for many noninteracting particles in a retarding dc field that are injected uniformly in time with a sinusoidally varying injection velocity $u = u_0(1 + 0.1 \sin \omega t)$. The total distance shown is the distance d, and the transit time across this distance of a particle traveling at u_0 is t_0. The modulation frequency is determined by the relation $\omega t_0 = 2\pi$. The number of particles per period $(2\pi/\omega)$ was 40.

(b) A distance–time diagram for many noninteracting particles as in (a) but with injection velocity $u = u_0(1 + 0.2 \sin \omega t)$. The modulation frequency is the same as before, but the number of particles per period $(2\pi/\omega)$ is 35.

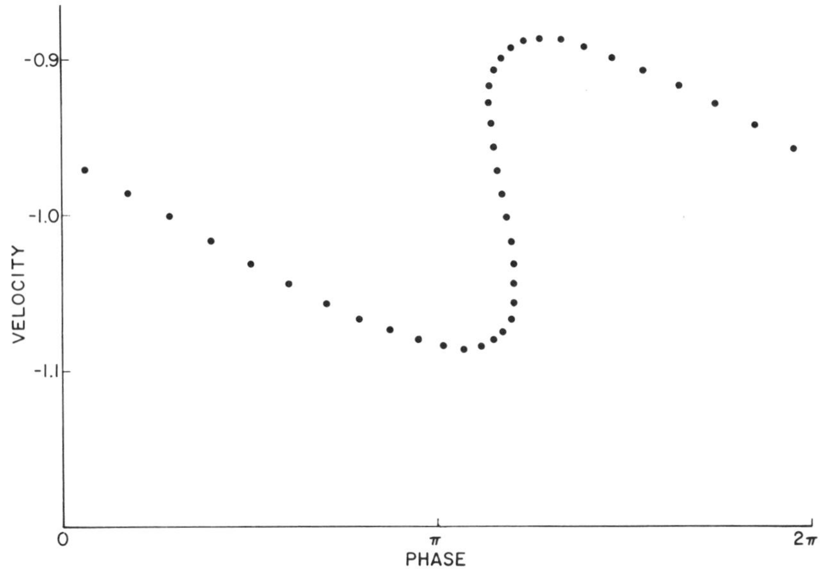

(a)

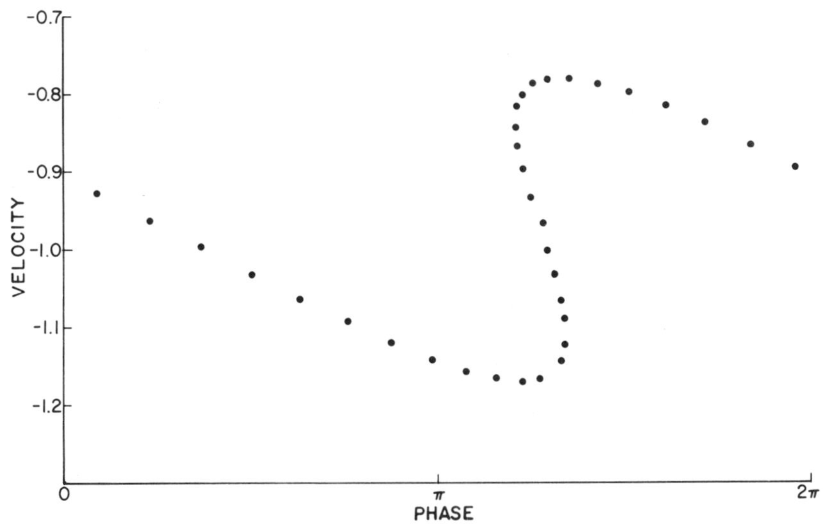

(b)

Fig. 5.6-8. (a) The velocity as a function of the arrival phase of a sequence of particles during one cycle of the modulation frequency, for the case of Fig. 5.6-7a at the position $z = 0$. For this value of velocity modulation the current is just beyond the single peak condition.

(b) The velocity as a function of arrival phase, for the case of Fig. 5.6-7b at the position $z = 0$. For this value of velocity modulation the current is clearly overbunched.

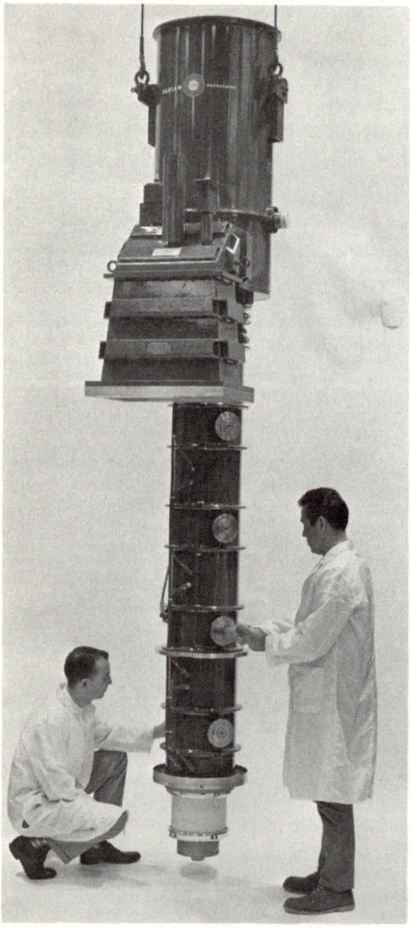

Fig. 5.6-9. A commercial klystron amplifier designed for use in radar applications. It provides 25 MW peak power, 300 kW average power in the 400–500 MHz band. [Courtesy Varian Associates.]

Figure 5.6-9 is a photograph of a multicavity klystron amplifier designed for high efficiency and high power output. The design of such tubes is based on the methods of this section and of Chapter 6, as modified by experiment.

Problem 5.6-1 Klystron Amplifier

A klystron amplifier has a sawtooth wave applied to the beam at the buncher "cavity" (which must now be a broadband circuit capable of handling the bandwidth of the sawtooth wave). The velocities of particles leaving the buncher are given by

$$u = u_0[1 + \alpha(t/t_0)], \qquad 0 < t < t_0$$

during each cycle of duration t_0. Then the velocity returns to u_0 to start the next cycle. Plot figures analogous to Figs. 5.6-3 and 5.6-4 for this case with $\alpha = 0.2$. Allow enough distance to get a crossover. Compare the crossover distance with that in Fig. 5.6-3. Explain the difference qualitatively.

PROBLEM 5.6-2 PARTICLE ACCELERATOR BUNCHER

A linear particle accelerator has a field that when viewed from the rest frame of a moving observer appears to vary as cos kz. Consider a set of 8 particles distributed initially uniformly every $(\pi/4)$ throughout a full wavelength and at rest in this frame of reference. Neglect relativistic effects and interparticle interaction. Calculate the trajectories of these particles as a function of z and t.

5.7 Electron Guns and Lenses

In this section we examine qualitatively some essentially two-dimensional effects of fundamental importance in electron tubes and particle accelerators. Many of these effects involve space-charge fields, i.e., the effects of the electrons or particles on the fields that affect them, which we treat for a number of one-dimensional problems in the next two chapters. The purpose of this section is simply to describe these effects and some approaches that can be used in their solution. The solution of these problems has been the subject of many books over the last 50 years that are referenced in the works listed at the end of this chapter.

In the preceding section drawings of klystron tubes showed electrons flowing smoothly out of electron guns without striking the gun electrodes or the cavity through which the electron beam must pass. It was implied there that the electron flow was smooth and laminar without crossovers. While it is possible to make electron guns that approximate this ideal performance, it is only possible to do so within certain ranges of the parameters of the guns. For example, only relatively low perveance designs meeting these conditions are possible, i.e., the perveance defined as $(I/V^{3/2})$ must be less than about 1×10^{-6} for electrons. When strip beams or hollow beams are used, the relevant comparable value of perveance is the value per square having sides equal to the beam thickness.

The basic concept of an electron gun dates back to the nineteenth century, and indeed it is very easy to produce a relatively low perveance nonlaminar beam by almost any crude arrangement of a cathode emitter opposite an accelerating anode with a hole in it. Higher current, more efficient guns were sought for cathode ray tubes in the 1920s and various attempts were made to shape electrodes outside the beam in an empirically optimum manner. However, it was found that the beam shape and laminarity of the current were extremely sensitive to the way that the electrodes were shaped.

An organized approach to this problem was provided by Pierce in the 1930s by a very simple and elegant line of reasoning. Laminar flow of electrons between concentric cylinders or spheres was known to be possible with straight line rectilinear trajectories. Pierce's idea was to take a segment of this spherical system and replace the electrons surrounding the segment desired with electrodes having shapes that would produce exactly the same field along the surface of the beam as the missing electrons [11].

Figures 5.7-1a and 5.7-1b illustrate this design approach. In the case of Fig. 5.7-1b the anode is imagined to be a conductor with a spherical dome exactly as in the original spherical flow system. This dome could be made substantially transparent to electrons by using a fine mesh grid, in which case an electron beam would travel on beyond the anode to a collector.

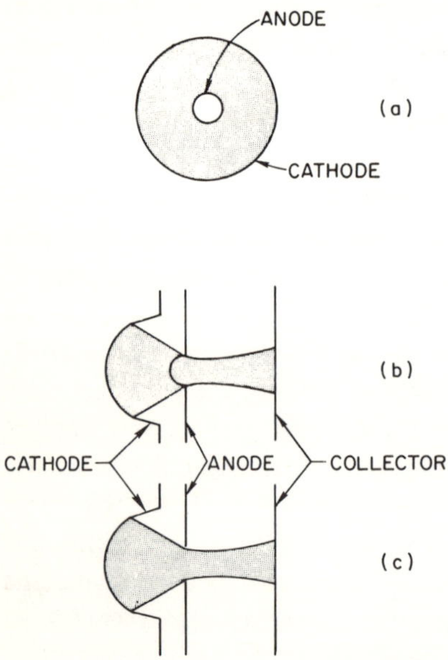

Fig. 5.7-1. (a) Rectilinear space-charge flow between concentric spheres. (b) A Pierce gun using a segment of the spherical flow pattern and a gridded anode. (c) A Pierce gun with an ungridded anode.

The cathodes in Fig. 5.7-1 consist of a spherical portion which is the emitting portion and conical "focus electrodes" which are thermally insulated from the emitting cathode and are intended to be nonemitting. Both the emitting cathode and the focus electrode are operated at the same potential. The emitting portion is typically operated at a temperature of 900°C if a BaO–SrO coated nickel cathode is used.

In some electron tubes, such as reflex klystrons, gridded guns of this type are used. However, in many tubes it is important to avoid wasting the power at the anode, in which case an ungridded aperture is used as shown in Fig. 5.7-1c. Obviously, such a gun is no longer a perfect Pierce gun, because flow is no longer between spherical electrodes. Nevertheless, excellent performance is obtained from ungridded guns in many performance ranges. Some criteria of gun performance are the perveance, the uniformity of the current density over the beam cross section, uniformity of velocity over the cross section, area convergence, and the percentage of the beam intercepted by the anode. A requirement arises in connection with ion propulsion beams to have very low current intercepted by the anode in order to avoid sputtering of the anode by the ions in the beam, typically cesium, which would over a long time period eat away the anode and change the performance of the gun. Guns with better than a part in 10^{-5} interception current have been built. A part in 10^{-3} is easily obtained.

Area convergence is important in guns for small diameter electron beams where very high current densities are required, because electron emitters are limited to a certain maximum value for a given emission life. Area convergences of 10 to 1 are easy and 100 to 1 or more are possible.

High beam perveance is usually desired, and it may be obtained by decreasing the cathode anode spacing up to the point at which the anode hole becomes important. If the anode is simply moved closer to the cathode, it will reach a point where it cannot do as good a job as the spherical dome that it replaced in drawing current off the center of the cathode. At this point, the current distribution off the cathode changes and approaches that of a hollow beam with very little emission coming from the center of the cathode [12]. A typical value of electron perveance at which this effect becomes important is 2×10^{-6}.

One of the most important effects of the anode hole is the lens effect that it produces [7, 10, 13]. Figure 5.7-2a is a sketch of an electron gun with a few field lines indicated from cathode to anode. In the anode hole the field lines must turn outward radially and electrons in the beam will be forced outward by these radial fields. Obviously, if we regard the anode hole as a lens, it is a divergent lens.

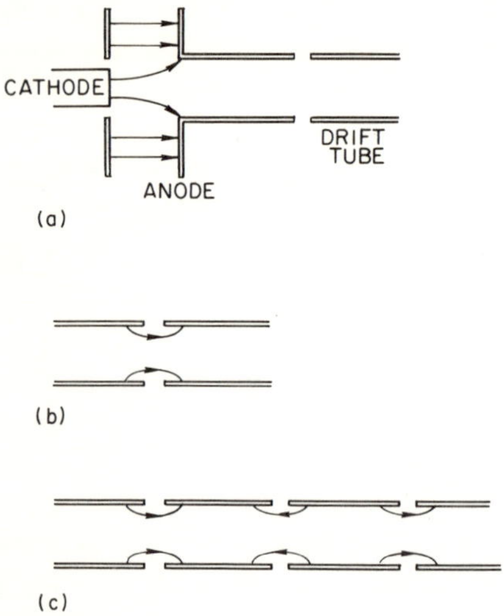

Fig. 5.7-2. (a) An electron gun. Lines showing the direction of force on an electron are shown. The drift tube and anode are at the same potential. (b) An electron lens consisting of a gap between two drift tubes operated at different potentials. Force on an electron is indicated by the lines with arrows. (c) A periodic lens system.

Many electron lenses are convergent. Figure 5.7-2b shows the field configuration between two cylinders at different potentials. The field is radially inward in the region where the electron is traveling more slowly and radially outward in the region where the electron is traveling more rapidly. Because the electron spends more time in the inward field, the net effect of such a lens is convergent. A very extensive literature on lenses exists which is useful for approximate calculations [7]. However, it is now possible, with a digital computer, to calculate exact trajectories of particles, given the electrode shapes, at extremely low cost. It is possible at a somewhat higher cost to calculate the effects of space charge on the trajectories, by means of a step-by-step process. One approach is given in the next chapter. Another is a relaxation method in which, iteratively, trajectories are calculated for the entire region of interest, fields are calculated from the trajectories and then new trajectories are found in a, hopefully, convergent calculation [12, 14].

In all of this discussion so far the presence of magnetic fields has not been mentioned. In most beam-type electron tubes, especially high-power tubes, an axial magnetic focusing field is used in order to prevent interception of the beam by the electrodes of the tube. In most linear particle accelerators focusing magnets are used for the same reason. The particular electron gun configuration illustrated in Fig. 5.7-2 with a flat cathode would produce a diverging beam which would be almost entirely intercepted by the anode, if nothing further were done to get the beam through the hole. This gun configuration is used extensively in electron tubes in which the entire gun can be immersed in a uniform axial magnetic field which confines the beam and keeps it focused in spite of the effects of space-charge or lens effect due to electric fields as illustrated in Fig. 5.7-2. Magnetic fields can also be used in a way that produces magnetic lens effects. Any variation in the axial intensity of the magnetic field will have this effect as we have seen in Section 5.3. In many systems, a good approximation to particle trajectories can often be found by considering the effects of electric and magnetic fields separately. For example, the electric lens effect at the gun anode can be calculated without considering the magnetic field and then a calculation of trajectories in the magnetic field can be made, using the electric lens effect as an initial condition. Again, computer solutions of problems of this type are now feasible, but a combination of space charge, static magnetic fields, and static electric fields can lead to costly computations in some cases.

An important additional factor in many high power microwave tubes and particle accelerators is the effect of strong time-varying fields on particle trajectories. The rf field configuration at the output gap of a klystron is typically like that of Fig. 5.7-2b and the axial-field strength at this gap is normally sufficient to almost stop most of the beam. Therefore, the radial field is also very strong and it is not surprising that a strong lens is created at such a gap, requiring a very intense magnetic field to contain the beam beyond this point. A particle accelerator will normally have a traveling-wave field with essentially this same configuration. In the moving frame of the particle beam, particles may experience only one lens effect during their entire travel, but this may be sufficient to cause problems.

One way that lens effects can be beneficial is when they are used in a periodic focusing system as suggested in Fig. 5.7-2c. Both periodic electric and periodic magnetic lens systems have been used successfully to transmit particle beams through long pipes, both in electron tubes and particle accelerators [7].

Problem 5.7-1 Lens Effects in a Magnetically Focused Beam

An electron gun is followed by a drift tube in the configuration shown in Fig. 5.7-2a. The whole system is immersed in a uniform axial focusing field. The lens effect at the anode produces a ripple in the beam diameter which is undesirable. Suppose that the drift tube can be operated at any desired potential relative to the anode potential. Can you see a way to reduce the beam ripple after it enters the drift tube? What are the parameters that would need to be adjusted? [See J. R. Pierce, *Bell Syst. Tech. J.* **30**, 825–829 (1951).]

REFERENCES

1. F. Rohrlich, "Classical Charged Particles." Addison-Wesley, Reading, Massachusetts, 1965.
2. R. P. Feynman, R. B. Leighton, and M. Sands, "The Feynman Lectures on Physics," Vols. I and II. Addison-Wesley, Reading, Massachusetts, 1964.
3. H. Alfvén and C. G. Falthammer, "Cosmical Electrodynamics." Oxford Univ. Press, London and New York, 1963.
4. C. L. Longmire, "Elementary Plasma Physics." Wiley (Interscience), New York, 1963.
5. T. G. Northrup, "The Adiabatic Motion of Charged Particles." Wiley (Interscience), New York, 1963.
6. B. Lehnert, "Dynamics of Charged Particles." Wiley (Interscience), New York, 1964.
7. J. R. Pierce, "Theory and Design of Electron Beams," 2nd ed., Van Nostrand-Reinhold, Princeton, New Jersey, 1954.
8. D. A. Dunn and R. E. Holaday, Electron trajectories in a nonuniform axially symmetric magnetic field, *J. Appl. Phys.* **32**, 1612–1620 (1961).
9. D. A. Dunn, R. P. Borghi, and G. Wada, A crossed-field multisegment collector for beam-type tubes, *IRE Trans. Electron Devices* **ED-7**, 262–267 (1960).
10. A. Arsenjewa-Heil and O. Heil, A new method for the generation of short, undamped, electromagnetic waves of high amplitude, *Z. Phys.* **95**, 762–764 (1935); Reprinted in English, *Electronics*, **16** (July), 164–167 (1943).
11. J. R. Pierce, Rectilinear flow in electron beams, *J. Appl. Phys.* **11**, 548–554 (1940).
12. P. T. Kirstein, G. S. Kino, and W. E. Waters, "Space Charge Flow." McGraw-Hill, New York, 1967.
13. O. Klemperer, "Electron Physics." Butterworth, London, 1959.
14. D. A. Dunn, V. Hamza, and J. A. Jolly, Computer designs for beam pentodes, *IEEE Trans. Electron Devices*, **ED-12**, 343–351 (1965).

CHAPTER 6

Motion of Many Interacting Particles—the Lagrangian Model

I think we may have more difficulty in exploring the full limits of the computer than we have had with earlier gadgets. I think there may be more danger in the period of trial and error than there has been with earlier devices. These earlier devices—looms, engines, generators—resisted at critical points human ignorance and stupidity. Overloaded, abused, they stopped work, stalled, broke down, blew up, and there was the end of it. Thus they set clear limits to man's ineptitudes. For the computer the limits, I believe, are not so obvious. Used in ignorance or stupidity, asked a foolish question, it does not collapse, it goes on to answer a fool according to his folly. And the questioner being a fool will go on to act on the reply.

E. E. Morison, in "Men, Machines, and Modern Times," p. 91. MIT Press, Cambridge, Massachusetts, 1966. Reprinted by permission.

6.0 Introduction

We now examine a set of problems in which the field experienced by a particle is dependent both on its position and on the positions of a large number of other particles. We do not deal with problems in which both the velocity and the position of the other particles are important, i.e., we do not include both electric and magnetic fields produced by the other particles, but only deal with one effect at a time. Omitting either electric or magnetic interactions may seem to be a drastic simplification, and it is, but there are many practical problems in which this approach is an excellent approximation and yields results that accurately reproduce the behavior of experimental systems. As seen in Chapter 4, these assumptions are equivalent to the assumption of quasistatics, and the error we make is of the order of $(v/c)^2$ where v is particle or wave velocity within the system.

We primarily deal with one-dimensional charge sheets or current sheets and their interactions within bounded regions of space. In the charge-sheet problems we will picture a diode, a region bounded by conducting walls separated by a distance d, into which charge sheets can be injected with prescribed initial velocities. Given the positions of all the charge sheets, the electric field at every point within the diode can be computed. Once this field is available, time is allowed to proceed for a short finite time increment and the motion of all the charges is computed. Then time is stopped and, using the new positions a new field is obtained. Then time is allowed to proceed for another increment, and so on. The mathematical technique that is needed for this procedure is that of finite difference equations, and the actual computations require a digital computer for all but the simplest cases. Setting up and programming this type of computation is quite easy, however, and interesting results can be obtained in many cases that require only a very small amount of computer time.

It is interesting that we are in fact solving a so-called "many-body" problem by this procedure, a type of problem in physics that is still insoluble in any very general form at the present time. The reason that we are able to do this at low cost and with great simplicity is that we take advantage of the symmetry inherent in many physical problems, and use a highly simplified model in which particles are constrained to move in only one dimension. We also treat in detail only the case of particles or charge sheets infinite in the transverse dimensions, and such particles obey a very simple force law. The force between two infinite charge sheets is independent of the distance between them. Disks or rods that move in one dimension have forces that vary with the separation and consequently calculations on such particles require much more extensive computations than are encountered in the infinite sheet problem. Problems involving two and three dimensional motion can also be treated by the same technique, but at the cost of both computer time per time step and the amount of computer storage required per particle. The result is that a much smaller number of particles can be treated on any given computer at reasonable cost, if more freedom of motion is allowed, i.e., there is a trade off between the number of particles and the number of degrees of freedom, as suggested in Fig. 6.0-1.

In a typical vintage 1970 computer about 10^4 charge sheets or disks restricted to one-dimensional motion can be handled or 10^3 rods in two-dimensional motion. The next generation of computers is expected to provide something close to an order of magnitude increase in these numbers at moderate cost. Even today, if one is willing to pay the price, 10^5 one-dimensional particles or 10^4 two-dimensional particles can be handled on a large computer.

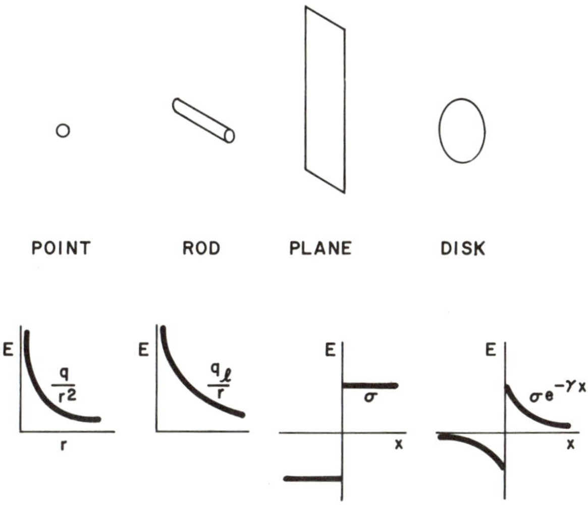

Fig. 6.0-1. Lagrangian models of plasmas can be designed around models of charged particles that are: point charges that can move in three dimensions, rod charges that can move in two dimensions, or plane or disk charges that can move in only one dimension with field variations as shown. There is a trade off between the number of particles that can be included in a problem and the number of dimensions in which motion is allowed.

For some problems it has been found that 10^4 to 10^5 particles are needed to get a valid simulation, and there are many important problems that require a three-dimensional simulation. Computer improvements can therefore be expected to play an important role in this area of research [1].

What is more important for the student is an understanding of some of the limitations and potential pitfalls associated with this analytical technique. Many formal aspects of the mathematics of this approach are only now receiving the attention they should have. However, empirically, one can often establish the fact that one's results are convergent and stable by varying the computational parameters, such as the duration of the time step and the number of particles, and demonstrating the insensitivity of the results to variations in these parameters. Some problems are very complex in this regard, and it is always unwise to assume that one's solution is a valid one without varying all of the parameters available over a considerable range. Perhaps the most dangerous feature of this approach and one that is inherent in the use of difference equations in place of differential equations, is the potential occurrence of numerical instabilities, i.e., instabilities that are not part of the physical problem being studied but that are introduced purely through the nature of the finite mathematical technique. Properly handled, difference equation techniques yield valid results, but it is essential to understand the ways that cumulative errors can creep in and to take precautions, both in the initial formulation of the problem and in the kinds of checks that are made on the results. Some examples of these precautionary techniques are given in the sections that follow that emphasize once again the importance of getting simple things straight before rushing off to solve complicated problems.

Many typical plasma and electron device problems have the general form suggested in Fig. 6.0-2, of a finite cylindrical region bounded on the ends by conducting walls. At the ends particles are injected or removed. Often

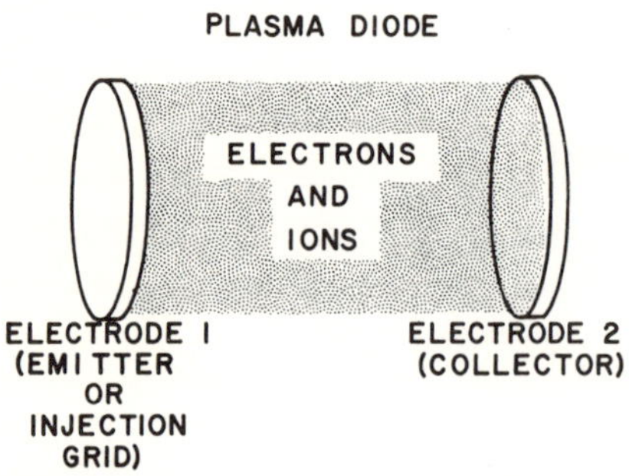

Fig. 6.0-2. A plasma diode with symmetry about an axis through the electrodes. The drawing can represent a dc gas discharge tube if electrode 1 is an emitter and electrode 2 is operated at a potential higher than electrode 1 so a current is drawn through the ionized gas or plasma consisting of electrons and ions plus neutral molecules. A thermionic energy converter can have a similar construction.

there is a high-frequency circuit coupled to the system either by connections to the end electrodes or in the form of an external circuit, not shown in the figure, that might take the form of one or more resonant cavities or a slow-wave circuit. A dc magnetic field may be present with varying degrees of symmetry.

While it is obvious that in a real physical system of this sort electrons, ions, and neutral atoms will undergo three-dimensional motion, there are many modes of operation in which we may be interested in only the axial motion of the particles. A typical example is one in which the essential energy exchange process takes place only with axially directed electric fields, as in a simple rectifier in which electrons are emitted from electrode 1 and collected on electrode 2 and a sinusoidally time-varying electric field exists between the two electrodes. In such a case we would be interested solely in modes of operation with axial symmetry. If the spacing between the electrodes were large in comparison with the diameter of the active region, it might be important to use disk particles in a computer simulation. If the electrode spacing were small in comparison with diameter, infinitely large charge-sheet particles would probably provide an accurate simulation of the physical effects.

Given a choice of allowable particle motion that seems to have a good chance of simulating the effect of interest, it is next important to consider the simulation of collisions between the charged particles in one and two-dimensional models. First, consider two zero-diameter charged rods with the same sign of charge on each rod. Particles of this type can never actually collide, because the repulsive forces between them go to infinity as the rods approach each other. If two such particles pass near each other, they undergo what is defined as a close Coulomb collision if their relative directions of motion are altered by more than 90°. If one is interested in phenomena involving primarily the effects of long-range Coulomb forces, as typically we are in studying the collective behavior and oscillations of aggregates of charges, it is important either to eliminate, or at least not to overemphasize, the effects of close collisions which act as a type of disordering or randomizing effect. It is possible in a rod model to set the force at small distances equal to zero for all rods, thus simulating the situation in which no collisions are close collisions. Some action of this sort is necessary, if the basic approach of simulating the motion of many physical particles with a much smaller number of computer particles is to be at all successful. We lump, say, 10^5 physical particles together into one computer particle, and we wish to simulate the behavior of, say, 10^{15} physical particles in a given physical region of space with 10^3 computer particles. Roughly speaking, the effects of close collisions on our very highly charged computer particles would be 10^5 times the true effect, if a force cutoff at small particle spacings were not used, as has been pointed out by Hockney [2, 16, 29] in connection with his study of a rod model.

Figure 6.0-3a illustrates the force between two rods which are constrained to travel in straight lines and which pass by each other at a distance. The axial force passes through zero as the rods pass each other and reaches a maximum when the rods are separated axially by a distance about equal to their spacing at closest approach. Let us suppose that we do not wish to allow any greater disturbance of trajectories than would occur in this case if the rods had been free to move in two dimensions. Then, for particles that would, in the absence of interaction, pass each other with a smaller separation (impact parameter) than that shown, the force must be reduced artificially. One model of a particle that accomplishes this result is a charged cloud in which the charge is distributed uniformly over a specified finite area.

Figure 6.0-3b shows the force between two charged clouds that pass directly through each other with zero impact parameter. A simple way of

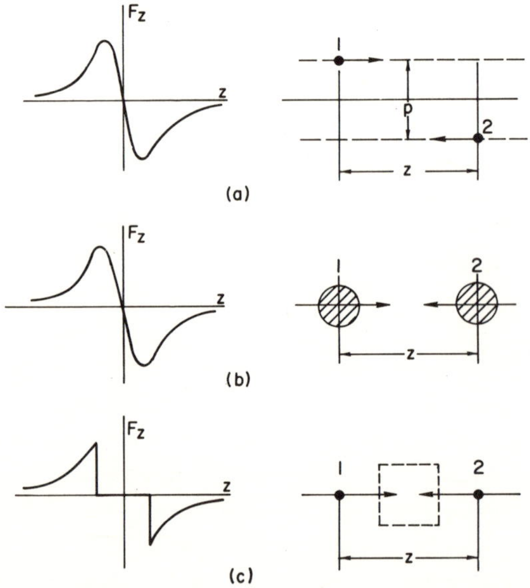

Fig. 6.0-3. Collisions between charged two-dimensional particles, 1 and 2, showing z-directed force F_z as a function of the separation z between particles.
(a) Zero-diameter rod particles constrained to move in straight lines separated by the distance p, the impact parameter. (b) Finite-diameter charged clouds approaching each other with zero impact parameter. (c) Zero-diameter charged rods with a force cutoff for small z when the rods occupy the same square cell, indicated by dashed lines. Close collisions with small impact parameters must be artificially suppressed in computer models in which a small number of charged rods is used to simulate a much larger number of physical particles. The models of (b) and (c) are typical approaches to this problem.

representing a charged cloud on the computer is to reduce the force to zero during the overlap period rather than reducing it gradually. This model is shown in Fig. 6.0-3c and is the one used by Hockney [2].

We will only be concerned with models of particles in one-dimensional motion and primarily we will be interested in the simplest type of particle which is infinite in the direction transverse to the direction of motion. In this case we cannot experience large angle collisions because no deflection is possible. There is still, however, the possibility of using other models of charged particles than the zero-thickness charge sheet shown in Fig. 6.0-4a. Here we can also consider a model which has a gradually changing force with distance during the time of overlap with another charge as shown in Fig. 6.0-4b. This type of model has been used by Mihran and Yu [3, 4]. However, it is expensive to use. Of considerable interest is the fact that a

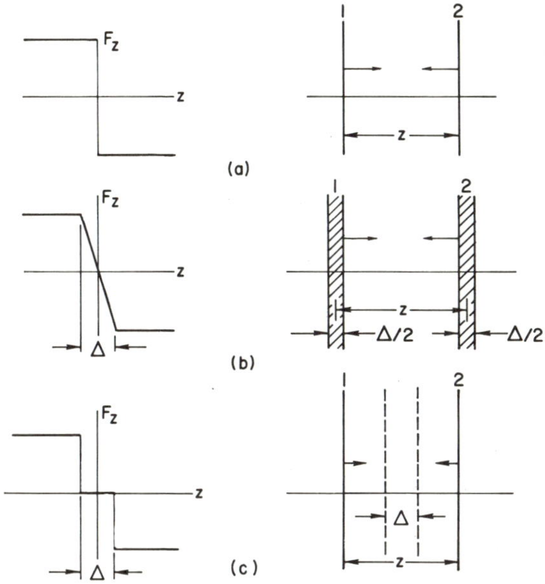

Fig. 6.0-4. Collisions between charged one-dimensional particles, 1 and 2, showing the z-directed force F_z as a function of the separation z between particles. (a) Zero-thickness charge sheets. (b) Finite-thickness charged clouds. (c) Zero-thickness charge sheets with a force cutoff when the charges occupy the same cell, indicated by dashed lines.

very similar smoothing effect can be introduced simply by coarse-graining the particle positions in configuration space for the purpose of making the field calculation. This type of coarse-graining is very important for other reasons and is actually introduced as a time-saving technique [5]. Thus it turns out that, although we will seemingly deal with zero-thickness charge and current sheets, the effective force law we will use in our final calculations will be cut off for charges that are very close together, if we use the coarse-graining technique discussed in Section 6.3. Figure 6.0-4c shows the force law for charge sheets in this type of simulation. The shape of the force versus distance curve is seen to be analogous to that of the Fig. 6.0-3c two-dimensional model.

6.1 Electric Field Calculations for Zero-Thickness Charge Sheets

Here we treat the problem of a parallel-plane diode having infinite transverse dimensions and containing many charge sheets, all having the same sign and magnitude of charge [6]. We calculate the field as a function of position within the diode, and leave to the next section the computation of particle motion. We start by considering the problem of a diode containing only a single charge sheet of finite thickness, as shown in Fig. 6.1-1.

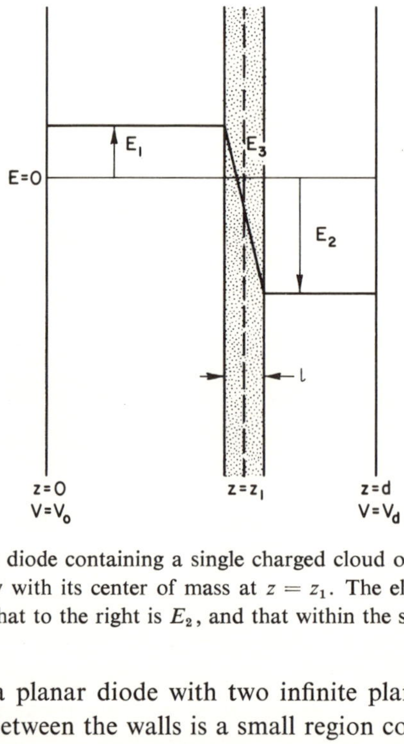

Fig. 6.1-1. A planar diode containing a single charged cloud of finite thickness l and uniform charge density with its center of mass at $z = z_1$. The electric field outside the slab to the left is E_1, that to the right is E_2, and that within the slab is E_3.

This figure shows a planar diode with two infinite plane parallel walls at $z = 0$ and $z = d$. Between the walls is a small region containing a uniform volume charge density ϱ_0. This region has a length l and its center is a distance z_1 from the plane at $z = 0$. The potential of the wall at $z = 0$ is V_0 and the potential of the wall at $z = d$ is V_d. We wish to solve for the electric field and potential within the diode.

Evidently from the differential form of Gauss's law in one dimenison

$$\partial E_z/\partial z = \varrho/\varepsilon_0 \tag{6.1.1}$$

the electric fields to the left and right of the charge slab will be constant, because the charge is zero in these regions, and the field within the charge slab will vary linearly with z, because the charge is uniformly distributed within the slab. Let the fields to the left and right of the slab be E_1 and E_2, respectively, as indicated in Fig. 6.1-1. We can then write an expression for the field within the slab, E_3, in terms of E_1 and E_2

$$E_3 = E_1 + (E_2 - E_1)\frac{[z - z_1 + (l/2)]}{l} \tag{6.1.2}$$

as required to give a linear variation within the slab and to give the proper values at $z = z_1 \pm (l/2)$.

If we now integrate over a volume with unit area perpendicular to the z-direction and length l coincident with the charge slab, we will get, from the integral form of Gauss's law of Eq. (3.1.2),

$$-E_1 + E_2 = (1/\varepsilon_0)\int_{z_1-(l/2)}^{z_1+(l/2)} \varrho \, dz = \varrho_0 l/\varepsilon_0 \tag{6.1.3}$$

We may now use the definition of static potential to obtain the integral of the field across the diode, and using Eq. (6.1.2) we get

$$\begin{aligned} V_d - V_0 &= -\int_0^d E_z \, dz = \int_0^{z_1-(l/2)} E_1 \, dz + \int_{z_1-(l/2)}^{z_1+(l/2)} E_3 \, dz + \int_{z_1+(l/2)}^{d} E_2 \, dz \\ &= E_1[z_1 - (l/2)] + E_2[d - z_1 - (l/2)] + E_1 l + (E_2 - E_1)(l/2) \\ &= E_1 z_1 + E_2(d - z_1) \end{aligned} \tag{6.1.4}$$

Now combining Eqs. (6.1.2)–(6.1.4) we can write the field in the three regions as

$$E_1 = \frac{V_d - V_0}{d} - \frac{\varrho_0 l}{\varepsilon_0}\left(1 - \frac{z_1}{d}\right), \qquad 0 < z < z_1 - \frac{l}{2} \tag{6.1.5}$$

$$E_3 = \frac{V_d - V_0}{d} - \frac{\varrho_0 l}{\varepsilon_0}\left(\frac{z_1}{l} - \frac{z}{l} - \frac{z_1}{d} + \frac{1}{2}\right), \quad z_1 - \frac{l}{2} < z < z_1 + \frac{l}{2} \tag{6.1.6}$$

$$E_2 = \frac{V_d - V_0}{d} + \frac{\varrho_0 l}{\varepsilon_0}\left(\frac{z_1}{d}\right), \qquad z_1 + \frac{l}{2} < z < d \tag{6.1.7}$$

We will be interested in the field acting at the center of mass of the slab which we obtain from Eq. (6.1.6) for $z = z_1$ as E_{3s}, the effective self-field acting on the charge slab

$$E_{3s} = \frac{V_d - V_0}{d} - \frac{\varrho_0 l}{\varepsilon_0}\left(\frac{1}{2} - \frac{z_1}{d}\right) \tag{6.1.8}$$

We see that for zero applied field, i.e., for $V_d = V_0$, the self-field for a positive charge is zero when the charge is at the center, is toward the left-hand plane for $z_1 < d/2$, and is toward the right-hand plane for $z_1 > d/2$. A single charge of either sign is thus in an unstable equilibrium at the center of the diode and will be pulled by its self-field to the diode wall in whichever direction it is initially displaced. If we wish to consider zero-thickness charge sheets with a surface charge density ϱ_s per unit area, we may use Eqs. (6.1.5)–(6.1.8) but we must replace $\varrho_0 l$ with ϱ_s.

Suppose now that there are k zero-thickness electron sheets present numbered consecutively from right to left, as in Fig. 6.1-2. We wish to

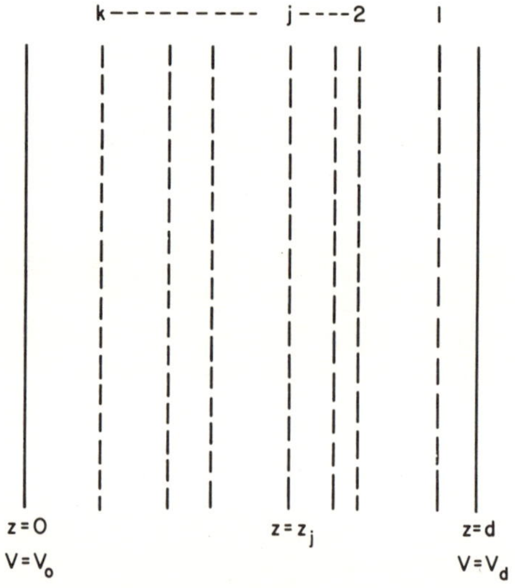

Fig. 6.1-2. A one-dimensional diode containing k zero-thickness charge sheets numbered consecutively from right to left. Distance is measured from left to right.

calculate the force acting on the jth sheet in the presence of the others. We compute the total field by taking the sum of the self-field plus the field due to charges to the left of the jth plus that due to charges to the right of the jth.

The self-field E_s is, from Eq. (6.1.8),

$$E_s = \frac{V_d - V_0}{d} - \frac{\varrho_s}{\varepsilon}\left(\frac{1}{2} - \frac{z_j}{d}\right) \tag{6.1.9}$$

The field at z_j due to charges to the left of the jth charge E_l is from Eq. (6.1.7)

$$E_l = \sum_{i=j+1}^{k} \frac{\varrho_s}{\varepsilon_0} \frac{z_i}{d} \tag{6.1.10}$$

The field at z_j due to charges to the right of the jth charge E_r is from Eq. (6.1.5)

$$E_r = \sum_{i=1}^{j-1} \frac{\varrho_s}{\varepsilon_0} \left(\frac{z_i}{d} - 1\right) \tag{6.1.11}$$

So the total field E_j acting on the jth charge sheet with $k - 1$ other charge sheets present in the space is

$$E_j = \frac{V_d - V_0}{d} + \frac{\varrho_s}{\varepsilon_0}\left[\frac{1}{2} - j + \sum_{i=1}^{k} \frac{z_i}{d}\right] \tag{6.1.12}$$

If the sheet has a surface mass density m_s per unit area, we may express the acceleration of the sheet a_j as

$$a_j = (\varrho_s/m_s)E_j \tag{6.1.13}$$

This concludes the first cycle of the computation that we wish to make in which the field acting on each charge is computed from the charge positions. We have computed the exact field as a function of position, with particular reference to the positions of the charge sheets. The field so calculated is the electrostatic field, i.e., time has not been considered in this calculation. Consistent with this so-called electrostatic approximation is the fact that the field is purely electric, i.e., the particle velocities do not produce magnetic fields in this model. The dual of this problem involves a system in which there is no charge separation and the only fields are magnetic.

Example VI A Problem of Many Interacting Particles That Can Be Solved Exactly—Plasma Oscillations

Here we obtain an exact solution to a problem that is of fundamental importance in electron beams and plasmas. We study the problem of a beam or set of initially stationary charge sheets as shown in Fig. 6.1-2, which are immersed in a uniformly distributed cloud of charge of the opposite sign which has the same total charge as that of the set of charge sheets [7–9]. If there are k charge sheets of charge ϱ_s per unit area, there is then a total charge of the opposite sign equal to $k\varrho_s$ per unit area distributed uniformly throughout the diode.

The electric field in the diode can be expressed, as in Eq. (6.1.12), at the position of the jth sheet, for the case of equal potentials on both walls as follows:

$$E_j = (\varrho_s/\varepsilon_0)\left[\tfrac{1}{2} - j + \sum_{i=1}^{k} (z_i/d)\right] - (k\varrho_s/\varepsilon_0)[(z_j/d) - \tfrac{1}{2}] \qquad \text{(VI.1)}$$

where the second term is the field due to the volume charge which we assume is fixed in space, maintaining a uniform distribution without regard to the motion of the charge sheets. Consider the case of a single charge sheet so that $j = k = 1$:

$$E_1 = (\varrho_s/\varepsilon_0)[\tfrac{1}{2} - 1 + (z_1/d)] - (\varrho_s/\varepsilon_0)[(z_1/d) - \tfrac{1}{2}] = 0 \qquad \text{(VI.2)}$$

It is seen that the neutralizing charge has completely canceled the self-field of the charge sheet, as we might expect.

Now consider a set of k charge sheets uniformly distributed over the diode at positions z_{0j} separated by a distance (d/k). We can write the positions when the sheets are perturbed from these positions in terms of δ_j, a normalized distance, as follows:

$$z_j/d = (z_{0j}/d) + \delta_j \qquad \text{(VI.3)}$$

We can allow for a displacement of the entire group of sheets from a symmetrical relation to the walls by defining η as the normalized displacement according to the relation

$$z_{0j}/d = 1 + \eta - (j/k), \qquad 0 < \eta < 1/k \qquad \text{(VI.4)}$$

When the system is perfectly symmetrical $\eta = 1/2k$. If we now define the plasma frequency ω_p as

$$\omega_p^2 = \varrho_s^2 k/m_s\varepsilon_0 d \qquad \text{(VI.5)}$$

where m_s is mass per unit area as used in Eq. (6.1.13), we can make use of Eq. (6.1.13) and substitute Eqs. (VI.3) and (VI.4) in (VI.1) to obtain

$$d^2\delta_j/dt^2 = -\omega_p^2\delta_j - (1/k)\sum_{i=1}^{k} \delta_i \qquad \text{(VI.6)}$$

which is the usual equation for simple harmonic motion if

$$\sum_{i=1}^{k} \delta_i = 0 \qquad \text{(VI.7)}$$

This is the case, if there is symmetry about the central plane, because then half the sheets have positive δ_i and half have negative δ_i of corresponding magnitudes. In deriving Eq. (VI.6) we made use of the fact that

$$\begin{aligned}\sum_{i=1}^{k} (z_i/d) &= \sum_{i=1}^{k} \delta_i + \sum_{i=1}^{k} (z_{0i}/d) \\ &= \sum_{i=1}^{k} \delta_i + \sum_{i=1}^{k} [1 + \eta - (i/k)] \\ &= \sum_{i=1}^{k} \delta_i + (k/2) - \tfrac{1}{2} + k\eta \end{aligned} \tag{VI.8}$$

We will examine only the case in which the charge sheets execute simple harmonic motion, i.e., a case of symmetrical excitation so that Eq. (VI.7) applies. In this case we can write, by inspection

$$z_j/d = (z_{0j}/d) + \delta_{0j} \cos \omega_\mathrm{p} t + u_{0j} \sin \omega_\mathrm{p} t \tag{VI.9}$$

for initial conditions

$$\left.\begin{aligned} z_j/d = z_{0j}/d + \delta_{0j} &= \text{normalized initial position} \\ u_{0j} &= \text{normalized initial velocity} \end{aligned}\right. \tag{VI.10}$$

assuming zero average displacement so that Eq. (VI.7) applies and zero average initial velocity, i.e., $(1/k) \sum_{i=1}^{k} u_{0i} = 0$.

Figure 6.1-3 is a set of trajectories plotted from this equation for a particular set of initial conditions. This figure shows a large amplitude oscillation at the plasma frequency ω_p. The frequency is not a function of amplitude and the motion is ordered even at this relatively large wave amplitude. If crossovers occur, this analysis is no longer directly applicable and we must resort to the methods to be described in the following sections.

This analysis can easily be extended to the case of an average initial velocity, but then particles are swept out of the diode. They can be replaced as they leave, however, and then we have a neutralized beam. Let t_j be the normalized times of injection and let δ_{0j} be the normalized initial perturbations of the trajectories and u_{0j} the initial velocities, as before. In addition let the normalized dc beam velocity be u_0 and t_{0j} be the initial times of injection that would have occurred, if the particles had been all injected at the dc beam velocity. In this case Eq. (VI.9) can be written

$$z_j/d = u_0(t - t_{0j}) - u_0(t_j - t_{0j}) \cos \omega_\mathrm{p}(t - t_j) + (u_{0j} - u_0) \sin \omega_\mathrm{p}(t - t_j) \tag{VI.11}$$

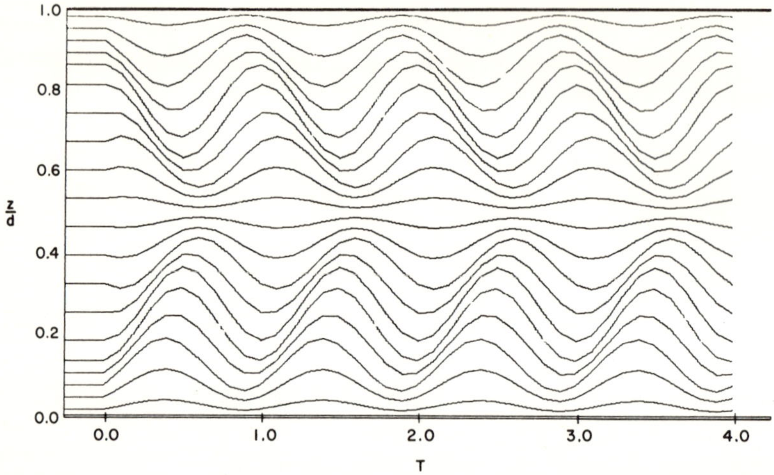

Fig. 6.1-3. Electron trajectories in a diode with conducting walls. An infinite-mass ion background providing dc neutralization is present. Sheets were given initial velocities u_{0j} and displacements δ_{0j} with $\delta_{0j\max} = 0.09$ and $u_{0j\max} = 0.25$ varying sinusoidally in z from zero at the center to the outer wall. Time is normalized to the plasma period $(\omega_p/2\pi)$ and distance to the diode spacing d.

where t is normalized time. For sinusoidal initial velocity and current modulation at frequency ω

$$u_{0j} = u_0 + u_1 \cos \omega t, \qquad J_{0j} = J_0 + J_1 \cos(\omega t + \varphi) \qquad \text{(VI.12)}$$

where u_1, J_{0j}, J_0, and J_1 are all normalized quantities. The ac current density is proportional to $d(t_j - t_{0j})/dt$, so we can write Eq. (VI.11) in the form

$$z_j/d = u_0(t - t_{0j}) + (J_1/J_0)(\omega_p/\omega)u_0 \sin(\omega t_j - \varphi) \cos \omega_p(t - t_j) + u_1 \cos \omega t_j \sin \omega_p(t - t_j) \qquad \text{(VI.13)}$$

Figure 6.1-4 is a trajectory plot from this equation for a specific set of conditions. Two beat patterns can be seen. One corresponds to the fast wave and one to the slow wave discussed in Section 7.4.

Problem 6.1-1 Field in a Diode

Five zero-thickness charge sheets lie within a one-dimensional diode with both walls at $V = 0$. The positions of the sheets are

$$z_1 = 0.70, \qquad z_2 = 0.65, \qquad z_3 = 0.60, \qquad z_4 = 0.35, \qquad z_5 = 0.20$$

Find the field and the potential within the diode as a function of z.

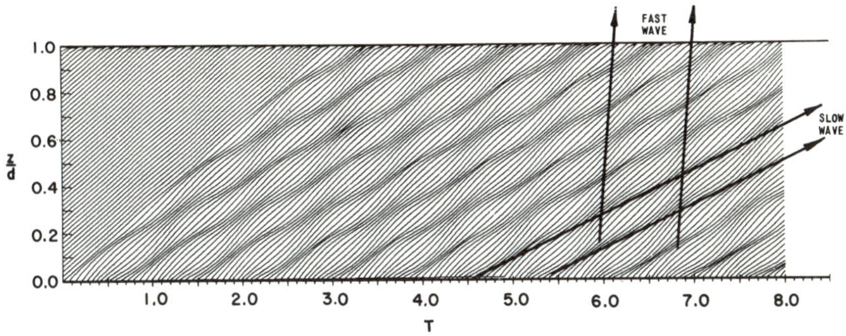

Fig. 6.1-4. Electron trajectories in a neutralized electron beam in a diode with conducting walls.

An infinite-mass ion background is present. Sheets were given an initial ac velocity $(u_1/u_0) = 0.33$ and an initial ac current density $(J_1/J_0) = 0.5$. The modulating frequency ω was 1.3 times the plasma frequency ω_p and the phase of the current modulation was $\varphi = 90°$ with respect to the velocity modulation. Time is normalized to $(\omega_p/2\pi)$ and distance to the diode spacing d. Both fast and slow space-charge waves are excited and their velocities are indicated.

Problem 6.1-2 Field near a Wall

A semiinfinite diode has infinite conducting walls at $z = 0$ and $z = \infty$. A number k of zero-thickness one-dimensional charge sheets with charge density ϱ_s per unit area are at distances z_j ($j = 1$ to $j = k$) from the wall at $z = 0$. Solve for and sketch the electric field and potential as functions of z for a single sheet, i.e., for $k = 1$. Write an expression for the field acting on the jth sheet, in the presence of the other $k - 1$ sheets using the step function S, where

$$\begin{aligned} S(z - z_j) &= 0, \qquad z < z_j \\ &= \tfrac{1}{2}, \qquad z = z_j \\ &= 1, \qquad z > z_j \end{aligned}$$

Problem 6.1-3 Particle Trajectory in a Diode

Solve for the trajectory (distance versus time) of a single charge sheet injected into a diode of spacing d with both walls held at the same potential V_0. Let the injection velocity be u_0, where $u_0^2 = 2\varrho_s V_0/m_s$, the charge per unit area be ϱ_s, and mass per unit area be m_s. For a given charge ϱ_s, what is the "escape" velocity u_0 for which the particle just makes it to the opposite wall of the diode?

6.2 Finite-Difference Solutions to Particle-Trajectory Equations

We have seen that the diode problem of the preceding section with many charge sheets results in a field equation that is simple to calculate for any particle within the diode. The complete solution of a problem of this type requires a separate field calculation at successive instants of time for each particle, because the field acting on each particle depends on the positions of every other particle in the diode as well as that of the particle of interest. There is no closed form solution to such a problem unless the motion is completely orderly with no crossovers. The basic method of solution to this type of problem makes use of finite-difference equations, in which time is quantized and all of the particles are allowed to move under the influence of the field computed from their previous positions. After a short time increment, preferably after only a few crossovers have occurred [10], all particles are stopped and a new field is calculated. Then they are all allowed to move a small distance under the influence of the new field, and so the process continues. An essential new point in this type of trajectory calculation is the use of difference equations. We will consider a uniform time step of duration δ and restrict our treatment to the use of the forward difference operator Δ. Its meaning is expressed by the equation

$$\Delta z(t_k) = z(t_{k+1}) - z(t_k) \tag{6.2.1}$$

where t_{k+1} is the time one time step δ after t_k. The velocity is then expressed not as dz/dt, but as

$$v(t_k) = \Delta z(t_k)/\delta \tag{6.2.2}$$

which, of course, has the limit dz/dt as $\delta \to 0$. Similarly acceleration is not d^2z/dt^2, but

$$a(t_k) = \Delta v(t_k)/\delta = \Delta^2 z(t_k)/\delta^2 \tag{6.2.3}$$

The solution of a difference equation is a relation that allows us to find the value of a specified variable such as z at any given time step, in terms of the number of the time step and the initial conditions. Uniqueness and existence theorems for linear difference equations are similar to those for differential equations [11].

Suppose now that we are following a particular particle in distance–time space. We know a series of past positions of the particle at time intervals δ apart up to time t_k as suggested in Fig. 6.2-1, and we wish to compute $z(t_{k+1})$. From Eqs. (6.2.1)–(6.2.3) we can obtain an expression for $z(t_{k+1})$

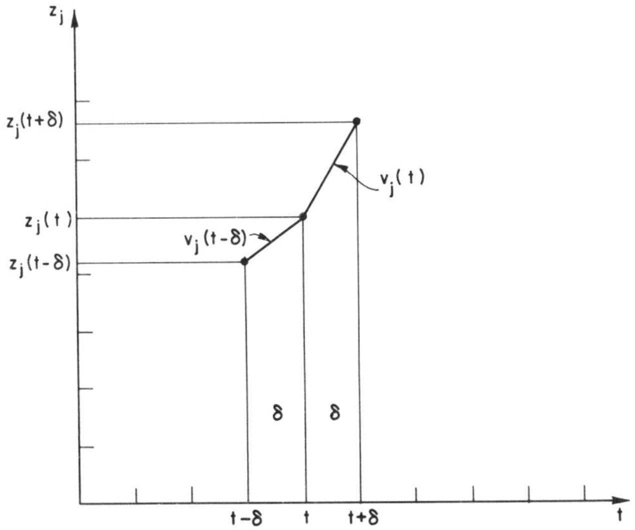

Fig. 6.2-1. The trajectory of the jth particle in the distance versus time plane using a quantized time scale with a time increment δ. The notation is that of a forward difference operator, such that the slope marked $v_j(t)$ refers to the forward difference $z_j(t + \delta) - z_j(t)$. In this notation $a_j(t - \delta)$ refers to the difference $v_j(t) - v_j(t - \delta)$.

in terms of the preceding positions, velocity, and acceleration

$$z(t_{k+1}) = z(t_k) + \delta v(t_{k-1}) + \delta^2 a(t_{k-1}) \tag{6.2.4}$$

Alternatively, this equation can be written in terms of the positions and acceleration only:

$$z(t_{k+1}) = 2z(t_k) - z(t_{k-1}) + \delta^2 a(t_{k-1}) \tag{6.2.5}$$

Note that, while $a(t_{k-1})$ and $v(t_{k-1})$ seem to be values of these quantities at time (t_{k-1}), according to the forward difference notation $v(t_{k-1})$ means δ times the difference between $z(t_k)$ and $z(t_{k-1})$ and $a(t_{k-1})$ means δ^2 times the difference between $v(t_k)$ and $v(t_{k-1})$. Thus both quantities inherently involve positions at more than one point in time. Velocity involves the difference between two positions and acceleration involves three positions. It is not appropriate, as in differential equations, to take the limit as δ approaches zero to find the values and meaning of these quantities at a particular instant of time. Both of these quantities are inherently spread out over more than one instant of time in difference notation.

We can understand the meaning of these equations in our context by first assigning some arbitrary values to acceleration and following through the solutions to the resulting difference equations. We here introduce the subscript notation in which the subscript refers to the time step, so that $z_k = z(t_k)$, $z_{k+1} = z(t_{k+1})$, etc. Using this notation Eq. (6.2.5) can be rewritten as

$$z_{k+1} = 2z_k - z_{k-1} + \delta^2 a_{k-1} \tag{6.2.6}$$

Now consider the simplest possible case of zero field and acceleration, for which $a_k = 0$ for all k. In this case the difference equation to be solved is

$$z_{k+1} = 2z_k - z_{k-1} \tag{6.2.7}$$

To solve this homogeneous linear second-order difference equation we must specify two initial conditions, just as in the case of a second-order differential equation. Let us set $z_0 = 0$ and $z_1 = 1$. Then we can substitute into Eq. (6.2.7) and calculate all successive values of z_k which leads to the solution

$$z_k = k \tag{6.2.8}$$

which expresses a steadily increasing position as time increases, corresponding to motion at a constant velocity equal to the initial velocity, here equal to δ^{-1}.

Let us now consider the next more complicated case of motion subject to constant acceleration. Let us set $\delta^2 a_k = A$ for all k. The equation to be solved is then

$$z_{k+1} = 2z_k - z_{k-1} + A \tag{6.2.9}$$

This is an inhomogeneous, linear, second-order difference equation. As in solving differential equations, standard procedure is to first solve the homogeneous equation and then to find a particular solution to the inhomogeneous equation and to add these solutions together to get a complete solution that matches the initial conditions. Although we have a solution to the homogeneous equation above, we will obtain the solution again by a general method, analogous to assuming an exponential solution to a differential equation. Here we assume a solution to the homogeneous equation of the form $z_k = m^k$ and substitute this solution into Eq. (6.2.7) which is the homogeneous form of Eq. (6.2.9) obtained from Eq. (6.2.9) by setting $A = 0$. We obtain

$$m^{k+1} - 2m^k + m^{k-1} = 0 \tag{6.2.10}$$

Dividing through by m^k we get

$$m^2 - 2m + 1 = 0 \tag{6.2.11}$$

which has as solution a double root at $m = +1$. As in differential equations, if the characteristic equation yields a double root the solution consists of a constant plus a linear term

$$z_k = c_1 + c_2 k \tag{6.2.12}$$

For the initial conditions $z_0 = 0$ and $z_1 = 1$ we obtain Eq. (6.2.8). The general solution to the homogeneous equation is Eq. (6.2.12). Let us now find a particular solution to the inhomogeneous equation, Eq. (6.2.9), using the standard method of undetermined coefficients, i.e., systematic guessing. If we try $z_k = Bk^2$ in Eq. (6.2.9) by analogy to the solution of the corresponding differential equation, we obtain

$$B(k + 1)^2 = 2Bk^2 - B(k - 1)^2 + A \tag{6.2.13}$$

which has the solution $A = 2B$, i.e., the undetermined coefficient is $\frac{1}{2}$ and a particular solution to the inhomogeneous equation is

$$z_k = \tfrac{1}{2}Ak^2 \tag{6.2.14}$$

The general solution is then the sum of Eqs. (6.2.12) and (6.2.14), i.e.,

$$z_k = c_1 + c_2 k + \tfrac{1}{2}Ak^2 \tag{6.2.15}$$

just as we might expect from experience with the differential equation for motion under constant acceleration.

Another interesting case is that of motion subject to a linear restoring force. The differential equation that would represent this case is

$$d^2z/dt^2 = -\omega_p{}^2 z \tag{6.2.16}$$

which leads to the equation of simple harmonic motion. In the case of the difference equation Eq. (6.2.6) we can use $a_{k-1} = -\omega_p{}^2 z_k$ or $a_{k-1} = -\omega_p{}^2 z_{k-1}$ or $a_{k-1} = -\omega_p{}^2 z_{k+1}$. Each of these choices leads to a different answer. We will here examine the case of $a_{k-1} = -\omega_p{}^2 z_k$ and leave an examination of the other two cases to a problem. For the case we wish to study, Eq. (6.2.6) can be written in the form

$$z_{k+1} - (2 - \omega_p{}^2\delta^2) z_k + z_{k-1} = 0 \tag{6.2.17}$$

which is a linear homogeneous equation, soluble as in the previous case by setting $z_k = m^k$ which leads to the characteristic equation

$$m^2 - (2 - \omega_p{}^2\delta^2) m + 1 = 0 \tag{6.2.18}$$

Solving for m yields the two complex conjugate roots

$$m = [1 - (\omega_p{}^2\delta^2/2)] \pm j\omega_p\delta[1 - (\omega_p{}^2\delta^2/4)]^{1/2} \qquad (6.2.19)$$

assuming $\omega_p\delta < 2$. If $\omega_p\delta > 2$ the roots are real, but this is a case of less interest to us because it corresponds to a very long time step that is comparable to the period of the oscillation. It is not surprising that we get an answer that is qualitatively different from the solution of the differential equation, Eq. (6.2.16), for such a coarse time step. Solving for z_k under the condition that z_k must be real, we obtain, for $\omega_p\delta < 2$, from Eq. (6.2.19)

$$z_k = A\cos(k\theta + \varphi) \qquad (6.2.20)$$

where A and φ are arbitrary constants determined by the initial conditions and where

$$\theta = \cos^{-1}[1 - (\omega_p{}^2\delta/2)] \qquad (6.2.21)$$

We have thus obtained a solution to the difference equation Eq. (6.2.17) that has the same form as the solution to the differential equation of simple harmonic motion. We have obtained this result by using $a_{k-1} = -\omega_p{}^2z_k$. It is shown in a problem that the other two alternatives, $a_{k-1} = -\omega_p{}^2z_{k+1}$ and $a_{k-1} = -\omega_p{}^2z_{k-1}$, lead to growing or decaying waves rather than constant amplitude harmonic waves.

The type of problem we are most interested in here is one in which we calculate the electric field from the charge positions at a given time step and then wish to calculate the motions of the charge sheets during the following time increment under the influence of the field, using Eq. (6.2.6). The question that remains is, do we interpret a_{k-1} as the value of the acceleration that is obtained from the electric field calculated from the most recent information on the positions of the charges, z_k? Or should we give some weight to the electric field calculated the previous time step from their previous positions, z_{k-1}? From what we have just calculated, the answer seems to be that one can only get a correct answer to problems such as that of an acceleration proportional to distance, if the most recent information is used. If old information is used, corresponding to the case $a_k = -\omega_p{}^2z_{k-1}$ in the example above, incorrect results will be obtained in the form of a "numerical instability." This interpretation also agrees with our understanding of a_{k-1} as the difference between the two slopes corresponding to the presently known velocity and the velocity to be calculated. If there is any "inherent" single time to which a_{k-1} might be said to refer it would be the kth time step rather than the $(k - 1)$th, in spite of the notation a_{k-1}, which is simply a result of using a forward difference operator rather than a central or backward difference operator in the definition of a_{k-1}. Finally, it is in accord with the idea that, since the field has changed from that at time $(k - 1)$ to that at k and the particles are subject to the field that now exists, it would be incorrect to use some previous, no longer existing, field to calculate their acceleration. In future calculations of this type we therefore use the following equation, obtained from Eq. (6.2.6) by substituting the acceleration from Eq. (6.1.13) based on the field at the most recent time step, $E(z_k)$:

$$z_{k+1} = 2z_k - z_{k-1} + \delta^2(\varrho_s/m_s)E(z_k) \tag{6.2.22}$$

If we wish to use the subscript j to refer to the number of the charge sheet, we can return to our previous notation and write Eq. (6.2.22) in the form

$$z_j(t_{k+1}) = 2z_j(t_k) - z_j(t_{k-1}) + \delta^2(\varrho_s/m_s)E_j[z_j(t_k)] \tag{6.2.23}$$

In some early work [6] in this field a coefficient of $\frac{1}{2}$ was used in front of the third term in this equation. This coefficient was justified incorrectly on the basis that the field is constant during a time step and, hence, by analogy to the form of the solution to the equation of motion under constant acceleration, the difference equation for arbitrary acceleration should have this form. We have seen that the correct form of the difference equation, Eq. (6.2.6), leads to a solution of the expected form, Eq. (6.2.15), when constant acceleration is assumed. Obviously an incorrect solution would have been obtained if a coefficient of $\frac{1}{2}$ had been used in Eq. (6.2.6). It may be shown that using a factor of $\frac{1}{2}$ in this way amounts to giving weight to the acceleration or field at the preceding time step and hence leads to numerical instabilities in some problems [12].

Problem 6.2-1 Motion of a Charge Sheet Subject to a Linear Restoring Force

The motion is to be computed in finite time increments of length δ by means of the linear second-order difference equation

$$z_{k+1} - 2z_k + z_{k-1} = -\omega_p{}^2\delta^2 z_{k-1}$$

where ω_p is real. Compute $z_k(k)$ for $z_0 = 0$ and $z_1 = 1$. Repeat for the case in which the restoring force is proportional to z_{k+1} rather than z_{k-1}, i.e., for

$$z_{k+1} - 2z_k + z_{k-1} = -\omega_p{}^2\delta^2 z_{k+1}$$

What is the essential difference in the result for these two cases?

6.3 Particle Ordering and Field Calculations Based on a Coarse-Grained Configuration Space

In this section we consider an approach to particle ordering and field calculations that is particularly suited to problems with large numbers of particles undergoing complex motions [13]. The need for this approach can be understood by examining a simple problem in which we introduce charge sheets into a diode according to some regular pattern as a function of time. As an example, consider the case of a single-velocity electron beam injected into a diode. The sheets representing the beam are injected uniformly spaced in time, with a given number of sheets per time step. These sheets are numbered on entry $j = 0, 1, 2, \ldots$ and the field may be calculated from Eq. (6.1.12) at each time step and the particles moved at each time step according to the trajectory equation, Eq. (6.2.23). Difficulties arise, however, if the sheets get out of the order that they were in when injected. Since Eq. (6.1.12) has the index j referring to charges ordered from right to left and the sheets are numbered according to the injection time, these two indices coincide only if the sheets stay ordered. Since most interesting problems of this type involve large numbers of sheets undergoing very complex motion, a systematic approach to this ordering problem is required. One approach is to reorder the sheets at every time step for the field calculations using an additional array, so that we always know the sequence of sheets from right to left and can continue to use field Eq. (6.1.12). This and other approaches based on the use of Eq. (6.1.12) fundamentally involve k ordering operations for every sheet and since there are k sheets the number of operations required for a field calculation is k^2. The approach to field calculation presented in this section requires only a constant times k operations to calculate the field. If the constant (which depends on the degree of coarse-graining) is less than k, this approach is preferable. For less than a hundred sheets and of the order of a thousand diode cells, there is little difference between the two methods. Orders of magnitude reduction in computation time are possible, however, as we increase the number of sheets to 1000 or more.

In this alternative method of computing the field, the diode is divided into a number of cells of equal length, as shown in Fig. 6.3-1. The field at the center of each cell is calculated as if all of the charges in the cell had their charge distributed uniformly over the length of the cell, regardless of their positions within the cell. The field change in a given cell can be considered to take place abruptly at the center of the cell or at the cell boundary, or to change linearly across the cell. We will consider the field

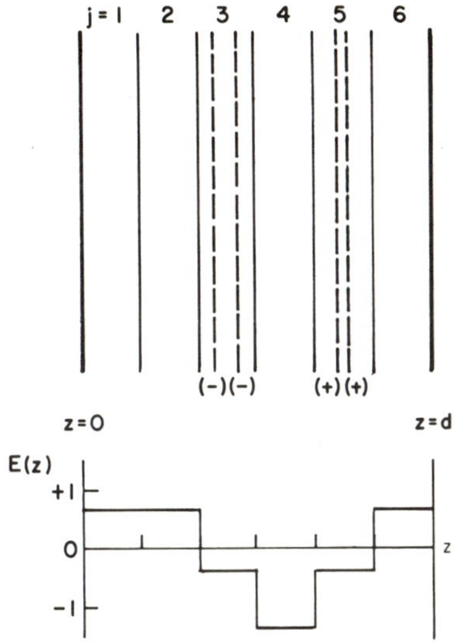

Fig. 6.3-1. The cell model of the diode with 6 equal length cells. Charge sheets located at arbitrary positions within a given cell are treated as if their charge were distributed uniformly over the length of the cell, for purposes of calculating the electric field at the center of each cell. The plot of electric field versus distance shown below the diode is for the case of two negative charge sheets in cell 3 at $z = 0.35$ and 0.4, two positive charge sheets in cell 5 at $z = 0.75$ and 0.8, and zero charge in all other cells. The field at the center of each cell has been calculated from Eqs. (6.3.2) and (6.3.4). The field is treated as if it were constant throughout each cell as shown, when particle trajectories are computed.

constant over the cell and to change at the boundary, as suggested in Fig. 6.3-1, but the other procedures would be equally valid. If 1000 field cells are used, this procedure is roughly equivalent to using only three-place accuracy in the location of each charge sheet for purposes of the field calculation. We continue to retain the precise location (typically to eight places) of each sheet for purposes of the trajectory calculation. The advantage of this procedure is not that lower accuracy can be used, but that the charges need not be reordered at each time step to calculate the field. We do have to record the number of charges in each cell at each time step, but we don't care which ones they are.

Once we have the amount of charge per cell, we can use a field equation equivalent to Eq. (6.1.12) to compute the field throughout the diode, but in this case the positions of the sheets stay fixed at the centers of the cells and the charge per cell fluctuates from time step to time step. To derive the new field equation, we note that in Eq. (6.1.12) the field changes by one unit each time we passed a charge sheet in moving across the diode. If

$\Delta E_j(t_k)$ is the amount of field change that occurs in crossing the jth cell in our present model in which there are J cells numbered 1, 2, . . . , J successively, starting with the first cell nearest to the left wall of the diode at $z = 0$, and each cell is of length (d/J)

$$\Delta E_j(t_k) = (\varrho_s/\varepsilon_0)m_j \tag{6.3.1}$$

where m_j is the number of charge sheets in the jth cell. Now if $E(0, t_k)$ is the field at $z = 0$ we can compute the field $E_j(t_k)$ at the center of the jth cell by starting at the left wall and adding ΔE_j each time we pass a cell as we move across the diode. Thus

$$E_j(t_k) = E(0, t_k) + \sum_{l=1}^{j} (1 - \tfrac{1}{2}\delta_{lj})\, \Delta E_l(t_k) \tag{6.3.2}$$

where $\delta_{lj} = 0$ for $l \neq j$ and $\delta_{lj} = 1$ for $l = j$. The potential V in the diode can be found at the boundaries of the cells at $z = (jd/J)$ from

$$V(jd/J, t_k) = -(d/J) \sum_{l=1}^{j} E_l(t_k) \tag{6.3.3}$$

The value of $E(0, t_k)$ is found by integrating the field times the distance across the diode. In the absence of an applied potential across the diode the field should integrate to zero, i.e., $V = 0$ at $z = 0$ and $z = d$, and $E(0, t_k)$ is the value of the field at the left wall of the diode that makes this integration come out correctly. From Eqs. (6.3.2) and (6.3.3) it may thus be shown that

$$E(0, t_k) = -(1/J) \sum_{j=1}^{J} (J - j + \tfrac{1}{2})\, \Delta E_j(t_k) \tag{6.3.4}$$

Of some interest is the fact that, although this method of field calculation was devised to save computer time by simplifying the reordering of the sheets from time step to time step, it also has the effect of changing the force law between sheets when the sheets are very close to each other. Two sheets in the same cell do not have any effect on each other, except insofar as they affect the self-field acting on both sheets. Thus a positively charged sheet and a negatively charged sheet in the same cell would produce zero field outside the cell and also would not act on each other as long as both sheets remain in the cell. Thus the general effect of this type of coarse-graining is to smooth the force law between sheets, as discussed in the general discussion of collisions in the introduction to this chapter in Section 6.0.

PROBLEM 6.3-1 FIELD IN A DIODE

Calculate the field and potential in a diode for the same configuration of charges as in Problem 6.1-1, but using the cell method with six cells, i.e., $J = 6$ in Eq. (6.3.4).

PROBLEM 6.3-2 FIELD IN A DIODE

Show that, for $V = 0$ at $z = 0$ and $z = d$, Eq. (6.3.4) follows from Eqs. (6.3.2) and (6.3.3).

6.4 Particle Injection in the Lagrangian Model—Specifying the Initial Conditions

In the preceding sections we have considered the field and trajectory calculations as two separate steps in the computation. An additional requirement for obtaining a complete solution is the specification of the initial conditions, i.e., the position and velocity of each particle when it is injected into the diode. If we were dealing with differential equations and infinitesimal time increments, we would simply specify the position and velocity of each particle at the instant of time that injection took place for that particle. However, as we have seen in Section 6.2, velocity and acceleration are quantities that inherently involve two or more positions and two or more times in a finite difference problem. Therefore, if we want to specify the "initial" velocity at the initial position, we are inherently involved not only with the initial position but also with the position one time step prior to the time of injection. In a diode problem this means that there is a first time t_k at which a given particle appears in the diode and starts to affect the field. We may call its position z_k at this time its initial position. However, clearly we must also specify a previous position z_{k-1} and the field at t_k, in order to calculate its motion during the next time step to find z_{k+1}. As usual there are several possible approaches to this problem. We will go through one procedure that has been used in the literature [13, 14] and that is easy to understand. This initialization procedure applies equally well to problems involving injection through a grid, injection from an emitting wall, and to volume ionization processes in which electron–ion pairs are created throughout a volume.

Figure 6.4-1 shows a portion of a particle trajectory during and preceding

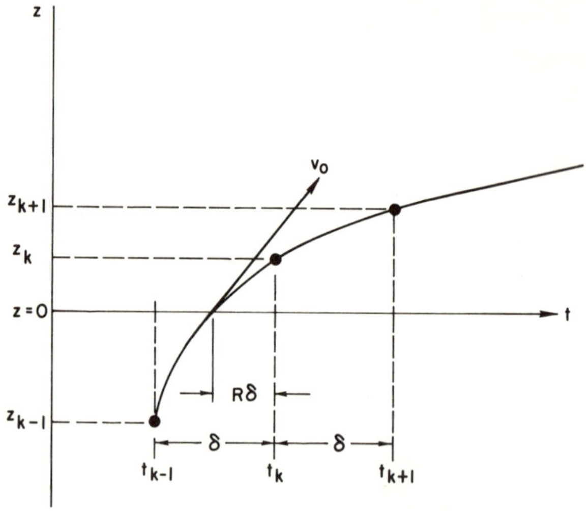

Fig. 6.4-1. Trajectory of a particle during and after injection into a region extending from $z = 0$ to large values of positive z. The initial position at which the particle appears in the diode is z_k. Its velocity at the plane $z = 0$ is v_0 and the time at which it passes this plane is $(t_k - R\delta)$.

the first time step that the particle appears in the space of interest. In this example we specifically picture injection from a wall at $z = 0$, simulating injection through a grid or as a result of emission from the wall. Particle positions z_{k-1}, z_k, and z_{k+1} are shown and a smooth line is shown connecting these positions that represents the trajectory. Of course, in a finite-difference model we actually only have information on particle position at the times t_{k-1}, t_k, etc. and not in between. The velocity is specified as v_0 at $z = 0$. The fraction R specifies the fraction of the time step δ that the particle spends in the space $z \geq 0$ during the injection time step between t_{k-1} and t_k.

In this initialization model we assume that a constant electric field exists within the space $z \geq 0$, and adjust our procedure to give an exact solution for this case. We could do this for zero electric field, but that would clearly not be making the best use of the data available which includes the value of the electric field at the plane $z = 0$ as well as throughout the diode. Making this assumption of a field that is independent of position, we can immediately write the exact value of z_k in terms of the acceleration a_0 due to the field at $z = 0$ at time t_{k-1}.

$$z_k = v_0 R\delta + \tfrac{1}{2}(R\delta)^2 a_0(t_{k-1}) \tag{6.4.1}$$

In this equation R must be specified in order to calculate the initial position

z_k. If we imagine injecting one particle per time step, we might appropriately allow each particle to see the field during half a time step, i.e., we would use $R = \frac{1}{2}$. In this case

$$z_k = \tfrac{1}{2}v_0\delta + \tfrac{1}{8}\delta^2 a_0(t_{k-1}) \tag{6.4.2}$$

If we introduce a number of particles per time step we should use a different value of R for each particle, so the first particles see the field longer than the last ones. Similarly, if we have a random distribution of particle entrance times, corresponding to a noise modulated injection current, R should be taken to be a random number uniformly distributed between zero and unity. Both a random velocity distribution and a random current may be simulated by randomly varying v_0 and R simultaneously [15].

PROBLEM 6.4-1 COMPLETE SOLUTION FOR PARTICLE TRAJECTORIES IN A DIODE

For the case of a dc electron beam injected into a diode, formulate an orderly computational procedure for computing the field and the particle trajectories as a function of time. Allow for the possibilities of crossovers and of particles being returned to the injection plane. Your procedure should be complete enough to allow a computer program to be written directly from it.

6.5 The Single-Species Diode with a dc Injection Velocity

In this section we consider the problem of a one-dimensional diode with a single velocity beam of particles of a single species injected at the left. Both walls of the diode are at the same potential, and the velocity of the injected particles corresponds to the wall potential, i.e., the particles can be thought of as having come from a zero potential emitter. The only physical variable is the beam current. For very small currents, the potential in the space is not significantly affected by the presence of the charges. As we increase the current, the potential is depressed (for negatively charged particles), until we reach a certain critical current at which some particles are returned to the injection plane and only a fraction of the current is transmitted. Below this critical current all the current is transmitted and the only solution to Poisson's equation is a static (non-time-varying) one described in Section 7.2. Above the critical current there can be more than one static solution and there can also be a time-varying solution which is solved for here. The Lagrangian model used here always selects the time-varying solution above critical current, even if it is started in one of the modes corresponding to a static solution. Below critical current both the Lagrangian and Eulerian models yield the same static solution and this solution agrees with physical experiments. Above critical current, physical experiments yield confusing results involving various apparently static states [6, 16]. The oscillations predicted by the Lagrangian model that we will study here have not been observed in experiments with electrons, but an experiment with an ion beam oscillated at about the right frequency [17]. Its oscillation may not, however, have been due to this cause.

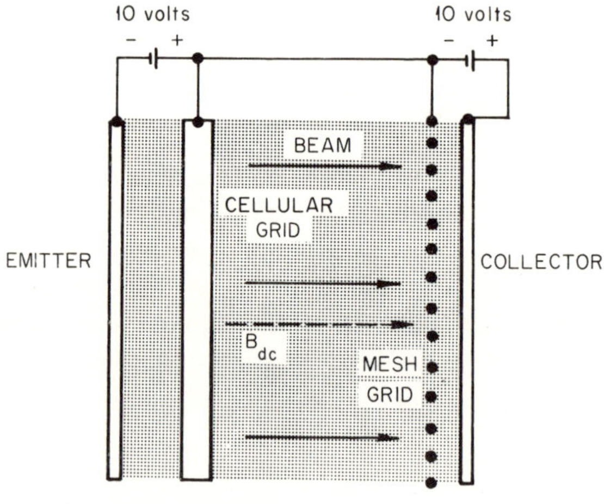

Fig. 6.5-1. A physical diode that might be expected to exhibit approximately the same behavior as the model of this section.

Electrons are injected into the diode at a potential of 10 V, below the ionization potential of any gas present, in order to prevent the buildup of positive-ion space charge. The diode spacing is much smaller than its diameter. The injection grid is cellular to prevent multiple particle transits and a collector electrode is operated positive with respect to the second grid in an attempt to prevent secondary electrons created at the second grid from drifting back into the diode. A dc magnetic field is applied in the direction of the particle motion to limit transverse motion.

Figure 6.5-1 shows a physical diode that might be expected to yield results in agreement with the theory presented here. However, diodes very similar to this one have been built with no evidence of oscillations of the type predicted here [18]. Figure 6.5-2 shows the one-dimensional model that we analyze. In the Lagrangian model we inject charge sheets at the injection plane at $z = 0$, all with the same velocity, and remove them from the computation whenever they reach either wall. For all charge sheets within the space we calculate the field and move the sheets at every time step using the techniques described in the preceding sections.

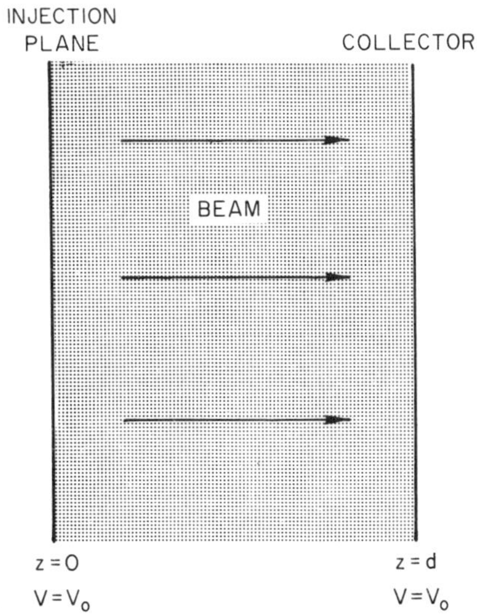

Fig. 6.5-2. A one-dimensional model of the diode of Fig. 6.5-1. Here the injection plane at $z = 0$ is considered to be transparent to electrons entering from the left and opaque to particles striking it from within the space. The wall at $z = d$ similarly collects all particles striking it, and there are no reemitted particles from either wall, i.e., there are neither slow nor elastic secondary electrons. Both walls are at potential V_0.

The first problem that arises is to select a number of particles to inject per unit time to represent the given physical current. This choice is strictly a computational one, because the same physical current can be represented by a very few charge sheets per unit time with a large charge per sheet or by a very large number of sheets with a small charge per sheet. In order to talk about this problem we need to set a time scale, and we normally find that the particle transit time at the injection velocity is a useful unit of time. We then have two interrelated variables, the time step and the number of charge sheets injected per time step. If we want a fairly fine grained representation of the problem, we might start with a time step of the order of 1/100th of a transit time. If we were to inject one particle per time step, this would lead to of the order of 100 particles in the diode at any given time. This might be enough for the representation of simple cases, but we may need more and there is no storage problem in typical computers even with 10^4 particles, so we may want to be able to inject many particles per time step. Similarly, we must choose the number of cells to divide the diode into for purposes of field calculations, if we use the cell model. A typical number is of the order of 1000.

The basic test that can be performed with each of these parameters, commonly referred to as coarse-graining parameters, such as the time step and the number of particles in the diode, is to coarsen the particular parameter of interest and to observe the results. If no significant change takes place, one can have some confidence that the result is insensitive to the degree of coarse-graining being used. The coarse-graining problem is one step more complex than this, however, because there may be a change in the noise level that results from changing the number of particles. Such a change in noise level does not necessarily signify a difficulty with the degree of coarse-graining. The basic noise phenomenon of interest here is that of shot noise, i.e., the noise due to the discrete nature of the charges. This source of noise is present in nature, because of the discrete charge of the electron. In the Eulerian model of the next chapter, this source of noise is eliminated by treating a large number of charges as if they were a continuous fluid. In the Lagrangian model used here, we retain the discrete nature of the charges, but reduce the number used to represent the physical charges in any given problem to the point at which artificial results begin to appear. As we reduce the number, i.e. as we increase the degree of coarse graining, we increase the level of shot noise, because the level of shot noise is proportional to the charge per particle. Schottky [19] showed in 1918 that emission from a temperature limited cathode results in a current with a mean square fluctuation $\bar{i}^2$ at the point of emission of

$$\bar{i}^2 = 2eI_0B \tag{6.5.1}$$

where e is the charge per particle, I_0 is the average current emitted, and B is the bandwidth of the measurement system. Evidently, if we increase the charge per particle by a factor of 10^6 we will have 10^6 times the amount of shot noise experienced with physical electrons. If we can still clearly see the results that we are seeking, then this increased noise level may not be harmful. It is possible to obtain a check on the fact that a given fluctuation is due to shot noise by increasing the number of particles and decreasing the charge per particle. If the noise current observed decreases proportional to the square root of the number of charges present, it is reasonable to attribute the source of noise to shot noise. If the noise level is unaltered by a change in the number of particles, it is reasonable to attribute the noise to a process that would be present in nature with very many more particles.

Given the above point of view, let us now see how this problem arises in the simple diode problem illustrated in Fig. 6.5-3. Charge sheets are shown within the diode in an orderly pattern, corresponding to operation below limiting current. The potential is depressed near the center and the

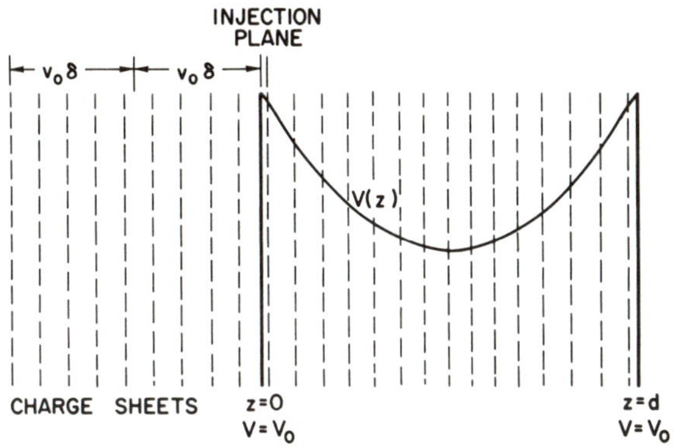

Fig. 6.5-3. The particles in the one-dimensional model prior to injection can be visualized as shown here, lined up to the left of the injection plane. The initial velocity of all particles is v_0 and the time step is δ, so $v_0\delta$ is the distance traveled by a particle in one time step in a field-free region. For the particular case shown, the number of particles entering the diode during the next time step will be 4 and during the time step following that, 5. The potential within the diode varies as a function of z with a minimum at the center.

sheets are closer together in this region. Sheets are also shown outside the diode, lined up prior to injection with a uniform spacing, although these sheets do not interact with each other or with those inside the diode. For a time step δ and an initial velocity v_0, the group of sheets lying within a distance $v_0\delta$ are injected during one time step. For the case shown, the number alternates between 4 and 5. It would also be easy to arrange to have the same integer number of sheets injected every time step. If this is the case, i.e., if the same number is injected every time step, and if the system is operating below limiting current in a static mode, the ordering and spacing of sheets within the diode is identical at every time step. The particle positions for this problem as a function of time are perfectly repetitive and when a computer is used, the output repeats.

However, when the number of injected sheets fluctuates, as in Fig. 6.5-3, between two or more numbers, there are at least two different sheet patterns allowable within the diode, and the potential and the current therefore fluctuate from time step to time step. This fluctuation is one simple form of shot noise. It is illustrated in Fig. 6.5-4 which is a plot of the minimum potential $V_{\min}$ within the diode and the total diode current I as a function of time for a number of different values of the computational parameters, defined as follows: N_0 is the number of sheets injected per transit time, and ΔT the time step in units of transit times. The small tics on the horizontal axis are separated by one transit time. The large vertical dashes

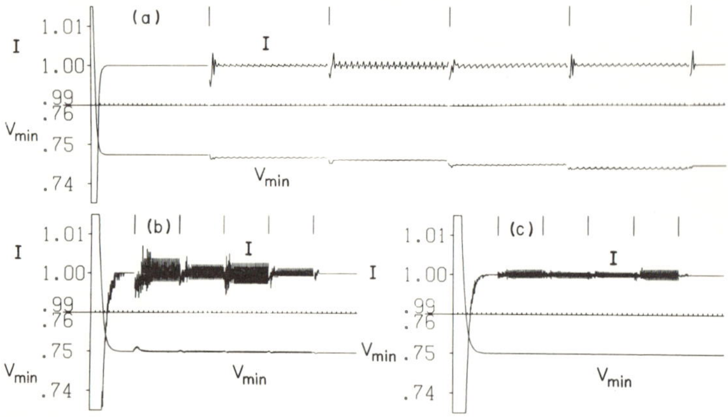

Fig. 6.5-4. Potential V_{min} at the minimum potential point within the diode and total diode current I as a function of time.

Each tic on the horizontal axis is one transit time at the injection velocity. For all cases the current is 0.714 times the limiting current. For each case the number of sheets injected per time step, $N_0 \Delta T$, is changed through six values with the first and last being an integer. (a) $N_0 = 32$, $N_0 \Delta T = 8.0, 8.8, 9.6, 10.4, 11.2, 12.0$. (b) $N_0 = 32$, $N_0 \Delta T = 1.0, 1.2, 1.4, 1.6, 1.8, 2.0$. (c) $N_0 = 256$, $N_0 \Delta T = 8.0, 8.2, 8.4, 8.6, 8.8, 9.0$.

above the curves separate cases with different values of the parameter $N_0 \Delta T$ which is the average number of sheets injected per time step. In each case the first and last cases are integer values of $N_0 \Delta T$, so there is no shot noise for these cases. The noise amplitude is also a function of the number of cells into which the diode is divided, being less for larger numbers of cells. This case of operation below limiting current is particularly simple, because the solution to Poisson's equation is static and there is no physical oscillation present. Shot noise in this case is superimposed on pure dc and is therefore clearly visible. For currents above limiting current, shot noise must be separated from an oscillatory solution [20].

Figure 6.5-5 shows a set of electron trajectories for a diode operating below limiting current that has just been switched to a current above limiting current. The parameter that specifies the injected current is (d/s), where d is the diode spacing of a diode with the injection plane at potential V_0, and s is the spacing of an emitter diode carrying a space charge limited current with the emitter at zero potential and the other wall of the diode at the potential V_0. The static theory presented in Section 7.2 predicts limiting current for $(d/s) = 2.8$. Higher injected currents correspond to closer spacings s of the emitter diode and hence to larger values of (d/s). For very large (d/s) all the injected current is returned to the injection plane.

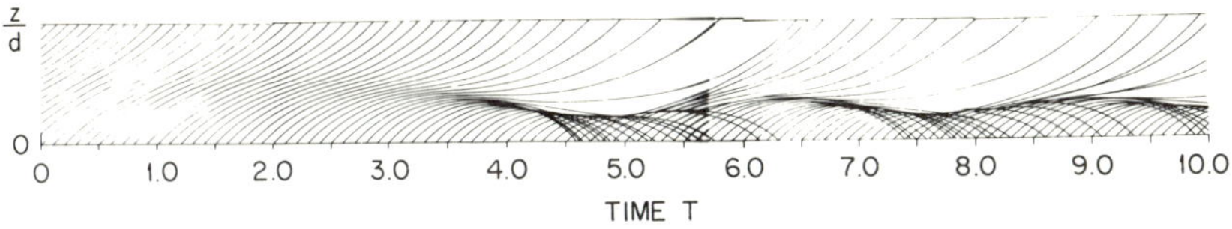

Fig. 6.5-5. Electron trajectories in a diode with single-velocity injection. Prior to $t = 0$ the diode was operating below limiting perveance. At $t = 0$ the charge on each sheet in the space and on all sheets injected after this time was increased from 7 units to 8.2 units. Limiting current is 8 units on this scale. Time is measured in transit times at the injection velocity. The frequency of oscillation after the transient has died out is equal to the plasma frequency corresponding to the charge density at the potential minimum. [from C. K. Birdsall and W. B. Bridges, "Electron Dynamics of Diode Regions." Academic Press, New York, 1966].

The turnaround point in this case is at $0.707s$, because the injected current is stopped and turned around at a point of zero potential and the problem is thus exactly the same as that of the emitter diode, except that the current is twice as great.

The basic feature of diodes with $(d/s) > 2.8$ is the fact that there is some current returned to the injection plane. In order to have a return current, the potential within the diode must drop to zero. But if a finite current comes to a stop at a point in space, there is an infinite charge density at that position. This condition is not theoretically impossible, and indeed is what is predicted by the static theory of Section 7.2. However, nature is not constrained to a static solution to Poisson's equation, and an infinite charge density is easily avoided, if the particles are allowed to turn around at different points at different times, as in Fig. 6.5-5. A very similar result is obtained in nature, if there is a nonzero velocity distribution in the beam initially. In that case the slower electrons turn around closer to the injection plane and the faster electrons either turn around further downstream or pass on to the other diode wall. In nature there is also the possibility of two-dimensional motion and of spatial variations in a direction normal to the injected beam velocity. Again by this additional means, it is possible to avoid an infinite charge density without time variations, simply by having the turnaround point oscillate in space rather than in time. It may be that it is for these reasons that the oscillations predicted by the one-dimensional model have not been observed.

The frequency of oscillation observed in the computer experiments is about equal to the electron plasma frequency (see Section 7.4) corresponding to the density of electrons at the oscillating turnaround point. Figure 6.5-6 is a plot of the potential at the position at which it has its minimum value within the diode, as a function of time. In this figure we show results for several different values of injected beam temperature. For this value of injected current, about 5% thermal spread is sufficient to completely remove the coherent oscillation. At this and higher values of beam temperature the current divides between the two walls on the basis of the particle injection velocity. The thermal spread was introduced into the beam in these computations by selecting the injection velocity v_0 in Eq. (6.4.1) from a table of

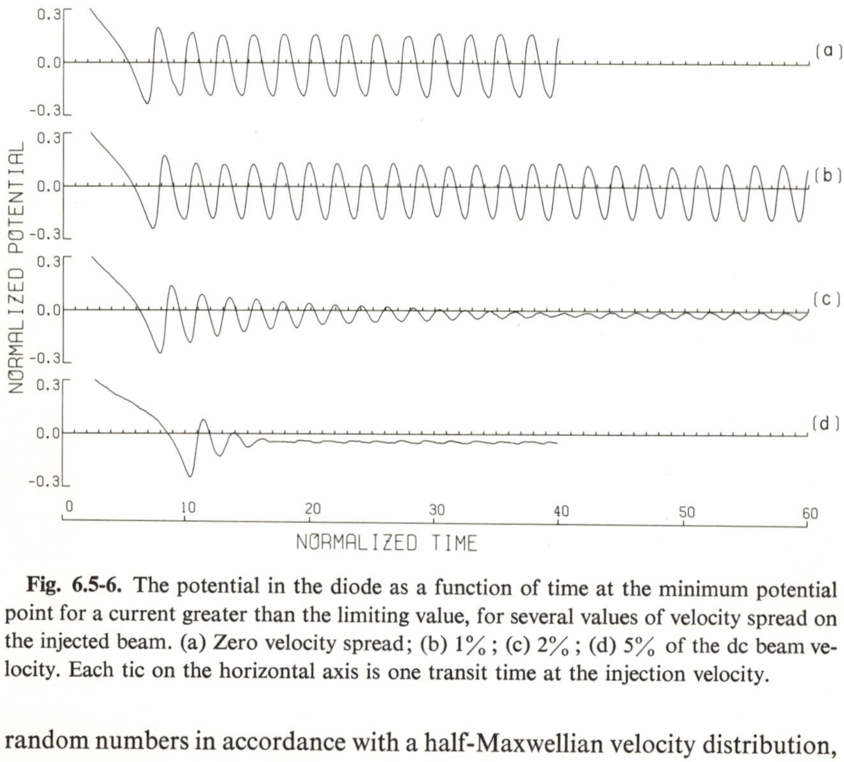

Fig. 6.5-6. The potential in the diode as a function of time at the minimum potential point for a current greater than the limiting value, for several values of velocity spread on the injected beam. (a) Zero velocity spread; (b) 1%; (c) 2%; (d) 5% of the dc beam velocity. Each tic on the horizontal axis is one transit time at the injection velocity.

random numbers in accordance with a half-Maxwellian velocity distribution, i.e., one with zero probability of any velocity less than the dc beam velocity and a Maxwellian distribution of velocities greater than the dc beam velocity [6]. The mean thermal velocity was 1, 2, or 5% of the dc beam velocity in Fig. 6.5-6. The current was not varied randomly, i.e., R in Eq. (6.4.1) was taken to be a sequence of uniformly spaced numbers corresponding to a dc beam. However, after a few time steps the particles are thoroughly scrambled as a result of the random injection velocity, so it is believed that a random current modulation would have negligible additional effect.

The basic result of increasing the thermal spread is what we might expect. However, an additional parameter that is apparently important in this problem is the number of particles that are available at the turnaround point for a given thermal spread. A small fluctuation in potential will suffice to sort out a large number of particles with a given velocity spread, whereas if there are relatively few particles and the same velocity spread, a large fluctuation is required. This point is illustrated by Fig. 6.5-7 where the same case is repeated using 1000 sheets and 3000 sheets. Only of the order of 100 sheets were used in the computation of Fig. 6.5-5 in the early

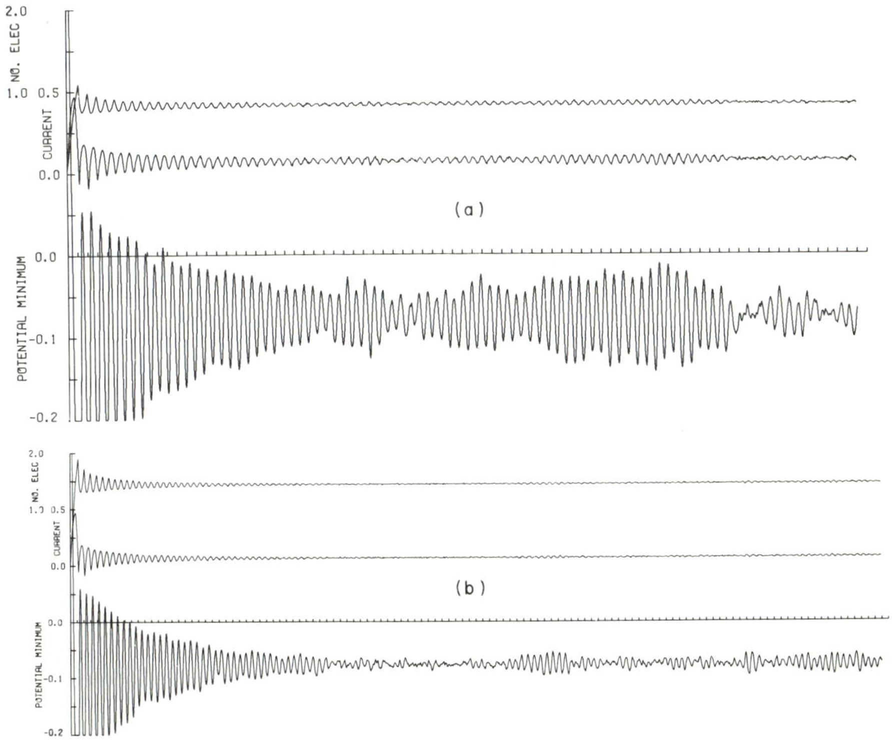

Fig. 6.5-7. Number of electrons, current, and value of the potential at the potential minimum as a function of time. Each tic on the horizontal axis is one transit time at the injection velocity. For both cases $(d/s) = 6.0$ and the velocity spread on the injected beam is 3% of the dc beam velocity. In case (a) there were approximately 1000 sheets in the diode at equilibrium. In case (b) there were approximately 3000 sheets.

work of Birdsall and Bridges [6]. The amplitude of the coherent oscillation is seen to be much greater for the 1000 sheet case. This result is evidently not due to the variation in shot noise produced by varying the number of sheets, as it appears that in both of these cases the shot noise is quite small. For smaller numbers of sheets, such as 100, the shot noise can become large enough to mask this effect and also so large as to confuse the picture with respect to oscillation amplitude as a function of thermal spread. However, once it is as small as it is for the 1000 sheet case of this figure, further increases in the number of sheets primarily affect the amplitude of the coherent oscillation.

The net result appears to be that, if enough sheets are used in combination with some finite but quite small initial beam temperature, plasma oscillations do not occur. The precise relationships are not available at this time, but it appears that at least some of the experimental results which have yielded no oscillations may be the result of an essentially one-dimensional effect as illustrated here. The work on the static theory of a planar diode by Fry and Langmuir [21] may thus turn out to be quite accurate for most physical conditions.

6.6 The Single-Species Diode with Time-Varying Injection Velocity—Klystron Amplifiers and Oscillators

It is a straightforward extension of the case of the preceding section to allow for a time-varying injection velocity. Here we do this and also consider the effects of an infinite-mass ion background which exactly neutralizes the dc space charge of the injected beam and allows ac space charge effects to be observed more clearly.

We consider an injection velocity u that varies sinusoidally with amplitude u_1 about a dc value u_0 with angular frequency ω:

$$u = u_0[1 + (u_1/u_0) \cos \omega t] \qquad (6.6.1)$$

It is convenient, as usual, to normalize with respect to the dc velocity and also the transit time t_0 at the unperturbed velocity u_0.

In Fig. 5.6-1 a two-cavity klystron amplifier was illustrated that could physically produce velocity modulation approximately as indicated by Eq. (5.6.3). We saw in Section 5.6 how such a device could be analyzed, if we neglected the interaction between the particles. We are now in a position to repeat that analysis including space charge. We now must specify the amount of dc current injected and we do so in terms of the parameter (d/s) discussed in the preceding section and also in Section 7.2. At limiting perveance for a dc unneutralized beam (d/s) has the value 2.8.

Figure 6.6-1 is a trajectory plot for a case below limiting perveance with $(d/s) = 1.8$, $(u_1/u_0) = 0.2$, and $\omega t_0 = 2\pi$. The number of particles injected per period $(2\pi/\omega)$ is $N_p = 40$. This case can be directly compared with

Fig. 5.6-3 which was for $(d/s) = 0$, but all other conditions the same. It can easily be verified that the trajectories in Fig. 6.6-1 are not straight lines as were the trajectories of Fig. 5.6-3, indicating that space charge is indeed having an effect. The expected effect of space charge would, at first glance, be to prevent as tight bunching as was observed without it and to cause the current maximum to occur further from the injection plane. Actually, the current maximum in Fig. 6.6-1 is clearly closer to the injection plane than it was in Fig. 5.6-3 and is, if anything, more tightly bunched than before. The difficulty with this comparison is the effect of the dc space charge, which is slowing the beam down and thereby lengthening the effective length of the drift space. We could pursue this comparison, but if we return to the physical device that we are modeling in Fig. 5.6-1, we see that it has a metallic drift-tube wall surrounding the beam throughout its length. Thus, in the physical system there can be very little potential variation along the length of the system, even though there may be a potential variation radially across the drift-tube cross section. In a one-dimensional model, we can't very well simulate this radial potential variation, but we can eliminate the axial potential variation by going to a neutralized beam, i.e., a beam with

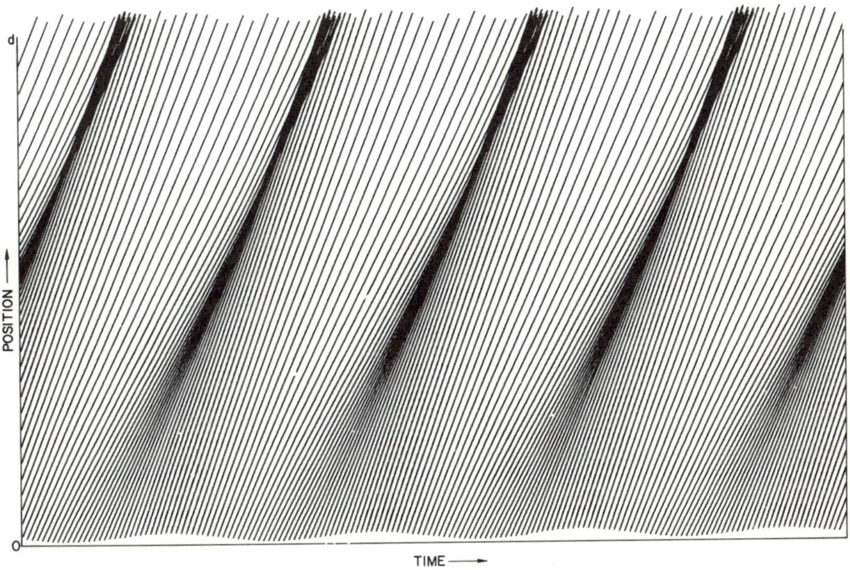

Fig. 6.6-1. A distance–time diagram for many interacting planar charge sheets injected into a diode with a sinusoidally varying injection velocity $u = u_0(1 + 0.2 \sin \omega t)$. The total distance shown is the distance d and the transit time across this distance of a particle traveling at u_0 is t_0. The modulation frequency is determined by the relation $\omega t_0 = 2\pi$. The number of particles per period N_p is 40. The injected current is below limiting current and is determined by the relation $(d/s) = 1.8$. Compare Fig. 5.6-3.

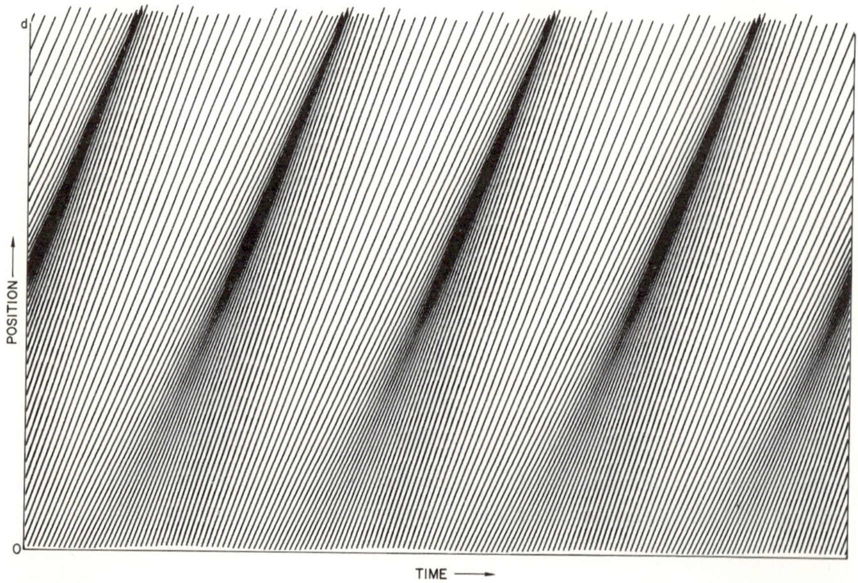

Fig. 6.6-2. A distance–time diagram for many interacting planar negative charge sheets injected into a diode with a uniformly distributed infinite-mass positive-ion background with a charge density equal to the charge density of the injected beam at the dc injection velocity. The injection velocity is $u = u_0(1 + 0.2 \sin \omega t)$, $\omega t_0 = 2\pi$, $N_p = 40$, $(d/s) = 1.8$. Compare Figs. 6.6-1 and 5.6-3.

no dc charge unbalance. The most convenient way of doing this that retains the possibility of ac space charge unbalance is through the use of an infinite-mass ion background. It is important that we are not trying to model a beam in the presence of a neutralizing ion background, but simply to model the effects of the conducting wall in Fig. 5.6-1. The amount of ion charge density that is to be postulated is the same as the density of the injected beam at the dc injection velocity.

Figure 6.6-2 is a trajectory diagram for a neutralized beam with the same parameters as the beam in Fig. 6.6-1. The trajectories are very like those of Fig. 5.6-3 and it would be difficult to make any very firm statements about the differences between these figures. We would therefore say that the amount of space charge represented by $(d/s) = 1.8$ is "small" in the sense that the results are not very different from the zero space-charge result of Fig. 5.6-3. As in Section 5.6 it is more illuminating to look at a velocity–phase diagram to get a detailed picture of the particle positions. Figure 6.6-3 is such a diagram at $(z/d) = 0.8$ and may be directly compared with Fig. 5.6-4a which is for the same position in the absence of space charge. Two rather clear differences can be noted. First, the current pulse is shifted in phase from about 0.8π to about 1.6π. Second, there is no overbunching in Fig. 6.6-3 whereas there is a small amount of overbunching without space charge. This is as expected; the bunch peak is occurring slightly further downstream with space charge than without it.

Figures 6.6-4 and 6.6-5 show plots for increasing values of injected current, and it is seen that for $(d/s) = 5$ or 6 we have reached a degree of space charge that is "large" in the sense that the results are quite different from the zero space-charge results of Fig. 5.6-3. At these values of (d/s) sharp

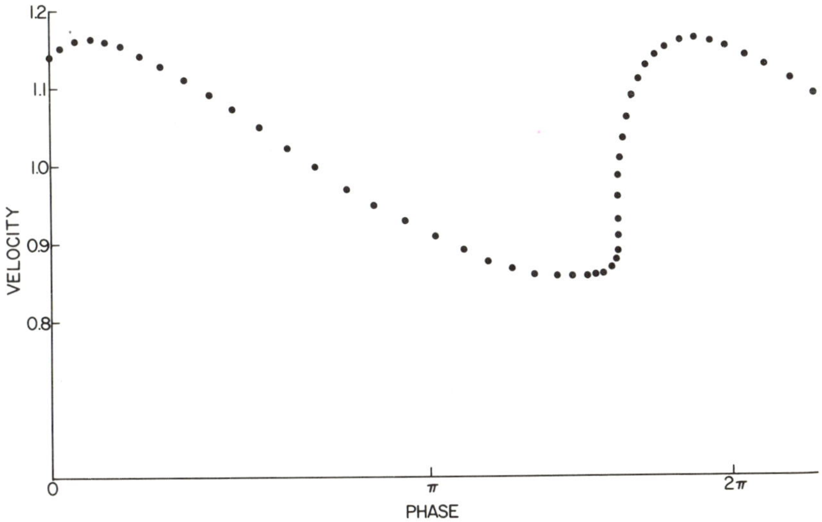

Fig. 6.6-3. A velocity–phase diagram for the beam of Fig. 6.6-2 at the position $(z/d) = 0.8$. The particles are nearly perfectly lined up to form a current pulse at about 1.6π. Compare Fig. 5.6-4a which is for zero space charge and which shows a greater degree of overbunching than this diagram.

bunches with crossovers no longer occur. Both of these large (d/s) cases would be beyond limiting perveance for an unneutralized beam, but going past 2.8 shows no significant dc space-charge effect when the beam is neutralized. Determining the limiting current for a neutralized beam is clearly an ac space-charge problem rather than a dc space-charge problem [22, 23].

A somewhat larger scale view of our beam is given in Figs. 6.6-6 and 6.6-7 which are trajectory plots for a neutralized beam with the same amount of velocity modulation, $(u_1/u_0) = 0.2$, but for twice the frequency, $\omega t_0 = 4\pi$, of the preceding figures. Raising the frequency has the effect of lengthening the drift space measured in terms of transit angle $(\omega z/u_0)$, so we can look at the beam after it reaches its first crossover. In these figures

Fig. 6.6-4. A distance–time diagram for a beam under the same conditions as Fig. 6.6-2, except that $(d/s) = 5.0$. As before $\omega t_0 = 2\pi$, $N_p = 40$, and $u = u_0(1 + 0.2 \sin \omega t)$.

Fig. 6.6-5. A distance–time diagram for a beam under the same conditions as Fig. 6.6-2, except that $(d/s) = 6.0$. As before $\omega t_0 = 2\pi$, $N_p = 40$, and $u = u_0(1 + 0.2 \sin \omega t)$.

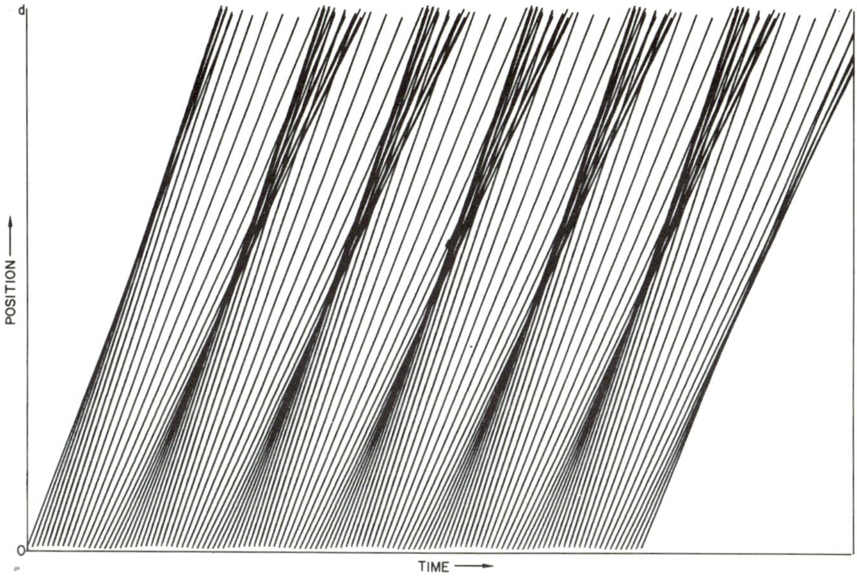

Fig. 6.6-6. A distance–time diagram for a beam under the same conditions as Fig. 6.6-2, except that the modulating frequency is twice as high, $\omega t_0 = 4\pi$. The other conditions are $(d/s) = 1.8$, $N_p = 20$, and $u = u_0(1 + 0.2 \sin \omega t)$. Note that $N_p = 20$ yields the same number of particles per unit length at this frequency as $N_p = 40$ in Figs. 6.6-2–6.6-5.

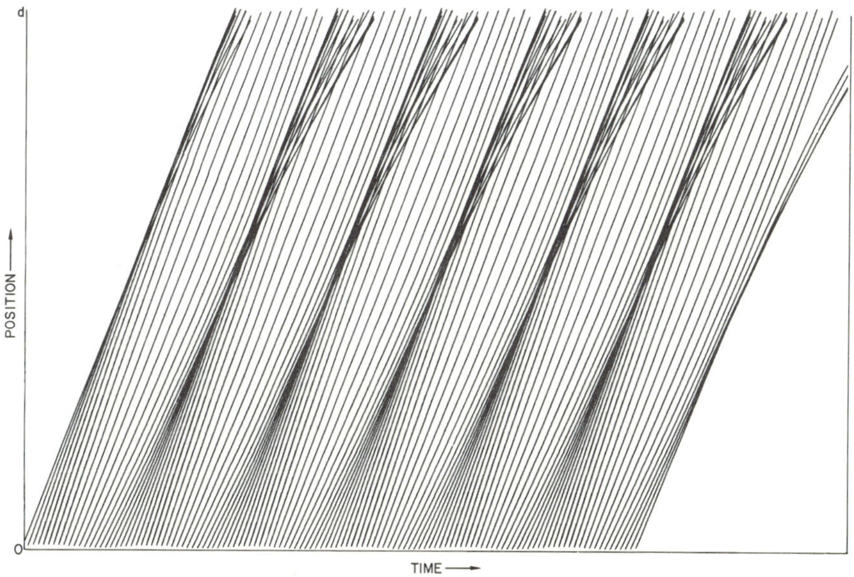

Fig. 6.6-7. A distance-time diagram for a beam under the same conditions as Fig. 6.6-6, except that there is a greater injected current, $(d/s) = 4.5$. As before $\omega t_0 = 4\pi$, $N_p = 20$, and $u = u_0(1 + 0.2 \sin \omega t)$.

two different values of (d/s) are considered, $(d/s) = 1.8$ and 4.5. It is seen that for this frequency we have not made the transition to a beam without crossovers, so we would regard $(d/s) = 4.5$ as a "small" value of space charge for this purpose, even though it is greater than the value for limiting current in the unneutralized case.

All of the results above were for the case of $(u_1/u_0) = 0.2$ which is a fairly large amplitude velocity modulation. If the results were repeated for a smaller modulation amplitude, the results regarding what are small and large values of space charge would have been different. For infinitesimal modulation amplitudes we finally arrive at a condition in which any non-zero space-charge results in the type of trajectories shown in Fig. 6.6-5, without crossovers. This small-signal theory can equally well be dealt with by means of the Eulerian model of the next chapter, and the material on space-charges waves in Section 7.4 is an introduction to this type of theory. The methods of this chapter can also be used for small-signal theory, but if only the small-signal results are of interest, it is much simpler to use the Eulerian model.

As we will see in Section 7.4, there are two waves that can be set up in a neutralized beam without crossovers, the slow and the fast space-charge wave. If pure velocity modulation is used, as in the cases considered thus far, both of these waves are excited. To excite only one wave, we must use a combination of velocity and current modulation. The trajectory diagram of Fig. 6.1-4 in Example VI illustrated both of these waves for a case with both velocity and current modulation.

Problem 6.6-1 Particle Accelerator Buncher

A linear particle accelerator has a field that when viewed from the rest frame of a moving observer appears to vary as $\cos kz$ in the absence of particles. Consider a set of particles distributed initially uniformly every $(\pi/4)$ throughout a full wavelength and at rest in this frame of reference. Neglect relativistic effects. Include interparticle interaction. Calculate the trajectories of these particles as a function of z and t. Take the boundary conditions at the end of the wavelength to be periodic, to simulate the presence of charges throughout all space repeating the motions of the charges in this wavelength. Choose the charge per particle to yield a self-field comparable to the field in the absence of particles. Compare your result with that for Problem 5.6-2.

6.7 The Two-Species Diode with Infinite Spacing—Ion Propulsion Beam Neutralization

In this section we allow the possibility of ion motion as well as electron motion. In order to keep the problem as simple as possible, we consider only dc injection and we further simplify the problem by taking one wall of the diode at infinity, i.e., we let $(d/s) \to \infty$. In this form, we have a rudimentary model of an ion propulsion engine [24].

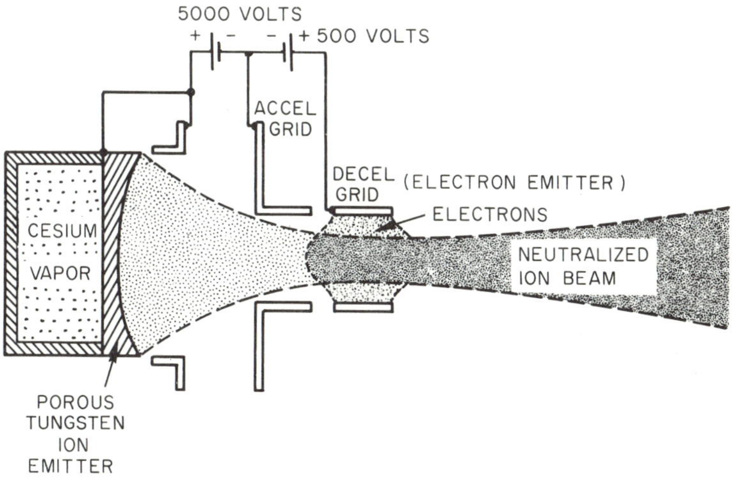

Fig. 6.7-1. Cross-section view of an ion propulsion system.

Ions are produced at the surface of the porous tungsten plug which is fed from a source of cesium vapor. The tungsten plug is operated at about 1100°C, well below the temperature for significant electron emission. The accel grid operates at about 5000 V negative with respect to the ion emitter and produces a well collimated ion beam with an intercepted ion current of less than 10^{-4} of the total ion current, when properly designed. The decel grid emits electrons with a mean thermal speed somewhat greater than the ion speed and is operated at a potential somewhat above that of the accel grid to prevent electrons from being drawn back into the ion emitter. The electrons eventually find their way into the ion beam in just the right number to neutralize the ion beam and to allow it to pass out into space, thus providing thrust to a space vehicle on which such a system would be mounted.

In practice such a system does not produce enough thrust when used as a single axially symmetric unit. It can be iterated in various ways: many long thin strip beams in parallel, many axially symmetric beams in two-dimensional arrays, many hollow cylindrical beams, etc. Even when the necessary paralleling has been done, the total thrust is typically only 10^{-4} *g*, a value suitable for very long missions at low thrust or for orbit correction systems. These devices can be built with very long life and high efficiency (>90%) and were well ahead of the technology of the nuclear power sources needed to drive them as of 1968.

Figure 6.7-1 is a picture of a physical ion propulsion system. Clearly it involves two-dimensional motion of the electrons which are allowed to enter the ion beam from electron emitters located out of the paths of the ion beams. The two-dimensional motion of electrons is not an essential feature of the neutralization process, although it is very important in obtaining long life to keep the electron emitters from being worn away by sputtering of the high energy ions.

Figure 6.7-2 shows a schematic and associated potential diagram of a one-dimensional ion propulsion system with the same elements as the system of Fig. 6.7-1. Our model is shown in Fig. 6.7-3. It simulates the region beyond the decel grid only, in that we consider electrons and ions with initial velocities in the positive z-direction only. We make the further assumption that our electrons are injected at a single dc velocity rather than with a thermal spread. Our model approximates the half-Maxwellian velocity distribution of the physical system with a beam having a single velocity equal to the mean thermal speed of the distribution.

If the ions are not neutralized, they will simply travel out from the decel grid or injection plane a distance approximately equal to 0.707 times the emitter to accel grid spacing s_i and return. Their trajectories in this case are very much like the trajectories of the particles that returned to the injection plane in the single-species diode beyond limiting perveance in Fig. 6.5-5. If we start with an initially empty region $z > 0$, the first few sheets injected will travel farther into the space than the subsequent ones.

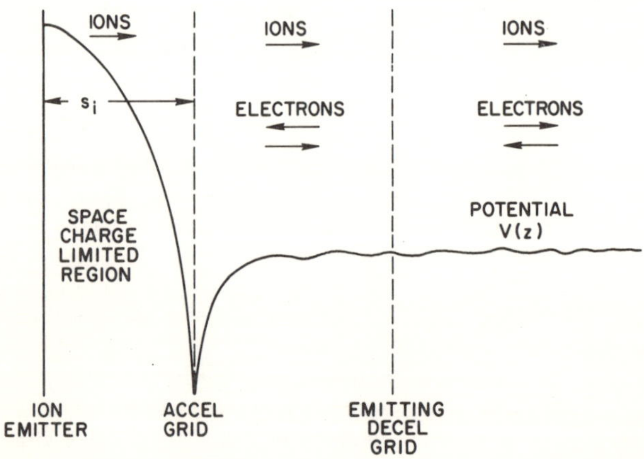

Fig. 6.7-2. A planar ion propulsion system and the associated potential diagram. Such a system would perform about the same as the system of Fig. 6.7-1, except that the grids would be worn down by ion sputtering and limit the life of the system.

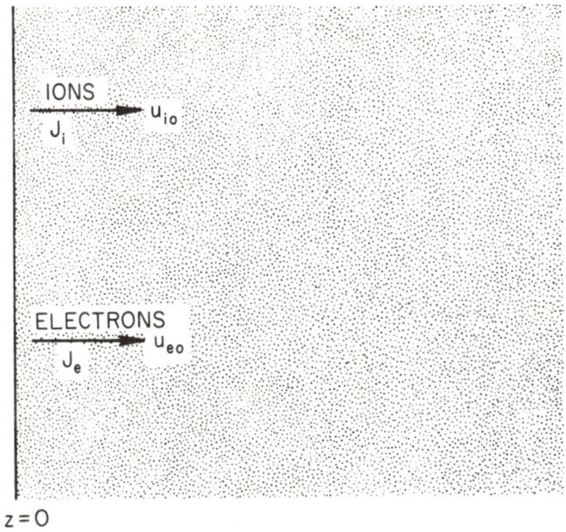

Fig. 6.7-3. One-dimensional model analyzed here has single velocity ion and electron beams injected at $z = 0$.

The injected currents are J_i and J_e and the injection velocities are u_{i0} and u_{e0}. The masses per particle per unit area are m_i and m_e. The injection plane has the property that it is transparent to electrons and ions injected in the positive z-direction, but it collects all particles that strike it traveling in the negative z-direction. The electron current density J_e is equal to the space-charge limited current that would be drawn in a diode of spacing s_e at the potential corresponding to u_{e0}. The corresponding ion-diode spacing is s_i, the spacing of the accel grid from the emitter in Fig. 6.7-2.

After this initial transient has died down we find trajectories as given in Fig. 6.7-4. It is seen that the average penetration is about $(z/s_i) = 0.707$. The pattern of sheets is almost periodic, but not quite, within the time shown. The sheets did not penetrate beyond $(z/s_i) = 1.4$ at later times; the fluctuation in the behavior of the sheets within this region is another example of shot noise.

When large numbers of excess electrons are supplied to the system, the ions travel in straight lines at constant velocity as shown by the dashed lines in Fig. 6.7-5. The solid lines in this figure are for the electrons, supplied in this case at a velocity 2.5 times the initial ion velocity. The behavior of the electrons is now analogous to the behavior of particles in a single-species diode beyond limiting perveance. Large numbers of electrons return to the injection plane, but some are transmitted on, randomly distributed in velocity around the ion velocity. The ion beam is thus supplied with whatever number of electrons are needed to maintain neutrality. If a slight excess of positive ions exists beyond the plane $(z/s_e) = 0.707$, a suitable number of additional electrons, which are available in large numbers at zero velocity at this plane, will be drawn out to neutralize the temporary charge un-

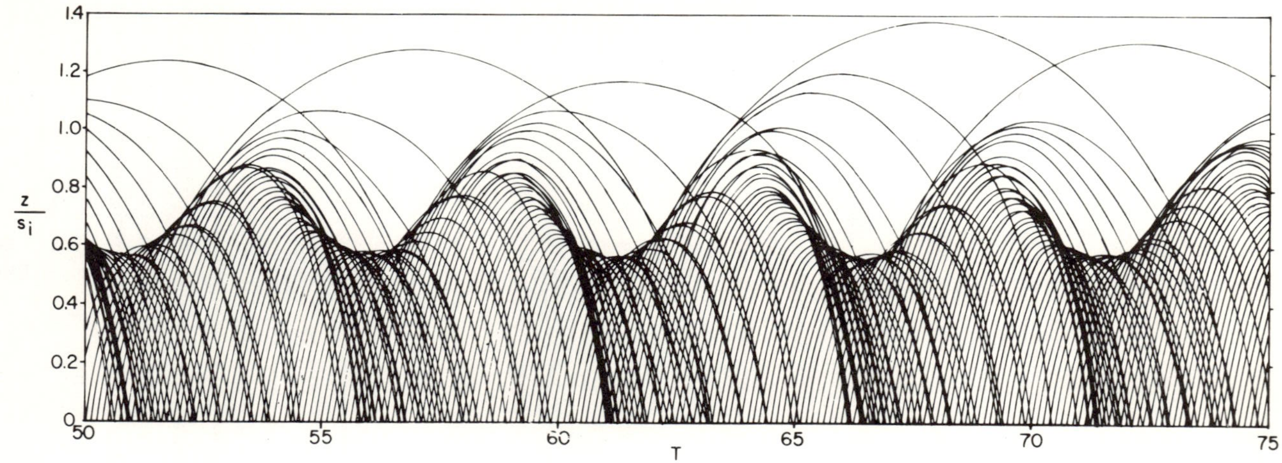

Fig. 6.7-4. Ion trajectories for a single-velocity ion beam injected into an infinite region after an initial transient has died out.

Distance is measured in units of the equivalent space-charge limited diode spacing s_i for the injected current density. Time is measured in ion transit times t_{i0} across the distance s_i at the injection velocity. The pattern repeats except for minor fluctuations due to shot noise. The time step here was $0.25t_0$, rather coarse by modern standards. This computation was done in 1962 with a small computer. For a still coarser time increment by a factor of two the pattern was substantially the same. One ion was introduced per time step [from D. A. Dunn and I. T. Ho, Computer experiments on ion beam neutralization with initially cold electrons, *AIAA J.* **1**, 2770–2777 (1963)].

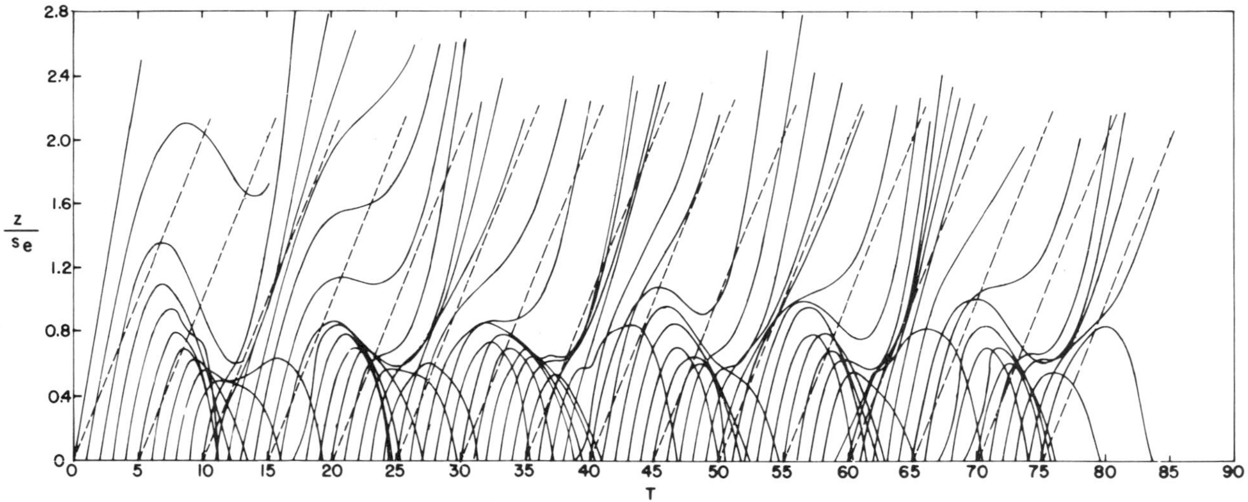

Fig. 6.7-5. Electron trajectories (solid) and ion trajectories (dashed) near the injection plane for a two-species case with $(u_{e0}/u_{i0}) = 2.5$, $(m_i/m_e) = 50$, and $(-J_e/J_i) = 2$.

The current ratio for this case is large enough to give complete neutralization by means of an oscillating electron sheath which supplies all the electron current needed for neutralization at velocities randomly distributed about the ion velocity. Distance is measured in units of s_e, the equivalent space-charge limited diode spacing corresponding to the injected electron current. Time is measured in units of the electron transit time t_{e0} across s_e [from D. A. Dunn and I. T. Ho, Computer experiments on ion beam neutralization with initially cold electrons, *AIAA J.* **1**, 2770–2777 (1963)].

balance. This is the fundamental principle of ion propulsion beam neutralization. It is clear that this is not a critical process and neutralization should equally well occur over a wide range of electron velocities and currents, provided only that the injected electron current is substantially greater than the injected ion current, so that enough low velocity electrons are available. The noncritical nature of this process should extend to electron sources that have large thermal velocity spreads and should be substantially independent of two-dimensional effects, if enough electrons are available [25]. The available experimental data [26] agrees with this simple result, and neutralization is found to be very easy to obtain under high electron current conditions.

In principle, neutralization can occur with the injected electron current just equal to the ion current. Once we go beyond the "oscillating sheath" of Fig. 6.7-5 the ion current must be exactly equal to the electron current if charge neutrality is to be maintained. Charge neutrality can only be obtained in the long run, if we eject the same number of electrons and ions from the system. Charge neutrality also requires that the electrons and ions travel with the same average velocity. If we eject exactly equal currents of both species at exactly equal velocities, we indeed get perfect neutralization with no regions of local charge unbalance. This case is the simplest one to understand. However, as noted in connection with Fig. 6.7-1, physical electron emitters emit particles with a wide velocity distribution and also with a mean thermal speed greater than the ion speed that is desired for typical missions. Hence, we must investigate cases of electron velocity greater than ion velocity. It is shown in Section 7.3 that, for single velocity beams, neutralization is obtained up to a velocity ratio of 2 with equal injected electron and ion currents without oscillations. Above that velocity ratio we find from the Lagrangian theory here that there are high-frequency oscillations both in the "sheath" near the wall as in Fig. 6.7-5 and further away that cause the electron trajectories to become randomized. Figures 6.7-6 and 6.7-7 show the electron and ion trajectories further away from the wall for a case with equal injected electron and ion currents. The electron trajectories for this case all start out penetrating deep into the space and bounce off the front of the ion beam. This basic trajectory pattern was predicted qualitatively and referred to as a plasma bottle (which is rapidly growing in length) [27]. For this case there is not enough electron current and eventually ions are returned, but the basic trajectory pattern for electrons is not significantly different for higher electron currents, except very near the injection plane where the electrons have a pattern like Fig. 6.7-5.

Figure 6.7-8 shows the conditions for neutralization for single velocity

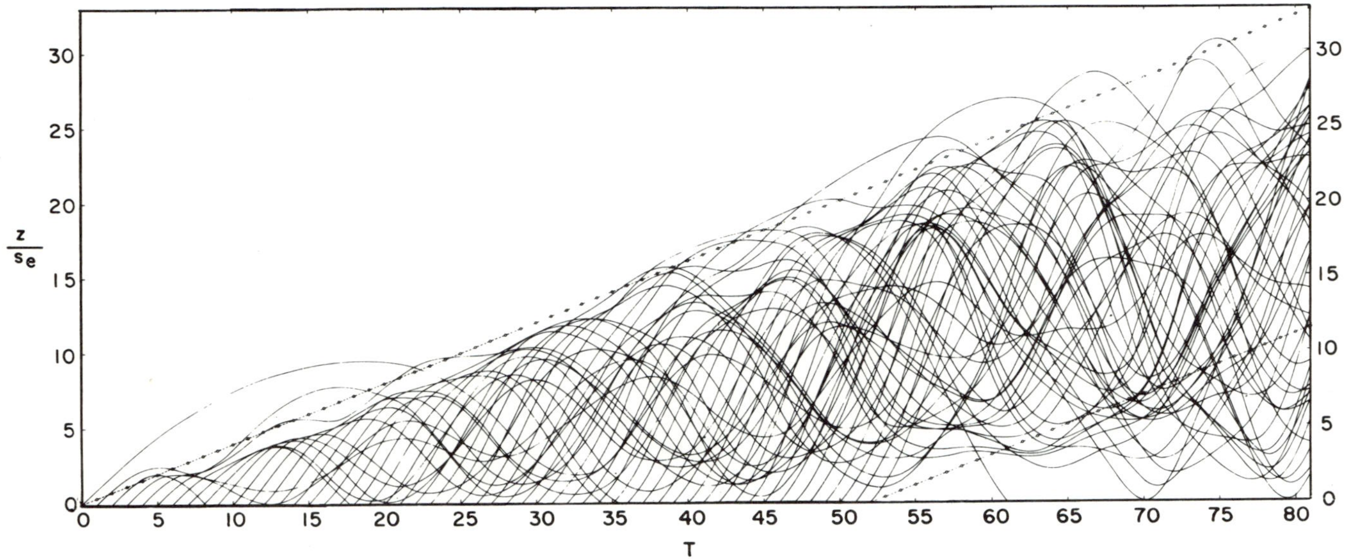

Fig. 6.7-6. Electron trajectories for a two species case with $(u_{e0}/u_{i0}) = 2.5$, $(m_i/m_e) = 1800$, and $(-J_e/J_i) = 1$. The current ratio here is insufficient to give full neutralization and some ions are eventually returned. Only every fifth trajectory out to $T = 55$ is shown, but in the computation electrons were injected uniformly beyond this time. Note that some electrons have already been returned to the injection plane at this time. The trajectory pattern here is the justification for regarding the beam as a "plasma bottle" [from D. A. Dunn and I. T. Ho, Computer experiments on ion beam neutralization with initially cold electrons, *AIAA J.* **1**, 2770–2777 (1963)].

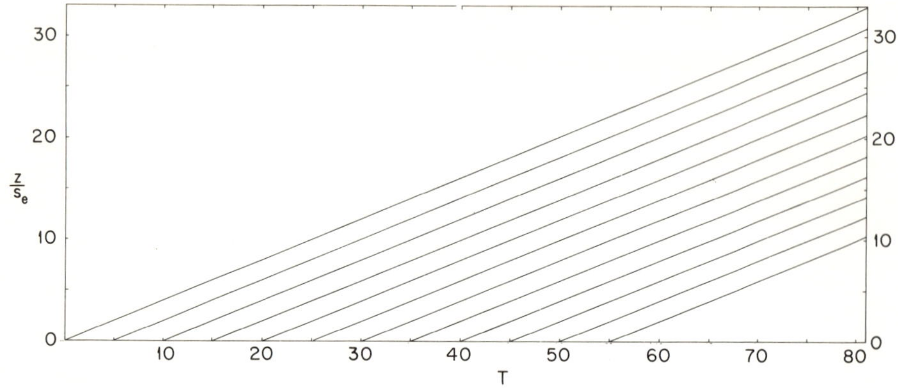

Fig. 6.7-7. Ion trajectories for the case of Fig. 6.7-6. At much greater times than that shown some ions turn back and return to the injection plane [from D. A. Dunn and I. T. Ho, Computer experiments on ion beam neutralization with initially cold electrons, *AIAA J.* **1**, 2770–2777 (1963)].

injection as a function of injected current and velocity ratios. This figure is based on five computer experiments which correspond to the points marked [24]. It tells us how much is enough electron current for neutralization as a function of the velocity ratio. Clearly, as we increase the velocity ratio, we require a greater injected electron current, as expected.

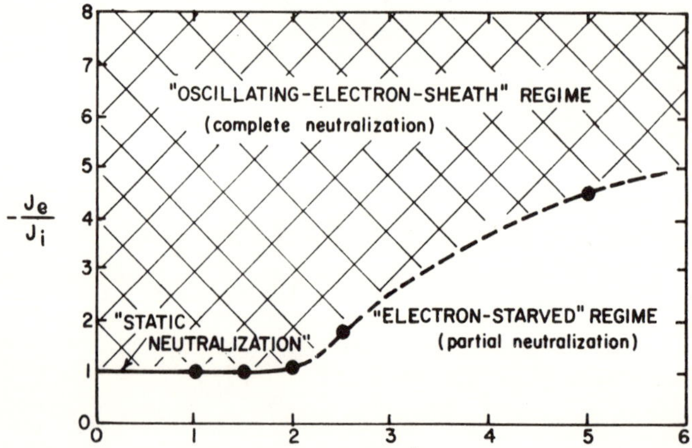

Fig. 6.7-8. Ratio of injected electron current to injected ion current, $(-J_e/J_i)$, for neutralization, as a function of the ratio of electron to ion velocity, (u_{e0}/u_{i0}) [from D. A. Dunn and I. T. Ho, Computer experiments on ion beam neutralization with initially cold electrons, *AIAA J.* **1**, 2770–2777 (1963)].

6.8 The Dual Problem—Interaction of Uncharged Current Sheets

The interaction of charge sheets of the same sign is always one of mutual repulsion. The interaction of current carrying wires or sheets, all carrying current in the same direction, is always one of mutual attraction. In both cases, the introduction of a second oppositely charged or oppositely directed species results in a combination of mutual attraction and repulsion which in the case of charged particles results in a plasma. There is an exact analog of current carrying wires or sheets in the case of gravitation which is always an attraction interaction. So far, no one has observed matter with a gravitational "charge" that has a sign opposite to that of conventional matter.

In this section we will introduce the problem of interaction among current carrying conductors. The approach and equations are equally applicable to gravitational interaction, and indeed this technique has been used to study both one- [28] and two-dimensional [29] models of galaxies. In the study of current carrying conductors, we are not actually interested in the way a group of wires interacts. Our real interest is in the behavior of plasmas carrying a unidirectional current, typically at very low pressures and low densities. Under these conditions collisions play only a small role and we can regard the plasma as essentially infinitely compressible. On the relatively slow time scale of interest here, the type of plasma oscillations we have been studying in previous sections is a very-high-frequency phenomenon. We will assume that any such oscillations have either damped out or that they continue at such a high frequency that we can ignore them. We simply regard the plasma as a type of neutral fluid conductor that is infinitely compressible.

The particular problem that we will discuss is that of a planar sheet plasma carrying a dc current in the x-direction. There are many modes that instabilities can take in such a system. However, we will eliminate consideration of most of them by starting our analysis with a sheet model in which the current is constrained to be uniform in the x- and y-directions, as suggested in Fig. 6.8-1. In our model the sheets stay flat, but can move freely in the z-direction. In an actual plasma there is nothing to prevent the plasma from squeezing together in periodic patterns in both x and y, and indeed it is usually found that a plasma is unstable to perturbations in both of these directions. The model here allows us to study the tendency of the plasma to pinch together and to derive its rate of pinching in terms of the amount of current, the size of the plasma, etc. In Section 7.9 we study this problem further using an Eulerian model and find that we can legit-

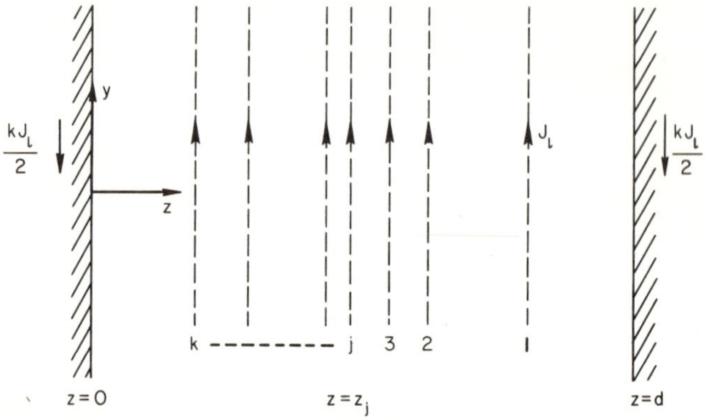

Fig. 6.8-1. A number k of infinite current sheets numbered in sequence from right to left. Each sheet carries a current J_l per unit width in the y-direction and has a mass per unit area m_s. The sheets are rigid planes that always lie in planes of constant z, but that can move freely in the z-direction. Conducting walls at $z = 0$ and $z = d$ carry the return currents of $(kJ_l/2)$ in each wall.

imately extend the results that we obtain here to a problem involving variations in y and z rather than t and z which is the form of our answer here.

In Fig. 6.8-1 we show k sheets, each carrying a current J_l per unit width in the y-direction. The total current per unit width is thus evidently kJ_l. As in the charge-sheet problem, we can calculate the field on either side of one of these sheets in the absence of the others and then sum the fields for a group of sheets to get the total field at any point. The force acting on any sheet is then J_l times the magnetic field at that sheet position.

Our basic equation for the magnetic field is, by analogy to the case with charge sheets, the quasi-static form of the curl equation, obtained by setting $(\partial\mathbf{D}/\partial t) = 0$ in Eq. (3.1.1).

$$\nabla \times \mathbf{H} = \mathbf{J} \tag{6.8.1}$$

With variations only in z and with J_y only this equation reduces to

$$\partial H_x/\partial z = J_y \tag{6.8.2}$$

This is directly analogous to Eq. (6.1.1) for charge sheets, and by the same reasoning the magnetic field is constant outside of a current sheet because the current is zero. Integrating Eq. (6.8.2) with respect to z yields for the field change across a current sheet with a current J_l per unit width

$$H_{x1} - H_{x2} = J_l \tag{6.8.3}$$

where H_{x1} is the field to the left and H_{x2} is the field to the right of the sheet analogous to Eq. (6.1.3). Here there is no analog of the image charge in the conducting walls for dc currents. However, there must be a means of carrying the return current. In Fig. 6.8-1 the return current is split equally between two fixed walls at $z = 0$ and $z = d$. These walls can be put at infinity, if desired, but if they are at finite z, the field outside these walls, for $z < 0$ and $z > d$, must be zero. Because there is no analog of the image charge here, we have a different and simpler set of equations for the field in this problem. From the Biot–Savart law, $\oint \mathbf{H} \cdot dl = \mathbf{I}$, and symmetry

$$H_{x1} = -H_{x2} = -J_l/2 \tag{6.8.4}$$

without regard to the position of the sheet within the space. Thus the equations analogous to Eqs. (6.1.5)–(6.1.7) are

$$H_{x1} = -J_l/2 \tag{6.8.5}$$

$$H_{x3} = 0 \tag{6.8.6}$$

$$H_{x2} = J_l/2 \tag{6.8.7}$$

where H_{x3} is the field at the center of the current sheet due to its own current. It is zero because there is no image current.

As in Section 6.1 we can find the field at z_j due to current sheets to the left of the jth sheet using Eq. (6.8.7):

$$H_{xl} = \sum_{i=j+1}^{k} J_l/2 = (k - j - 1)(J_l/2) \tag{6.8.8}$$

The field at z_j due to sheets to the right of the jth sheet is from Eq. (6.8.5)

$$H_{xr} = -\sum_{i=1}^{j-1} J_l/2 = (-j + 1)(J_l/2) \tag{6.8.9}$$

The total field at z_j is then

$$H_{xj} = (k - 2j)J_l/2 \tag{6.8.10}$$

which is, as we would expect, equal to zero for $j = k/2$, i.e., for a sheet halfway across. It is $(-kJ_l/2)$ for $j = k$ and $(kJ_l/2)$ for $j = 0$.

The force F_s per unit area on the jth sheet is, of course,

$$F_{sj} = J_l B_x = (\mu_0 J_l^2/2)(k - 2j) \tag{6.8.11}$$

and the acceleration of the jth sheet is

$$a_j = F_{sj}/m_s = (\mu_0 J_l^2/2m_s)(k - 2j) \tag{6.8.12}$$

If the sheets stay ordered, this problem is obviously very simple because the acceleration is constant, leading to parabolic trajectories. Even when the sheets get mixed up, the computation is still extremely simple, because the acceleration is only a function of ordering and not of position.

PROBLEM 6.8-1 MAGNETIC PINCH EFFECT

For a planar infinitely compressible plasma described by the model of this section calculate the trajectories of all sheets for $k = 4$, assuming an initially uniform spacing and zero initial velocity for all sheets.

REFERENCES

1. O. Buneman and D. A. Dunn, Computer experiments in plasma physics, *Sci. J.* **2**, 34–43 (1966).
2. R. W. Hockney, Computer simulation of anomalous plasma diffusion and numerical solution of Poisson's equation, *Phys. Fluids* **9**, 1826–1835 (1966).
3. S. P. Yu and T. G. Mihran, Nonlinear rf behavior of electron beams with velocity distribution, I. General analysis, *J. Appl. Phys.* **34**, 2972–2975 (1963).
4. T. G. Mihran and S. P. Yu, Nonlinear rf behavior of electron beams with velocity distribution, II. Application to rectangular velocity distribution, *J. Appl. Phys.* **34**, 2976–2983 (1963).
5. P. Burger, Zero-field solutions and their stability in the one-dimensional low-pressure cesium diode, *J. Appl. Phys.* **38**, 3360–3368 (1967).
6. C. K. Birdsall and W. B. Bridges, Space charge instabilities in electron diodes and plasma converters, *J. Appl. Phys.* **32**, 2611–2618 (1961). [Also "Electron Dynamics of Diode Regions." Academic Press, New York, 1966.]
7. A. S. Halsted and D. A. Dunn, *Quart. Res. Rev.* No. 5, pp. I-34–I-42, Stanford Electronics Laboratories, Stanford University, Stanford California, 1963.
8. C. K. Birdsall, "Plasma oscillations and particle trajectories in a drifting electron stream; Palmer diagrams," Rept. 458, Ser. 60, Electron. Res. Lab., Univ. of Cal., Berkeley, 1962.
9. H. Derfler, The frequency of nonlinear plasma oscillations, *J. Electron. Contr.* **11**, 189–192 (1961).
10. J. Dawson, One-dimensional plasma model, *Phys. Fluids* **5**, 445–458 (1962).
11. S. Goldberg, "Difference Equations." Wiley, New York, 1958.
12. A. S. Halsted, "Equilibrium conditions and beam focusing effects in a beam-generated plasma," Rept. SUIPR 10, Inst. Plasma Res., Stanford Univ., 1965.
13. P. Burger, Theory of large-amplitude oscillations in the one-dimensional low-pressure cesium thermionic converter, *J. Appl. Phys.* **36**, 1938–1943 (1965).
14. P. Burger, D. A. Dunn, and A. S. Halsted, Computer experiments on the randomization of electrons in a collisionless plasma, *Phys. Fluids* **8**, 2263–2272 (1965).

15. P. K. Tien and J. Moshman, Monte Carlo calculation of noise near the potential minimum of a high-frequency diode, *J. Appl. Phys.* **27**, 1067–1078 (1956).
16. R. W. Hockney, Formation and stability of virtual electrodes in a cylinder, *J. Appl. Phys.* **39**, 4166–4170 (1968).
17. J. M. Sellen and H. Shelton, Transient and steady state behavior in cesium ion beams, *in* "Electrostatic Propulsion" (D. B. Langmuir, E. Stuhlinger, and J. M. Sellen, eds.), pp. 305–356. Academic Press, New York, 1961.
18. H. H. Atkinson, Limiting currents and neutralization in cylindrical electron beams, *in Proc. Int. Congr. Microwave Tubes, Scheveningen, Netherlands, Pt. K*, no. 6, pp. 559–562, 1962.
19. W. Schottky, Spontaneous current fluctuations in various conductors, *Ann. Phys. (Leipzig)* **57**, 541–567 (1918).
20. C. W. Barnes and D. A. Dunn, Shot noise in the Lagrangian one-dimensional model, *Proc. Symp. Computer Simulation of Plasma and Many-Body Problems, Williamsburg, Va.*, NASA SP-153, pp. 31-38; April, 1967.
21. T. C. Fry, The thermionic current between parallel plane electrodes: velocities distributed according to Maxwell's law, *Phys. Rev.* **17**, 441 (1921). [Also I. Langmuir, The effect of space charge and initial velocities on the potential distribution and thermionic current between parallel electrodes, *ibid.* **21**, 419–435 (1923).]
22. J. R. Pierce, Limiting stable current in electron beams in the presence of ions, *J. Appl. Phys.* **15**, 721–726 (1944).
23. J. E. Faulkner and A. A. Ware, Effect of finite ion mass on stability of a space-charge neutralized electron beam, *J. Appl. Phys.* **40**, 366–368 (1969).
24. D. A. Dunn and I. T. Ho, Computer experiments on ion beam neutralization with initially cold electrons, *AIAA J.* **1**, 2770–2777 (1963). [Also O. Buneman and G. Kooyers, Computer simulation of the electron mixing mechanism in ion propulsion, *ibid.* **1**, 2525–2528 (1963).]
25. R. P. Wadhwa, O. Buneman, and D. F. Branch, Two-dimensional computer experiments on ion-beam neutralization, *AIAA J.* **3**, 1076–1081 (1965).
26. D. B. Langmuir, "Electrostatic Propulsion." E. Stuhlinger, and J. M. Sellen, eds., Academic Press, New York, 1961.
27. H. Mirels, On ion rocket neutralization, *in* "Electrostatic Propulsion" (D. B. Langmuir, E. Stuhlinger, and J. M. Sellen, eds.). Academic Press, New York, 1961.
28. F. Hohl and M. Feix, A one-dimensional model to study collective effects in stellar dynamics, *Phys. Lett.* **22**, 432–434 (1966).
29. R. W. Hockney, Gravitational experiments with a cylindrical galaxy, *Astrophys. J.* **150**, 797–806 (1967).

CHAPTER 7

Motion of Charged and Conducting Fluids—the Eulerian Model

It would seem an obvious thought to use a transmission line comprising a coiled-up wire inside a hollow conductor. (I must emphasize that I did not then think of the helix by itself. "Obviously" there had to be a shield, an outer conductor. This shows just how wrong one can be.) Anyway, these difficulties afforded me an incentive to compute the impedance and to make some estimate of what the phase velocity would be. It looked as if it were not impossible to make a circuit with a wave velocity much smaller than that of light.

In the group at the University of Birmingham, one man was considered to be the expert on transmission lines. I went to him and told him what I should like to try; what did he think of it? He said he did not think that such a line would propagate anything above 10 megacycles. Well, it was such an easy thing to do, so I took some copper wire and a steel rod, put the steel rod into the lathe and wound the helix on it; the turns were touching and I pulled them out a bit, to prevent shorts. Then I took out the co-ax inner conductor of a standing-wave detector, substituted the helix for it, and just took readings along it when a 10-cm signal passed through—and lo and behold, I got a standing wave, a very ragged one to be sure, but when I measured the distances between peaks and took the average, I found the velocity was down by a factor of 7.

This was an encouraging result. Incidentally, I confirmed that this factor of 7 was pretty near equal to the pitch of the wire. I was amazed. I did not expect that nature would be that simple—there it was, a real discovery. I had not predicted that result. As a matter of fact, if you take the distributed capacity of a helix into account you get a very messy sort of equation that I could not even understand; but nature is essentially simple.

R. Kompfner, "Invention of the Traveling-Wave Tube." San Francisco Press, San Francisco, California, 1964. Reprinted by permission.

7.0 Introduction

We can model an electron beam or a plasma containing many interacting particles using either the Lagrangian discrete particle model of the preceding chapter or the Eulerian fluid model that we present here. In the Lagrangian model that we have examined, there was no restriction on particle cross-overs, and there was no limitation on the number of different particle velocities that was allowed in any given region of space, other than the limitation imposed by the choice of the number of particles to be used in the model. Thus the Lagrangian model is well suited to studies of large-amplitude wave problems in which the waves "break" and in other problems involving order–disorder transitions.

The Eulerian fluid model is normally useful in quite different contexts. It allows, in a single fluid model, only a single drift velocity to be associated with a single point in space. If two fluids are allowed to interpenetrate, each may have a different velocity at a given point. In addition, random velocity distributions of prescribed form can be superimposed on the drift velocity of each fluid without introducing excessive mathematical complexity, but the temperature of each fluid is a parameter that must be assigned or determined through an equation of state. The basic importance of the Eulerian fluid model lies in its ability to predict static states of a system and to determine the stability of static states to various small-amplitude perturbations and to explore near-equilibrium states generally. In many problems it is much more important to know that a given static state is unstable and to know the growth rate of the instability, than it is to know the final state of the system. Thus, the two analytical techniques presented in these two chapters complement each other and are normally used in conjunction to study a new problem. For example, in the study of a microwave beam-type oscillation, a small-signal Eulerian model can be used to predict the conditions for start oscillation, but a Lagrangian model is needed to predict the large-amplitude limit of signal level and so the power output and efficiency of the device. In some types of electron beam and plasma problems a large number of static states of a system are predicted from the Eulerian model. A complete understanding of the problem requires a knowledge of these states and also of an additional set of essentially time-varying states that sometimes exist and that can only be obtained from a Lagrangian model. Often the time-varying states can best be understood by comparing them with static states, as will be seen.

There are many aspects of fluid behavior that we do not explore here at all. Our fluids are infinitely compressible and totally nonviscous. We thereby completely omit the effects of collisions between neutral particles and between charged particles and neutrals. The regime in which our results are correct is in low-pressure highly ionized media such as occur in electron tubes, gas tubes, thermionic energy converters, interplanetary space, fusion reactors, etc. There are many texts [1, 2] that are concerned with other pressure–ionization regimes that are of interest in MHD and related device applications.

7.1 Fundamental Equations for Charged and Conducting Fluids

Just as in the Lagrangian theory, there are two fundamental equations required in Eulerian analysis, a field equation from which we can compute the field due to the charges or currents, and a trajectory equation from which we can compute the motions of the charges or current elements due to the action of the field. In Lagrangian theory we made our computations self-consistent by going back and forth between these two equations, step by step in time. In Eulerian theory it is usually possible to simply equate quantities appearing in both equations and to force self-consistency in this manner.

The field equation for charged fluids is simply Poisson's equation which for one-dimensional variations of potential can be written as

$$\partial^2 V/\partial z^2 = -\varrho/\varepsilon \tag{7.1.1}$$

using $\mathbf{E} = -\nabla V$ and $\mathbf{D} = \varepsilon\mathbf{E}$ in the first Maxwell equation, $\nabla \cdot \mathbf{D} = \varrho$. Just as we used the approximation that the fields were determined only by the positions of the charges in our Lagrangian theory, and not by their velocities, so here we make the corresponding quasi-static approximation that $\mathbf{E} = -\nabla V$ which implies $\nabla \times \mathbf{E} = 0$. Hence we here ignore the effects of time-varying magnetic fields produced by the motions of the charges.

The corresponding quasi-static field equation for currents in uncharged conducting fluids is obtained from $\nabla \times \mathbf{H} = \mathbf{J}$, setting $\partial\mathbf{D}/\partial t$ equal to zero. If we consider only one-dimensional variations in J and H, we may picture J_y producing H_x with variations of both quantities only in z. Then

$$\partial H_x/\partial z = J_y \tag{7.1.2}$$

is the equation giving $\mathbf{H}$ in terms of a known $\mathbf{J}$.

Both Eqs. (7.1.1) and (7.1.2) can be used for truly static problems and for time-varying problems in which all wave and particle velocities are much less than the velocity of light in free space, which includes a very large class of interesting device problems.

We now wish to examine the equations of motion for a charged fluid that allow us to compute charge and current distributions from fields and initial conditions. Neglecting relativistic effects we may write the force equation in the form

$$q\mathbf{E} = m\, d\mathbf{u}/dt \tag{7.1.3}$$

Now we want to evaluate the total derivative of velocity for a particular piece of fluid that moves under the influence of the field $\mathbf{E}$. Let $\mathbf{u}$ be the velocity at $P(x, y, z)$ at time t and $\mathbf{u} + d\mathbf{u}$ be the velocity at a neighboring point $Q(x + dx, y + dy, z + dz)$ at time $t + dt$. In the Eulerian model only a single velocity is allowed at any point such as P or Q at any instant of time. Then $d\mathbf{u}$ is

$$\begin{aligned} d\mathbf{u} &= \frac{\partial \mathbf{u}}{\partial t}\, dt + \frac{\partial \mathbf{u}}{\partial x}\, dx + \frac{\partial \mathbf{u}}{\partial y}\, dy + \frac{\partial \mathbf{u}}{\partial z}\, dz \\ &= \frac{\partial \mathbf{u}}{\partial t}\, dt + (d\mathbf{r} \cdot \nabla)\mathbf{u} \end{aligned} \tag{7.1.4}$$

where $d\mathbf{r} = \mathbf{1}_x\, dx + \mathbf{1}_y\, dy + \mathbf{1}_z\, dz$ is the vector extending from P to Q and

$$d\mathbf{r} = \mathbf{u}\, dt \tag{7.1.5}$$

so that

$$d\mathbf{u} = (\partial \mathbf{u}/\partial t)\, dt + (\mathbf{u} \cdot \nabla)\mathbf{u}\, dt \tag{7.1.6}$$

and finally the force equation becomes

$$(q/m)\mathbf{E} = d\mathbf{u}/dt = (\partial \mathbf{u}/\partial t) + (\mathbf{u} \cdot \nabla)\mathbf{u} \tag{7.1.7}$$

What we now have is the total time derivative that we wanted, which is not the way the velocity changes at a particular point in space, but the way the velocity changes if we follow a particular piece of fluid as it moves along a trajectory in both time and space, often referred to as the substantial derivative and written in the form $D\mathbf{u}/Dt$.

Let us examine the form of the terms of $(\mathbf{u} \cdot \nabla)\mathbf{u}$ in Cartesian coordinates.

$$\begin{aligned} [(\mathbf{u} \cdot \nabla)\mathbf{u}]_x &= u_x(\partial u_x/\partial x) + u_y(\partial u_x/\partial y) + u_z(\partial u_x/\partial z) \\ [(\mathbf{u} \cdot \nabla)\mathbf{u}]_y &= u_x(\partial u_y/\partial x) + u_y(\partial u_y/\partial y) + u_z(\partial u_y/\partial z) \\ [(\mathbf{u} \cdot \nabla)\mathbf{u}]_z &= u_x(\partial u_z/\partial x) + u_y(\partial u_z/\partial y) + u_z(\partial u_z/\partial z) \end{aligned} \tag{7.1.8}$$

For one-dimensional fluid motion in the z-direction, only the electric field in this direction need be considered, so Eq. (7.1.7) can be written in the form

$$(q/m)E_z = (\partial u_z/\partial t) + u_z(\partial u_z/\partial z) \tag{7.1.9}$$

This is one basic equation we need. For an uncharged conducting fluid subject only to $\mathbf{J} \times \mathbf{B}$ forces we would have in place of E_z in Eq. (7.1.9) the z-component of $\mathbf{J} \times \mathbf{B}$ which for y-directed currents and x-directed magnetic field would give us, if $\mathbf{B} = \mu\mathbf{H}$,

$$-\mu J_y H_x = (\partial u_z/\partial t) + u_z(\partial u_z/\partial z) \tag{7.1.10}$$

We now need to relate $\mathbf{J}$, $\mathbf{u}$, and ϱ for charged fluids because we are often able to specify the current, but we may know neither the charge nor the velocity explicitly. The continuity equation for charge obtained by taking the divergence of the fourth Maxwell equation is

$$\nabla \cdot \mathbf{J} = -\partial\varrho/\partial t \tag{7.1.11}$$

which in one dimension becomes

$$\partial J_z/\partial z = -\partial\varrho/\partial t \tag{7.1.12}$$

The definition of current is

$$\mathbf{J} = \varrho\mathbf{u} \tag{7.1.13}$$

and in one dimension is

$$J_z = \varrho u_z \tag{7.1.14}$$

7.2 Static Theory of a Charged Fluid in a Planar Diode—Single Velocity Injection of a Prescribed Current

Here we picture the problem of an electron beam injected into a diode and solve for the dc states of the system with only one charge species present as in Fig. 7.2-1. As in our previous model of this problem, conducting walls at $z = 0$ and $z = d$ exist, and the single-velocity beam is injected at $z = 0$ with a velocity u_0 corresponding to the potential V_0 of the wall at $z = 0$, i.e., $u_0^2 = 2(q/m)V_0$. The wall at $z = d$ has a potential V_d, and we wish to solve for the potential $V(z)$. If all the injected current J is transmitted to the wall at $z = d$, J is constant throughout the space. If some or all of the current is returned to the injection plane, we must seek a self-consistent determination of the fraction returned. If current is returned, we consider it to be collected on the injection plane.

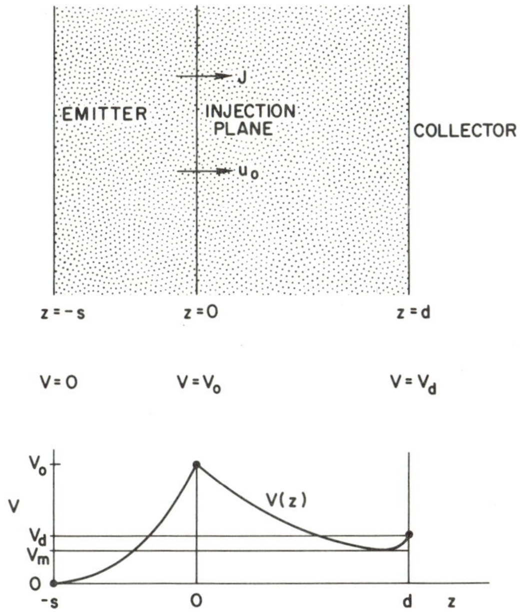

Fig. 7.2-1. A single species diode with walls at $z = 0$ and $z = d$ has a current density J injected at velocity u_0 at $z = 0$. A hypothetical emitter at $z = -s$ would yield the space-charge limited current density J, if the potential V_0 were applied. The corresponding potential diagram is given below. The potential $V(z)$ for $0 < z < d$ is sought. The potential at the position within the space $z = 0$ to $z = d$ at which the field is zero is V_m. The minimum potential within this space is V_{min}, which may or may not be the same as V_m.

We start with Eqs. (7.1.1) and (7.1.14) which for a constant J can be written in terms of $D = 2.33 \times 10^{-6}$, using $u_z^2(z) = 2(q/m)V(z)$

$$d^2V/dz^2 = (4J/9D)V^{-1/2} \tag{7.2.1}$$

where the negative sign in the charge q has been introduced, and where it has been assumed that V is non-time-varying. The first integration of Eq. (7.2.1) is carried out by multiplying both sides by dV giving

$$(dV/dz)\, d(dV/dz) = (4J/9D)(dV/V^{1/2}) \tag{7.2.2}$$

which can be integrated to give [3, 4, 5]

$$dV/dz = \pm[(16J/9D)V^{1/2} + C_1]^{1/2} \tag{7.2.3}$$

where C_1 is evidently the square of the electric field at the position at which $V = 0$, if any such position exists.

Here C_1 may be taken as defining a potential V_m as follows:

$$C_1 = -(16J/9D)V_m^{1/2} \tag{7.2.4}$$

and then Eq. (7.2.3) can be written

$$dV/dz = \pm[(16J/9D)(V^{1/2} - V_m^{1/2})] \tag{7.2.5}$$

We may have several types of potential distributions depending on the sign of $V_m^{1/2}$ and the sign of (dV/dz) at the potential minimum V_{min} wherever it may occur within the diode (Table 7.2-1). If we were dealing with an ion beam, we would be interested in potential maxima.

TABLE 7.2-1

TYPES OF POTENTIAL DISTRIBUTION IN A DIODE

Type	$V_m^{1/2}$	dV/dz (at $V = V_{min}$)	Fraction of current transmitted
A[a]	−	−	None
B	0	0	Part
C	+	0	All
D[a]	+ or −	+ or −	All

[a] Note that in cases A and D, V_{min} is not necessarily equal to V_m.

The standard case of Child's law is obtained by setting $V_m^{1/2}$ equal to zero, then (dV/dz) is zero at $V = V_m$ which in this case must also be V_{min}, and we have a type B distribution. In this case Eq. (7.2.5) takes the form

$$\int (dV/V^{1/4}) = \pm(\tfrac{4}{3})(J/D)^{1/2} \int dz \tag{7.2.6}$$

which may be integrated to give

$$\tfrac{4}{3}V^{3/4} = \pm(\tfrac{4}{3})(J/D)^{1/2}z + C_2 \tag{7.2.7}$$

Although we normally wish to use $z = 0$ at the injection plane, here let us choose $z = 0$ at the potential minimum $V = 0$, so that $C_2 = 0$ and we have Child's law in standard form

$$Jz^2/V^{3/2} = D = 2.33 \times 10^{-6} \tag{7.2.8}$$

Although we have here pictured the current as injected at an initial velocity u_0 and then reduced to zero velocity at a virtual cathode at $z = 0$, the problem is exactly the same, if we picture the current as starting at zero velocity at an emitter and accelerated to velocity u_0 by a potential applied to the diode, as in the usual derivation of Child's law. The constant D may be understood in terms of a diode carrying a space-charge limited current J_0 with a potential applied between a zero-potential emitter and a wall spaced a distance s from the emitter. The diode may be pictured as being made up of many cubes s on a side, each carrying a current $I_0 = J_0 s^2$ and having a perveance S_0 where

$$S_0 = I_0/V_0^{3/2} = J_0 s^2/V_0^{3/2} = D = 2.33 \times 10^{-6} \tag{7.2.9}$$

Thus the diode constant D is the perveance of such a cubical portion of an infinite parallel plane diode. The distance s specifies the current density for a given current.

It is also convenient to use the parameter (d/s) to specify the current density J injected into a diode of length d in terms of the Child's law current density J_0 that would be drawn in a diode with the same spacing d, but with one wall at zero potential and the other wall at the potential of the injection plane V_0. Thus

$$d/s = (J/J_0)^{1/2} \tag{7.2.10}$$

where here s is used to refer to the length of the diode that would produce the actual injected current density J. Thus we can picture an injection diode of spacing s to the left of the diode we wish to study, of spacing d, that is supplying a space-charge limited current J through the injection plane. If $(d/s) = 1$, for example, and if $V_d = 0$, we would have a completely symmetrical problem with the beam being accelerated from zero potential to V_0 at the injection plane a distance s from the emitter of the injection diode, and then decelerated back to zero potential at the collector, also a distance s from the injection plane. The solution to Poisson's equation would be identical in both regions, but with the potential as a function of distance simply reversed.

Let us now examine another type of solution to this diode problem, one for which $V_{\mathrm{m}}^{1/2}$ is positive. In this case we cannot simplify Eq. (7.2.5), but it can still be integrated directly, if we write it in the form

$$\pm \frac{4}{3}\left(\frac{J}{D}\right)^{1/2} z = \int \frac{2V^{1/2}\, d(V^{1/2})}{(V^{1/2} - V_{\mathrm{m}}^{1/2})^{1/2}}$$

$$= \frac{4}{3}(V^{1/2} - V_{\mathrm{m}}^{1/2})^{1/2}(V^{1/2} + 2V_{\mathrm{m}}^{1/2}) + C_2 \qquad (7.2.11)$$

which can be rearranged in the form of Eq. (7.2.12), if we set $z = 0$ at $V = V_0$ and introduce $\varphi = V/V_0$ and $\alpha = V_{\mathrm{m}}/V_0$.

$$z/s = \pm(\varphi^{1/2} + 2\alpha^{1/2})(\varphi^{1/2} - \alpha^{1/2})^{1/2} \mp (1 + 2\alpha^{1/2})(1 - \alpha^{1/2})^{1/2} \qquad (7.2.12)$$

The solution corresponding to the upper pair of signs gives a potential that rises as (z/s) increases from zero. This is one kind of type D distribution. The solution corresponding to the lower sign decreases with increasing (z/s) to a minimum at

$$(z/s)_{\min} = (1 + 2\alpha^{1/2})(1 - \alpha^{1/2})^{1/2} \qquad (7.2.13)$$

The equation to the right of the minimum for this solution is given by

$$(z/s)_{+} = (\varphi^{1/2} + 2\alpha^{1/2})(\varphi^{1/2} - \alpha^{1/2})^{1/2} + (1 + 2\alpha^{1/2})(1 - \alpha^{1/2}) \qquad (7.2.14)$$

We have a type C distribution if the potential reaches a minimum and increases again following Eq. (7.2.14).

Figure 7.2-2 gives a number of potential distributions plotted from Eqs. (7.2.12) and (7.2.14). To interpret this figure, picture a diode with boundary planes at $(z/s) = 0$ and $(z/s) = (d/s)$. We may select any value of (d/s) for any potential curve in the figure. If the curve rises steadily or falls steadily with no minimum within the selected distance, it is a type D distribution. Any curve starting at the left can thus be a type D distribution. If a minimum is encountered, a type C distribution is said to exist. If the potential on the

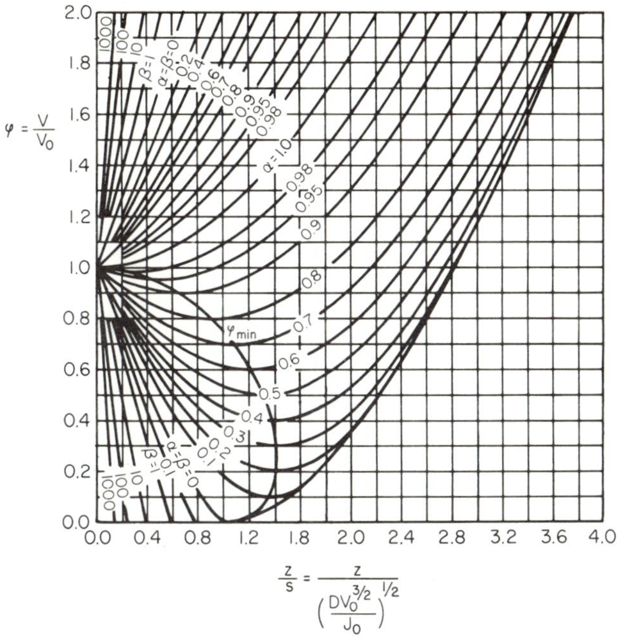

Fig. 7.2-2. Normalized potential, $\varphi = (V/V_0)$, as a function of normalized distance (z/s) for the electron diode with single velocity injection.

These curves apply to type C potential distributions which have a potential minimum within the space and also to type D distributions which have the potential minimum at one wall. For these curves all of the injected current is transmitted. The curve marked $\varphi_{\min}$ passes through the minima of the type C distributions and for all of these type C curves the value of α is equal to the minimum value of φ [from C. E. Fay, A. L. Samuel, and W. Shockley, On the theory of space charge between parallel plane electrodes, *Bell Syst. Tech. J.* **17**, 49–79 (1938)].

wall at $z = d$ is known, this uniquely specifies the curve for the specified value of (d/s). For example, take $V_d = V_0$. We can draw a straight line parallel to the abscissa that crosses all of the possible potential curves. If we now select a particular value of (d/s), we then determine which of these curves is to be used. Starting with a small (d/s) and increasing this parameter amounts to increasing the injected perveance which decreases the value of the potential at the center of the diode, because more negative space charge is present in the diode. This statement is correct up to the point at which $d/s = 2.8$ and (V_m/V_0) is reduced to 0.25, but interestingly enough a further reduction in potential can only be obtained at lower values of injected perveance. Poisson's equation thus has two solutions in this region, one with a potential minimum below $(V_m/V_0) = 0.25$ and one with a higher minimum. It is thus unlike Laplace's equation which has only one unique solution for any given set of boundary conditions. In the preceding chapter we saw that there is, in addition to these two static solutions to Poisson's equation, at least one time-varying solution. Attempts to inject a higher value of perveance than the limiting value predicted in Fig. 7.2-2 result in this time-varying solution. If the injected perveance is decreased after having been increased either just to the maximum or beyond the maximum, the upper portion of the curve of Fig. 7.2-2 is always followed, so the lower portion of this curve with static potentials less than a quarter of the wall potential is apparently not physically accessible.

Type D distributions can also exist when the sign of the constant in Eq. (7.2.3) is positive rather than negative. We may see this by rewriting Eq. (7.2.5) in the form

$$dV/dz = \pm[(16J/9D)(V^{1/2} + \beta^{1/2}V_0^{1/2})] \tag{7.2.15}$$

which leads to an equation analogous to Eqs. (7.1.12) and (7.2.14) of the form

$$z/s = (\varphi^{1/2} - 2\beta^{1/2})(\varphi^{1/2} + \beta^{1/2})^{1/2} - (1 - 2\beta^{1/2})(1 + \beta^{1/2})^{1/2} \tag{7.2.16}$$

which applies for $\varphi > 1$ and to

$$z/s = -(\varphi^{1/2} - 2\beta^{1/2})(\varphi^{1/2} + \beta^{1/2})^{1/2} + (1 - 2\beta^{1/2})(1 + \beta^{1/2})^{1/2} \tag{7.2.17}$$

which applies for $\varphi < 1$. Equations (7.2.16) and (7.2.17) are plotted in Fig. 7.2-2 and marked with values of β appropriately.

Type A distributions are identical to type D distributions with $\varphi < 1$, as given by Eq. (7.2.16). However, in type A distributions all of the current is reflected, so the current J in s must be replaced by $2J$ in Eq. (7.2.16). This simply requires that the rhs of Eq. (7.2.16) be multiplied by $2^{-1/2}$ to give type A distributions. The notation of Fay, Samuel, and Shockley [4] has been used in all of the final equations and in Fig. 7.2-2. Curves for types A and B and another type C distribution are given in this reference.

A critical issue in the late 1930s [3, 4, 5] was the design of efficient triodes, tetrodes, and pentodes. Efficiency could only be obtained at low plate voltages and it was important to know how to design the grid to plate region of such tubes. If we let the $z = 0$ plane represent the screen grid of a tetrode, and the $z = d$ plane be the plate, it is evident that too large a screen to plate spacing will result in a $(d/s) > 2.8$, thus exceeding limiting perveance and causing some current to return to the screen. These calculations tell us precisely what spacing is required to avoid this condition, for any given plate voltage. In high-power tube design there is a requirement for a certain minimum spacing to avoid voltage breakdown during the time that maximum voltage is applied to the tube. The design must proceed by considering these two requirements of the maximum spacing to avoid space-charge blocking and the minimum spacing to avoid breakdown, when arriving at a maximum efficiency design for a given power output. Similar considerations govern the design of the suppressor grid configuration in pentodes. Obviously real tubes with real grids also involve two-dimensional effects of considerable importance. Some calculations have been made of electron trajectories in such high-space-charge two-dimensional problems, using digital computers [6].

PROBLEM 7.2-1 LIMITING CURRENT IN AN ELECTRON TUBE

A high-power tetrode is required to have a 1-in. spacing between screen grid and anode, in order to prevent voltage breakdown when anode voltage is at its maximum value of 100 kV. At minimum anode voltage, maximum current is drawn. If the minimum anode voltage is equal to the screen grid voltage which is 1 kV, what is the maximum allowable current density if the perveance is to be below limiting perveance? What beam area is required to produce 100 amp?

7.3 Static Theory of a Neutralized Beam

Many of the phenomena exhibited in the single-species problem discussed in the preceding section are altered if the space charge of the injected beam is neutralized by the presence of charges of the opposite sign. There are several methods of physically introducing ions into an electron beam. Two of the most commonly used in device applications are: (1) the introduction of a neutral gas that is ionized by the high-energy-beam electrons, and (2) thermal ionization of a neutral alkali vapor at a hot electrode immersed in the neutral gas, as in thermionic energy converters. Often the same electrode is used for ionizing the vapor and for emitting electrons. In real physical energy converters of this type there is some volume ionization as well as ionization of the vapor at the emitter. A full treatment of such a two-fluid system tends to be complex. However, one simplification that can yield accurate answers in certain cases is the treatment of one of the fluids, usually that representing the ions, as if it had infinite mass. Then the positively charged background behaves as if it were part of a solid, and only the motion of the electrons is studied. This approach, however, omits interactions between electrons and ions that can be important, as we will see in Section 7.5 where we treat both fluids in a two-fluid problem.

Here we treat the problem of ion propulsion beam neutralization, in which an ion beam is ejected from a space vehicle at a fixed dc velocity, subject to the assumption of infinite ion mass insofar as changes in ion velocity are concerned [7]. Electrons are ejected from the vehicle at a dc velocity that can be different from but comparable to that of the ions. In the real physical problem, electrons have a large thermal spread of velocities and typically the mean thermal speed is larger than the injection velocity of the ions. We simulate this situation with a single-velocity electron beam having a velocity equal to the mean thermal speed. There is no volume ionization or recombination in our model.

Let us consider the specific configuration of a semiinfinite diode with one wall at $z = 0$ and one wall at $z = \infty$ as shown in Fig. 7.3-1. A current density J_e of electrons is injected at $z = 0$ with velocity u_{e0}, where $u_{e0}^2 = 2(e/m_e)V_e$ and m_e is electron mass. The wall at $z = 0$ is at $V = 0$ so the electrons must have come from an emitter at $-V_e$. A current density of positive ions J_i is also injected into the space with a velocity u_{i0} at $z = 0$ such that, with m_i the ion mass, $u_{i0}^2 = 2(e/m_i)V_i$, so that the ions must have come from an emitter at a potential $+V_i$ with respect to the wall at $z = 0$. Actually, since we treat the ions as if they had infinite mass in their interaction with the electrons, we mean here that the ion mass is m_i only in the sense that the charge density and current of the ion beam is known in terms of m_i, V_i, etc. We further restrict the problem to the case $J_i = -J_e$ so there is "current neutrality," i.e., the net total charge per unit time ejected from the wall is zero.

As in the case of the single-species diode we may obtain the static theory for this case from Poisson's equation with no time variations. Here

$$d^2V/dz^2 = -\varrho/\varepsilon_0 = -(J_i/\varepsilon_0 u_i)[1 - (u_i/u_e)] \tag{7.3.1}$$

where ϱ, V, u_i, and u_e are total charge density, potential, ion velocity, and electron velocity, respectively, at arbitrary values of z.

Now if $m_i \gg m_e$ and also u_{e0} is nearly equal to u_{i0}, potential fluctuations of the order of V_e or less that allow all of the ions to be transmitted will not significantly alter the ion velocity u_i. Therefore we can solve for

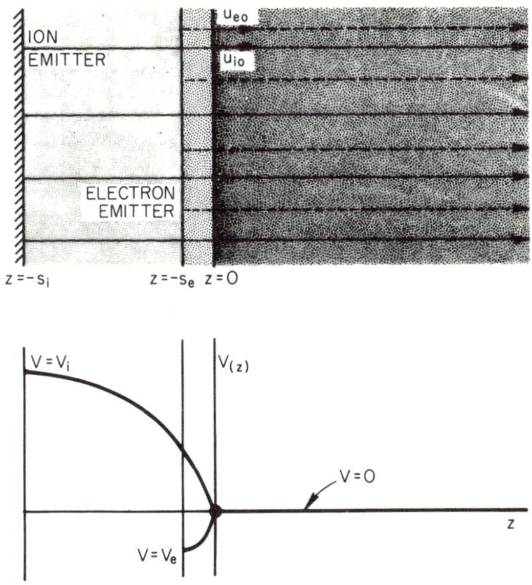

Fig. 7.3-1. A semiinfinite diode with one wall at $z = 0$ and one wall at $z = \infty$. Electron and ion beams are injected at $z = 0$ with constant dc velocities u_{e0} and u_{i0} and current densities J_e and J_i. In this model the ions are treated as if they had infinite mass, insofar as any perturbation of their injection velocity is concerned. Current neutrality is assumed, i.e., $J_e = -J_i$. The corresponding potential diagram is given below for the special case of equal ion and electron velocities in which case $V = 0$ for all $z > 0$. Hypothetical electron and ion emitters at $-s_e$ and $-s_i$ at potentials V_e and V_i are shown. In a physical system these emitters must be arranged in some two-dimensional configuration such as that shown in Fig. 6.7-1.

the special case of constant ion velocity and obtain some information that will apply under these conditions. The assumption of constant ion velocity is equivalent to an assumption of infinite mass ions or of a fixed background of ion space charge. We let $u_i = u_{i0}$ and define $\gamma_0 = u_{e0}/u_{i0}$. We also restrict our solution to the case in which all of the injected current, both of ions and electrons is trasmitted. Subject to all of these assumptions we can then rewrite Eq. (7.3.1) as

$$d^2V/dz^2 = -(J_i/\varepsilon_0 u_{i0})[1 - (u_{i0}/u_e)] \tag{7.3.2}$$

Now we may express u_e as

$$u_e = [2(e/m_e)(V_e + V)]^{1/2} \tag{7.3.3}$$

And defining γ as

$$\gamma = \gamma_0[1 + (V/V_e)]^{1/2} \tag{7.3.4}$$

we may write Eq. (7.3.2) in the form

$$d^2V/dz^2 = (J_i/\varepsilon_0 u_{i0})[(1/\gamma) - 1] \tag{7.3.5}$$

As in the solution of the single species diode this equation can be integrated once by multiplying through by (dV/dz) obtaining in this case

$$\frac{1}{2}\left(\frac{dV}{dz}\right)^2 = \frac{J_i}{\varepsilon_0 u_{i0}}\left[\frac{2V_e}{\gamma_0}\left(1+\frac{V}{V_e}\right)^{1/2} - V\right] + C_1 \tag{7.3.6}$$

The constant C_1 can be evaluated by the following argument. We have assumed that all the current is transmitted and that $J_i = -J_e$. Therefore, the net charge must average to zero when the average is taken over all space, even though u_{i0} is not equal to u_{e0}. If this condition is met, there must be zero field at $z = 0$, i.e., $(dV/dz) = 0$ at $z = 0$, because if it did not, there would have to be a corresponding net nonzero charge to the right of $z = 0$ for the field lines to end on. So, if $(dV/dz) = 0$ at $z = 0$,

$$\left(\frac{dV}{dz}\right)^2 = \frac{2J_i}{\varepsilon_0 u_{i0}}\left\{\frac{2V_e}{\gamma_0}\left[\left(1+\frac{V}{V_e}\right)^{1/2} - 1\right] - V\right\} \tag{7.3.7}$$

Equation (7.3.7) can be integrated [7] to give

$$z = \frac{2}{\gamma_0}\left(\frac{\varepsilon_0 u_{i0} V_e}{2J_i}\right)^{1/2}\left\{\cos^{-1}\frac{1-\gamma_0[1+(V/V_e)]^{1/2}}{1-\gamma_0} \pm \frac{\gamma[(dV/dz)^2]^{1/2}}{(2J_i V_e/\varepsilon_0 u_{i0})^{1/2}}\right\} \qquad 0 < z < \frac{\lambda}{2} \tag{7.3.8}$$

where (dV/dz) is given in Eq. (7.3.7) and where the upper sign is taken for $\gamma_0 < 1$ and the lower sign for $\gamma_0 > 1$. The spatial period λ is defined by

$$\lambda = \frac{4\pi(m_e/m_i)^{1/2}V_i^{1/2}}{(2J_i/\varepsilon_0 u_{i0})^{1/2}} = \frac{2\pi u_{i0}}{\omega_{pe}} \tag{7.3.9}$$

where $\omega_{pe}^2 = (eJ_i/\varepsilon_0 u_{i0} m_e)$ is essentially the plasma frequency of the electrons. Thus λ is independent of the initial electron velocity and is of the order of the spacing through which the electrons are accelerated rather than the much larger spacing through which the ions are accelerated for equal currents and approximately equal velocities. It is interesting that if we move through the potential diagram of Fig. 7.3-2 at the ion velocity, we see potential variations at essentially the electron plasma frequency, as seen from Eq. (7.3.9).

It is found that Eq. (7.3.8) only applies for $\gamma_0 \leq 2$. For larger γ_0 there is a reflected electron current and a different relation comes out of the static theory. The form of Eq. (7.3.8) is indicated in Fig. 7.3-2 for several values of γ_0 between 0 and 2. For $\gamma_0 = 1$ there is no variation in potential at all, corresponding to complete charge neutrality. For all the other cases, there are perturbations in potential, in a sense to accelerate electrons on the

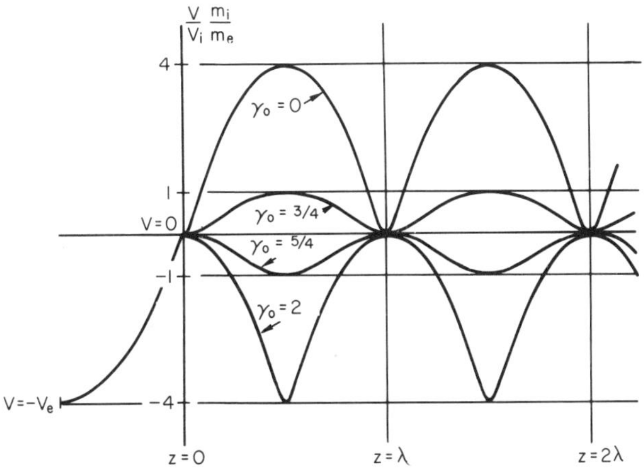

Fig. 7.3-2. Normalized potential (Vm_i/V_im_e) as a function of z for several values of $\gamma_0 = u_{e0}/u_{i0}$ between 0 and 2. For $\gamma_0 > 2$ this type of static theory yields solutions in which current loops of electrons are required to give charge neutrality. Such solutions are not physically realizable [from Ramo-Wooldridge Staff, Electrostatic propulsion, *Proc. IRE* **48**, 447–491 (1960)].

average, if they are injected at a velocity less than that of the ions, and in a sense to decelerate electrons on the average, if they are injected at a velocity greater than that of the ions. Thus in all cases there is space-average charge neutrality, and it results from periodic potential hills or valleys that, on the average, accelerate or decelerate the electron beam as needed to achieve this state. At $\gamma_0 = 2$ the beam is brought to rest at each potential minimum, which in this case are virtual cathodes. Thus for $\gamma_0 > 2$ it is not possible to slow the electrons further and a different mechanism must exist for charge neutralization, if it occurs at all. Static theory allows such a mechanism, consisting of recirculating current loops that flow between adjacent virtual cathodes, and with the potential diagram essentially the same as for $\gamma_0 = 2$. However, it can be shown by the techniques of the last chapter that such states are not established when an electron and an ion beam are injected into a semiinfinite space as postulated in the static theory. It is also possible to gain some understanding of this problem by considering perturbations of static states [8]. If the state is unstable to small perturbations, it cannot be expected to exist in nature. From these studies it has been concluded that states with current loops are not physically realizable in any ordinary system.

PROBLEM 7.3-1 ION PROPULSION BEAM

Using Eq. (7.3.7) establish that the extreme values of the potential diagrams of Fig. 7.3-2 are correct. What are the corresponding values of (V/V_e) as a function of γ_0?

7.4 Time-Varying Perturbations in a Neutralized Beam—Space Charge Waves

Here we take, as a special case of the neutralized electron beam with an infinite mass ion background, the case of complete local charge neutrality for which the static theory yields simply a constant potential everywhere and equal number densities of electrons and ions. Here we picture an electron beam neutralized by stationary ions, rather than by an accompanying ion beam, as in the preceding section, although in our present model, as before, the ions have infinite mass and cannot interact with the electrons, so the analysis applies to either case. Physically, a neutralized beam with ions at rest but with a small thermal spread can be obtained by passing an electron beam through a gas at low pressure. As a result of volume ionization, if the beam has an energy greater than the ionization potential of the background gas, a background of ions and electrons with no net drift velocity is created. A neutralized beam is one form of plasma, i.e., an electrically neutral assemblage of electrons and ions, in which plasma waves and oscillations can occur. If the beam velocity is zero, we have a plasma at rest. We here solve for the type of plasma wave that is carried with the electron beam and is often referred to as a space-charge wave. This wave is a slow wave with the electric field in the direction of wave propagation, so the electrostatic approximation that we have used both in the preceding chapter and this chapter is valid. High-velocity electromagnetic waves with electric and magnetic fields transverse to the direction of prop-

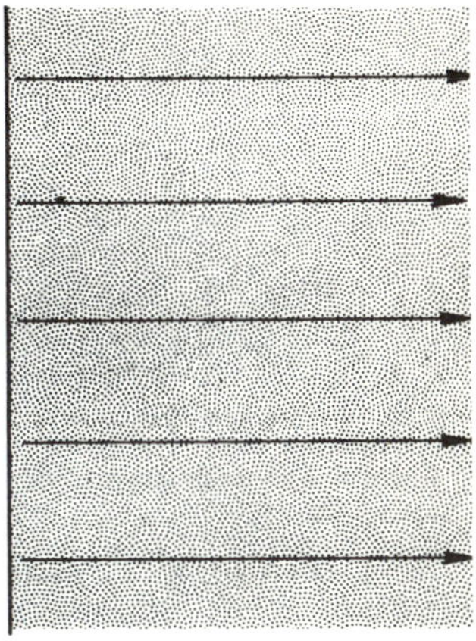

Fig. 7.4-1. A neutralized electron beam traveling through a semiinfinite diode extending from $z = 0$ to $z = \infty$. The electrons have a drift velocity u_{e0} which is constant throughout the space in the absence of perturbations. The ions have zero drift velocity and infinite mass. An initial ac velocity $u_{e1} = u_g \cos \omega t$ can be introduced at $z = 0$ and the propagation of space-charge waves in the z-direction can thereby be initiated.

agation that occur in ionospheric propagation through the plasma of space require a different treatment.

Here we solve for time-varying perturbations in a cold plasma which consists of a mixture of electrons and ions in equal numbers with negligible random velocities, but including a possible drift velocity of electrons relative to the ions as shown in Fig. 7.4-1. We first restrict the analysis to the case of an unbounded system. The static theory for such a plasma is very simple and is obtained from Poisson's equation, which may be written in the form

$$\partial^2 V/\partial z^2 = -(\varrho_i + \varrho_e)/\varepsilon_0 \tag{7.4.1}$$

for a region with variations only in the z-direction. If ϱ_i is uniform throughout all space, ϱ_e may either be uniform or a periodic function of distance, as seen in the preceding section, but in either case the space-average net charge density will be zero. Here we will consider only the simplest static mode in which the net charge density is zero and uniform throughout all space, i.e., $\varrho_i + \varrho_e = 0$ everywhere. However, now we are interested in perturbations of this static mode that can be both time and space varying. We will restrict the analysis to small signals in order to obtain a simple solution. In the preceding section, we derived solutions to Poisson's equation that could be regarded as spatially varying perturbations of the solution with zero density variation. Those solutions were exact and did not involve the small-signal approximation we make here. In order to excite these perturbed solutions all that was necessary was to inject a beam with either a higher or lower dc velocity than that corresponding to complete local charge neutrality. We will see that a time-varying injection velocity can be used to excite the waves we solve for here.

We will use subscripts 0 and 1 to refer to dc and ac perturbation quantities, respectively. We will write the electron current density **J** as

$$\mathbf{J}_e = \varrho_e \mathbf{u}_e \tag{7.4.2}$$

where $\mathbf{u}_e$ is electron velocity and ϱ_e is electron charge density. For one-dimensional flow the scalar equation is

$$J_e = \varrho_e u_e \tag{7.4.3}$$

and in terms of dc and ac quantities

$$\begin{aligned} J_e &= J_{e0} + J_{e1} = (\varrho_{e0} + \varrho_{e1})(u_{e0} + u_{e1}) \\ &= \varrho_{e0}u_{e0} + \varrho_{e0}u_{e1} + \varrho_{e1}u_{e0} + \varrho_{e1}u_{e1} \end{aligned} \tag{7.4.4}$$

Now

$$J_{e0} = \varrho_{e0}u_{e0} \tag{7.4.5}$$

and if we consider very small signals, the term $\varrho_{e1}u_{e1}$ is very much smaller than the other terms in Eq. (7.4.4), so we may write, to this approximation

$$J_{e1} = \varrho_{e0}u_{e1} + \varrho_{e1}u_{e0} \tag{7.4.6}$$

Now, velocity can be written

$$u_z = u_{z0} + u_{z1} \tag{7.4.7}$$

so the force equation can be written in the following form, from Eqs. (7.4.7) and (7.1.9), for small signals with E_{z1} being the perturbation electric field, there being zero dc electric field in this problem.

$$(q/m)E_{z1} = (\partial u_{z1}/\partial t) + u_{z0}(\partial u_{z1}/\partial z) \tag{7.4.8}$$

or, dropping the subscript z and writing the equation for electrons, we have in terms of V, where V is the ac potential corresponding to E_{z1},

$$-(q/m)(\partial V/\partial z) = (\partial u_{e1}/\partial t) + u_{e0}(\partial u_{e1}/\partial z) \tag{7.4.9}$$

which may now be combined with Eqs. (7.4.6) and (7.1.12) to obtain the complete trajectory equation. We assume all time-varying quantities vary as $\exp j\omega t$ and obtain equations that are functions only of z.

$$j\omega u_{e1} + u_{e0}(\partial u_{e1}/\partial z) = -(q/m)(\partial V/\partial z) \tag{7.4.10}$$

$$\varrho_{e0}(\partial u_{e1}/\partial z) + u_{e0}(\partial \varrho_{e1}/\partial z) = -j\omega \varrho_{e1} \tag{7.4.11}$$

These may be combined to yield the desired trajectory equation, first differentiating Eq. (7.4.10) with respect to z, as follows,

$$(\partial^2 \varrho_{e1}/\partial z^2) + 2j\beta_e(\partial \varrho_{e1}/\partial z) - \beta_e^2 \varrho_{e1} = \varepsilon_0 \beta_{pe}^2 (\partial^2 V/\partial z^2) \qquad (7.4.12)$$

where

$$\beta_e = \omega/u_{0e} = \text{beam wave number for electrons}$$
$$\beta_{pe} = \omega_{pe}/u_{0e} = \text{electron plasma wave number}$$
$$\omega_{pe} = (|q| \varrho_{e0}/m\varepsilon_0)^{1/2} = \text{electron plasma frequency}$$

The trajectory equation, Eq. (7.4.12), can now be combined with the field equation, Eq. (7.4.1), which can be rewritten in the form

$$\partial^2 V/\partial z^2 = -\varrho_{e1}/\varepsilon_0 \qquad (7.4.13)$$

to yield the combined equation for charge

$$(\partial^2 \varrho_{e1}/\partial z^2) + 2j\beta_e(\partial \varrho_{e1}/\partial z) - (\beta_e^2 - \beta_{pe}^2)\varrho_{e1} = 0 \qquad (7.4.14)$$

which may now be solved as an ordinary differential equation. We assume all quantities vary as $\exp(-jkz)$ and obtain the dispersion equation that relates frequency and wave velocity:

$$\frac{\beta_{pe}^2}{(k - \beta_e)^2} = 1 \qquad (7.4.15)$$

Equation (7.4.15) is a quadratic with the two solutions

$$-jk_{f,s} = -j\beta_e \pm j\beta_{pe} \qquad (7.4.16)$$

corresponding to the fast and slow space-charge waves, the fast wave, faster than the beam velocity, and the slow wave, slower than the beam velocity. Their velocities, v_f and v_s, may be found directly from Eq. (7.4.16), since

$$-jk_{f,s} = -j\frac{\omega}{u_{e0}} \pm j\frac{\omega_{pe}}{u_{e0}} = -j\frac{\omega}{u_{e0}[\omega/(\omega \pm \omega_{pe})]} \qquad (7.4.17)$$

so

$$v_f = u_{e0}\left[\frac{1}{1 - (\omega_{pe}/\omega)}\right], \qquad v_s = u_{e0}\left[\frac{1}{1 + (\omega_{pe}/\omega)}\right] \qquad (7.4.18)$$

This result is plotted in the ω–k diagram of Fig. 7.4-2 where the line $k = \beta_e$ and also the lines corresponding to the slow and fast space-charge waves of Eqs. (7.4.16) and (7.4.17) are shown. This is the same type of diagram used typically to illustrate the behavior of a waveguide. The velocity is the slope of a radius vector from the origin. Lines corresponding to waves at the velocity of light are indicated. Our analysis thus far has neglected time-

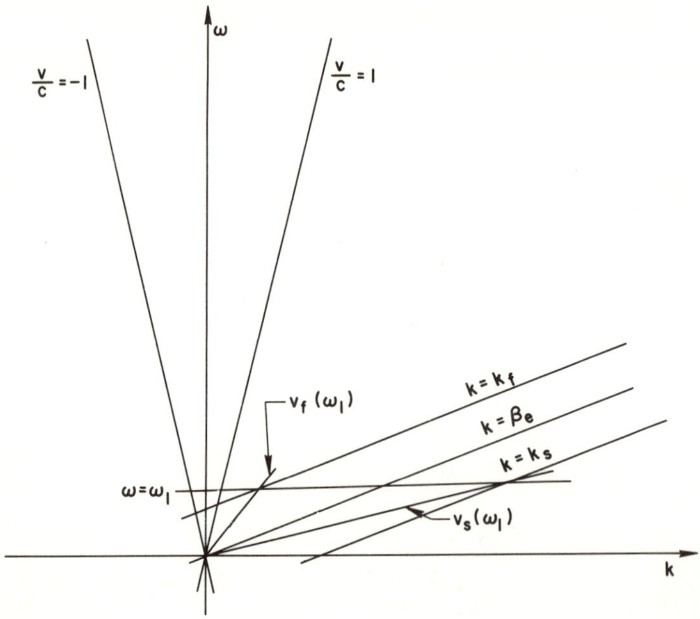

Fig. 7.4-2. The ω–k diagram for space-charge waves in a one-dimensional drifting plasma (neutralized electron beam) with a drift velocity $u_{e0} = \omega/\beta_e$. The phase velocity of the fast wave at frequency ω_1 is the slope of the line $v_f(\omega_1)$, and the phase velocity of the slow wave is the slope of the line marked $v_s(\omega_1)$. Infinite-mass ions are assumed as in Fig. 7.4-1.

varying magnetic fields and is thus invalid at velocities comparable to the velocity of light. We therefore must not attach any physical significance to the intercept of the k_f line with the ω-axis. A similar difficulty applies to the k_s intercept on the k-axis, because of the symmetry of the two waves. Neither of these intercepts occurs in the complete two-dimensional analysis, in which case the two curves for $k_{f,s}$ both bend in and arrive at the origin at zero frequency.

The velocity of each of these waves is seen to be near that of the drift velocity u_{e0} and tending to approach this velocity more and more as the frequency increases. The spread in velocity is seen from Eq. (7.4.18) to be set by the parameter (ω_{pe}/ω). This parameter is typically less than unity for electron beams produced from conventional electron guns. Higher beam perveances $(I/V^{3/2})$ can be obtained in plasma devices, and hence larger values of (ω_{pe}/ω) can be obtained, including values greater than unity.

Of considerable interest is the special case of zero drift velocity corresponding to a stationary cold plasma. The result here may be obtained directly from Eq. (7.4.15) for the case of zero u_{e0}, first dividing numerator and denominator by $(j\beta_e)^2$ to obtain

$$\frac{\omega_{pe}^2/\omega^2}{[(k/\beta_e) - 1]^2} = 1 \tag{7.4.19}$$

Now if β_e approaches infinity in Eq. (7.4.19) it is evident that the solution becomes

$$\omega = \omega_{pe} \tag{7.4.20}$$

for all k. The ω–k diagram for this case is simply a resonance line at $\omega = \omega_{pe}$ as in Fig. 7.4-3. The interpretation of this line, again ignoring the portions of the line corresponding to velocities greater than c, is that of a positively traveling wave for positive ω and positive k and a negatively traveling wave for positive ω and negative k. This statement may be checked by finding ω/k. The wave velocity in each of the two regions may have any value. In the figure two examples are given, that corresponding to the wave number $k = k_0$ being a forward wave and that corresponding to $k = -k_1$ being a backward wave. In this case the entire wave, for both positive and negative k, corresponds to what was the fast wave in the drifting plasma,

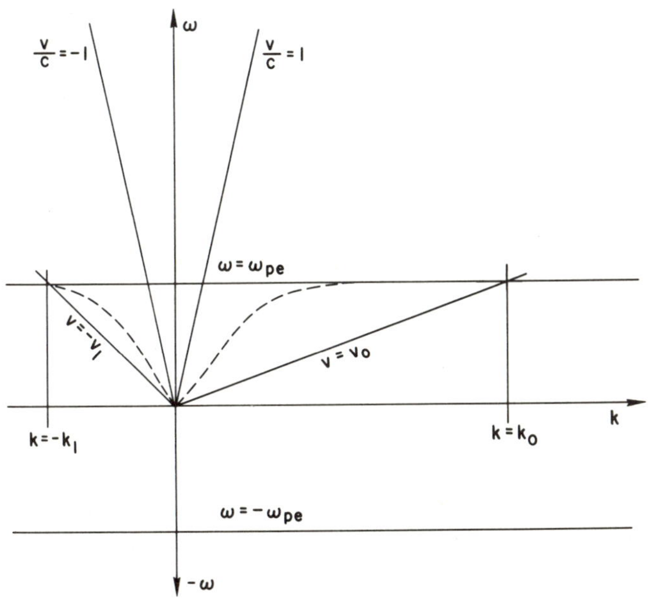

Fig. 7.4-3. The ω–k diagram for space-charge waves in a one-dimensional stationary plasma. The dashed lines are what would be obtained from a more complete two-dimensional theory. Infinite-mass ions are assumed as in Fig. 7.4-1.

while what was the slow wave now occurs at negative frequency and has no physical significance. When the effects of finite cross section are taken into account, the curves are modified for small k and take the form indicated by the dashed lines, thus allowing propagation at all frequencies below ω_{pe}. The asymptote in that case for large k is the value given by the analysis here.

An understanding of the dispersion relation does not, by itself, give us an understanding of any particular physical system. What it does is give us un understanding of the waves that can propagate in the system. Next we need to impose particular boundary conditions to fit the problem of interest and determine what amounts of what waves are excited and examine the variation in distance and time of the complete solution. Here we will carry out this calculation for a particularly simple case that also has importance in the theory of the klystron amplifier [9].

We return to the case of a nonzero drift velocity and picture at $z = 0$ a pair of grids with a spacing between them that is small in comparison with any wavelength of interest and with a constant amplitude sinusoidal voltage applied between them. The electron beam receives an approximately sinusoidal velocity modulation as it passes through this pair of grids. Because velocity is proportional to voltage to the one-half power, the velocity modulation is not sinusoidal if the rf voltage is sinusoidal, but it is nearly so for small signals, which is the case of interest here. Since all of the quantities of interest vary with z and t as $\exp(j\omega t - jkz)$, we may write the ac velocity in the form

$$u_{e1} = u_{e1s} + u_{e1f} = A \exp[j(\omega t - k_s z)] + B \exp[j(\omega t - k_f z)] \quad (7.4.21)$$

where A and B are arbitrary constants whose values we wish to solve for. Now if the velocity at $z = 0$ is sinusoidal as a result of passing the beam through the grids, i.e., if

$$u_{e1}(0) = u_g \cos \omega t \quad (7.4.22)$$

Then, using the usual convention with respect to taking the real part of complex quantities

$$u_g \exp(j\omega t) = A \exp(j\omega t) + B \exp(j\omega t) \quad (7.4.23)$$

and so we have

$$u_g = A + B \quad (7.4.24)$$

We need another equation if we wish to get the values of A and B. The other boundary condition at $z = 0$ is $J_{e1} = 0$. The grids can alter the velocities of particles passing through them but cannot change the current, because if the grid spacing is very small in comparison with the distance traversed by a particle during a cycle, the number of particles per unit time leaving the grids must be equal to the number per unit time entering. The charge density, however, can be altered suddenly, just as velocity can be changed, and in fact it must be the case that, if velocity is increased the charge density must decrease in order to keep the number leaving the grids per unit time constant, and vice-versa. The ac velocity and charge density are directly related, for each wave, through Eq. (7.4.11), which for waves varying as $\exp(-jkz)$ can be written

$$\frac{\varrho_{e1}}{\varrho_{e0}} = \frac{k}{\beta_e - k} \frac{u_{e1}}{u_{e0}} \tag{7.4.25}$$

Or, substituting Eq. (7.4.16) in Eq. (7.4.25), we obtain for the fast and slow waves individually

$$\frac{\varrho_{e1f}}{\varrho_{e0}} = \frac{\omega - \omega_{pe}}{\omega_{pe}} \frac{u_{e1f}}{u_{e0}}, \qquad \frac{\varrho_{e1s}}{\varrho_{e0}} = -\frac{\omega + \omega_{pe}}{\omega_{pe}} \frac{u_{e1s}}{u_{e0}} \tag{7.4.26}$$

Now the zero current condition at $z = 0$ requires, from Eq. (7.4.6), that the local ac charge density and velocity be related as follows at $z = 0$.

$$\varrho_{e1}(0)/\varrho_{e0} = -u_{e1}(0)/u_{e0} \tag{7.4.27}$$

So, if we substitute the individual wave relations of Eq. (7.4.26) into Eq. (7.4.27) we can solve for u_{e1f} in terms of u_{e1s}, yielding

$$B/A = u_{e1f}/u_{e1s} = 1 \tag{7.4.28}$$

Thus the two velocity waves are excited with equal amplitude by a short gap. We are now in a position to write the complete equations for velocity, charge, and current, using Eqs. (7.4.21), (7.4.26), (7.4.16), and (7.4.6)

$$u_{e1} = \frac{u_g}{2} \exp j[\omega t - (\beta_e + \beta_{pe})z] + \frac{u_g}{2} \exp j[\omega t - (\beta_e - \beta_{pe})z]$$

$$\begin{aligned} \varrho_{e1} = &-\frac{\varrho_{e0} u_g}{2u_{e0}} \left(\frac{\omega + \omega_{pe}}{\omega_{pe}}\right) \exp j[\omega t - (\beta_e + \beta_{pe})z] \\ &+ \frac{\varrho_{e0} u_g}{2u_{e0}} \left(\frac{\omega - \omega_{pe}}{\omega_{pe}}\right) \exp j[\omega t - (\beta_e - \beta_{pe})z] \\ J_{e1} = &-\frac{\varrho_{e0} u_g}{2} \left(\frac{\omega}{\omega_{pe}}\right) \exp j[\omega t - (\beta_e + \beta_{pe})z] \\ &+ \frac{\varrho_{e0} u_g}{2} \left(\frac{\omega}{\omega_{pe}}\right) \exp j[\omega t - (\beta_e - \beta_{pe})z] \end{aligned} \tag{7.4.29}$$

These quantities can be written in the form of standing waves as follows [9, 11]:

$$\begin{aligned} u_{e1} &= u_g \cos(\beta_{pe}z) \exp j(\omega t - \beta_e z) \\ \varrho_{e1} &= -(\varrho_{e0}u_g/u_{e0})[\cos(\beta_{pe}z) - j\omega/\omega_{pe} \sin(\beta_{pe}z)] \exp j(\omega t - \beta_e z) \quad (7.4.30) \\ J_{e1} &= j\varrho_{e0}u_g(\omega/\omega_{pe}) \sin(\beta_{pe}z) \exp j(\omega t - \beta_e z) \end{aligned}$$

The waves thus form a pattern with the electron plasma wavelength as a stationary envelope, and inside this envelope there are traveling waves with exactly the velocity of the beam electrons. This is the type of beating wave pattern that is always exhibited by two waves of equal amplitude traveling in the same direction at slightly different velocities. The beat wavelength in this case is the electron-plasma wavelength. Figure 7.4-4

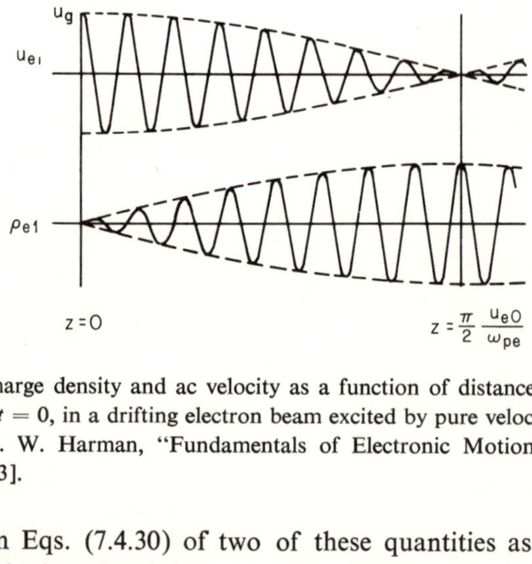

Fig. 7.4-4. Charge density and ac velocity as a function of distance, z, for a specific instant of time, $t = 0$, in a drifting electron beam excited by pure velocity modulation at $z = 0$ [from W. W. Harman, "Fundamentals of Electronic Motion." McGraw-Hill, New York, 1953].

is a plot from Eqs. (7.4.30) of two of these quantities as a function of distance, for the instant of time at which the rf gap voltage and the ac velocity at the gap are at their maxima. Figure 7.4-5 is a similar plot of the charge density and velocity from Eqs. (7.4.29) for the fast and slow waves, plotted separately. With the requisite ac velocity and current at the injection plane it is, in principle, possible to excite only one wave at a time as shown in these figures.

This analysis tells us how to select the drift length for a simple two-cavity klystron amplifier. We start with an input gap driven by a resonant cavity at frequency ω and generate ac velocity modulation at the exit of the gap, which converts to ac current a quarter electron-plasma wavelength downstream. For maximum small-signal gain, we thus should select our drift length to be this length, because this is the maximum-gain point. In practice, there are many complicating features. Of course, a finite-beam theory is required to get the correct wavelength for a finite beam. But beyond this, one is normally not interested only in maximum small-signal gain, but in a reasonable gain consistent with high efficiency. Thus, a large-signal theory of the sort discussed in the preceding chapter is needed. In practice, this compromise results in too low a gain (<10 dB) for most uses in a two-cavity klystron, and so almost all klystrons have from three

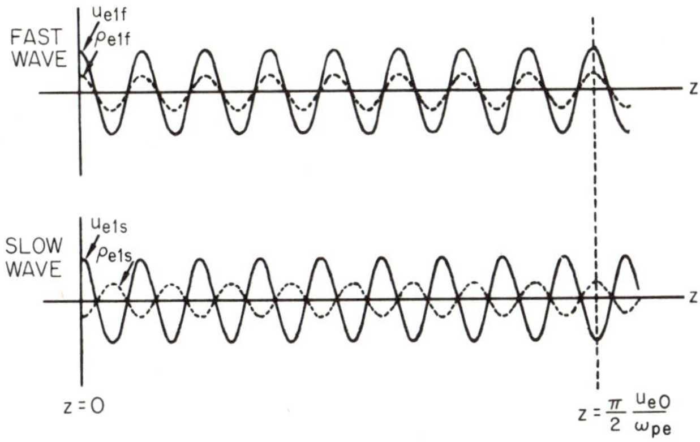

Fig. 7.4-5. Charge density and ac velocity as a function of distance, z, for a specific instant of time, $t = 0$, in a drifting electron beam. The upper graph shows the fast-wave pattern and the lower graph shows the slow wave. If a beam is excited with pure velocity modulation at $z = 0$, a combination of these waves is excited. For the specific excitations shown, the ac velocities of the two waves are in phase at $z = 0$ and 180° out of phase at $z = (\pi u_{e0}/2\omega_{pe})$, while the situation is reversed for the charge densities [from W. W. Harman, "Fundamentals of Electronic Motion." McGraw-Hill, New York, 1953].

to six cavities to give higher gain than is obtainable with two cavities. In multicavity klystrons the type of space-charge wave analysis presented here can be used to design all of the drift lengths except the last, because the signal amplitude ordinarily only becomes large enough to cause crossovers in the last section of such a tube.

This particular theory is useful in any beam-type microwave amplifier in which there is a drift region. It can readily be adapted to the study of noise induced waves as well as coherent signals. It is readily extended to the analysis of more complex devices in which the space-charge waves are coupled to waves on a distributed circuit, as in a traveling-wave amplifier, or to waves on another beam or plasma, as we shall see.

PROBLEM 7.4-1 SPACE-CHARGE WAVES

What ratio of velocity and current must be excited at $z = 0$ in a drifting neutralized beam in order to excite only the slow space-charge wave? What ratio will excite only the fast wave?

7.5 A Two-Fluid Problem—Growing Waves in a Plasma with Finite-Mass Ions

Here we repeat the problem of the preceding section, but remove the restriction to infinite-mass ions as shown in Fig. 7.5-1. There is no change in the field equation, Eq. (7.4.1) and no change in the trajectory equation for electrons, Eq. (7.4.12), each of which was derived without reference

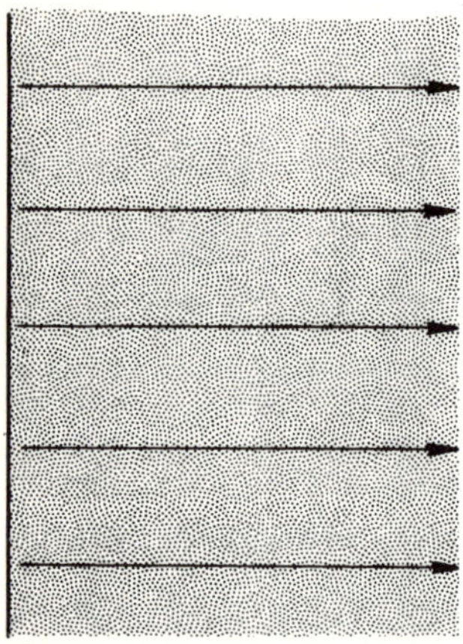

Fig. 7.5-1. A neutralized electron beam traveling through a semiinfinite diode extending from $z = 0$ to $z = \infty$. In this case the background ions are considered to have finite mass. Waves propagating can be initiated by an ac velocity modulation at $z = 0$.

to the ion motion. By analogy to Eq. (7.4.12), there must be an ion trajectory equation of the same form as Eq. (7.4.12), namely

$$(\partial^2 \varrho_{i1}/\partial z^2) + 2j\beta_i(\partial \varrho_{i1}/\partial z) - \beta_i^2 \varrho_{i1} = \varepsilon_0 \beta_{pi}^2 (\partial^2 V/\partial z^2) \qquad (7.5.1)$$

where

$\beta_i = \omega/u_{0i}$ = beam wave number for ions

$\beta_{pi} = \omega_{pi}/u_{0i}$ = ion plasma wave number

$\omega_{pi} = [(|q| \varrho_{i0}/m_e\varepsilon_0)(m_e/m_i)]^{1/2}$ = ion plasma frequency where we have assumed, as for the electrons, that the ions are allowed to move only in the z-direction, and where m_e and m_i are the electron and ion masses, respectively.

We now must obtain a solution to the three coupled differential equations (7.4.1), (7.4.12), and (7.5.1). This may be accomplished as in the preceding section by assuming all quantities vary as $\exp(-jkz)$, arriving at the dispersion equation, here written in the form of Eq. (7.4.19).

$$\frac{\omega_{pe}^2/\omega^2}{[(k/\beta_e)-1]^2}+\frac{\omega_{pi}^2/\omega^2}{[(k/\beta_i)-1]^2}=1 \tag{7.5.2}$$

This is a quartic in k and is only easily solved for various special conditions.

One interesting special case is that of an electron beam moving through an initially stationary ion background, i.e., there is zero ion drift velocity, but a nonzero ac ion velocity is allowed. Here we let u_{i0} approach zero so that β_i approaches infinity, and we have the equation

$$\frac{\omega_{pe}^2/\omega^2}{[(k/\beta_e)-1]^2}+\frac{\omega_{pi}^2}{\omega^2}=1 \tag{7.5.3}$$

We can solve Eq. (7.5.3) directly for k as follows:

$$k=\beta_e\left[1\pm\frac{\omega_{pe}}{\omega}\left(1-\frac{\omega_{pi}^2}{\omega^2}\right)^{-1/2}\right] \tag{7.5.4}$$

First let us examine the value of k for real ω and $\omega \ll \omega_{pi}$. In this frequency range, it is a good approximation to write

$$k \cong \beta_e[1 \pm j(\omega_{pe}/\omega_{pi})] \tag{7.5.5}$$

and, since we have assumed a neutralized beam so that $\varrho_{i0}=\varrho_{e0}$,

$$k \cong \beta_e[1 \pm j(m_i/m_e)^{1/2}] \tag{7.5.6}$$

Now since $m_i \gg m_e$, it is evident that the imaginary part of k is much greater than the real part, implying very rapid growth in distance, in fact so rapid that in one wavelength there is a very large change in the amplitude of the wave. As ω increases and becomes very nearly equal to ω_{pi}, the imaginary part of k evidently increases even more, approaching infinity as ω approaches ω_{pi}. This is a fairly spectacular result and shows that the infinite-mass approximation used in the analysis of the preceding section omitted an important effect. In an actual system of this type, of finite cross section, this infinite-gain condition at $\omega = \omega_{pi}$ changes to a finite-gain peak. Also, collisions and temperature effects further reduce the gain somewhat. Nevertheless, the gain due to this two-stream interaction is a very large effect and plays an important role in many electron tubes, plasma devices, and astrophysical phenomena. Figure 7.5-2 is an ω–k diagram for the system described by Eq. (7.5.4) for real ω and for $(m_i/m_e) = 16$.

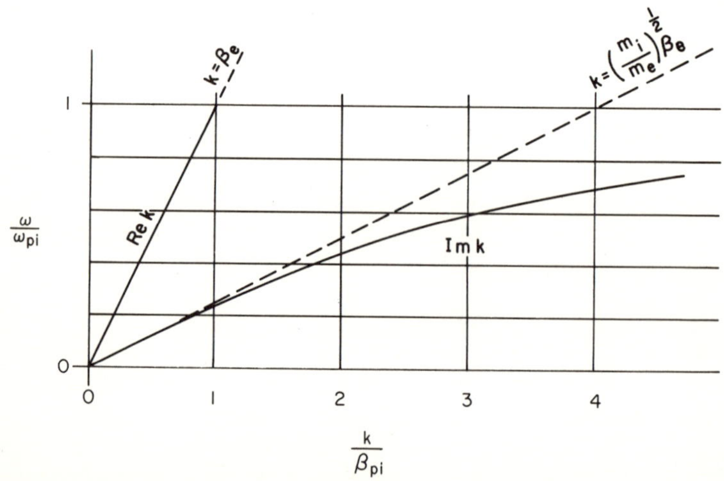

Fig. 7.5-2. An ω–k diagram for a one-dimensional electron beam traveling through a finite-mass ion background at rest. The mass ratio, (m_i/m_e), for this figure is 16. The real part of k is exactly equal to β_e for $\omega < \omega_{pi}$. The imaginary part of k which represents wave growth is approximately four times the value of Re k for $(\omega/\omega_{pi}) \ll 1$ but approaches infinity as (ω/ω_{pi}) approaches unity.

A way of explaining this result physically is as follows. What we have calculated is the mutual interaction between space-charge waves on two coupled charged fluids. Because the fluids are perfectly intermixed, the coupling between them can be very large. In the frequency range below the ion-plasma frequency the sense of the coupling between the waves is such as to enhance any initial perturbation. Thus the sinusoidal perturbation of charge density in each fluid is so phased as to enhance the perturbation in the other, resulting in an exponential increase of the perturbation in charge density in both fluids with distance. In a physical device, such a mechanism will cause small noise signals to grow to very large amplitudes. There is no necessity to supply an external signal or perturbation to get things started. In a physical device also, there are always ends or walls that produce reflected waves which can cause feedback, and hence oscillations at frequencies where the feedback signal has the right phasing. Thus the physically observed result of a two-stream mechanism of this type will often be noisy oscillations with peaks in the noise output at a number of frequencies where the length of the system is correct, all in the vicinity of the frequency for maximum gain, here ω_{pi}.

The same effect occurs when an electron beam is passed through a stationary plasma consisting of ions and electrons, but the main interaction occurs near the plasma frequency of the plasma electrons. This is an easy configuration to set up in the laboratory. The analysis involves three rather than two fluids, one for the beam, one for the electrons of the plasma, and one for the ions of the plasma. Again one can adopt the point of view that one fluid has infinite mass, but as we have seen, this approach must be used with some care. It is reasonable to take this approach for the ions at high frequencies, far above the ion plasma frequency, but it is not reasonable to neglect ion motion at lower frequencies without justification. In a problem it will be shown that there is justification for also neglecting ion motion at low frequencies in this case of the beam penetrating a plasma. Physically this is because oscillations at the ion plasma frequency are damped out by the motion of the much more mobile plasma electrons. Plasma electrons neutralize any charge unbalance due to ion motion and prevent such a disturbance from growing, unlike the situation where there are only two species present. The physical situation corresponding to only two species with one at rest and one in motion is very difficult to produce, although it is often postulated as existing for the purposes of analysis.

Problem 7.5-1 Two-Stream Instability

Compute and sketch the portion of the ω–k diagram for $\omega > \omega_{pi}$ that properly goes with the diagram of Fig. 7.5-2.

Problem 7.5-2 Two-Stream Instability

A cold one-dimensional plasma with plasma-ion density n_i and plasma-electron density n_e is interpenetrated by an electron beam of density n_b. The system is neutral, i.e., $n_i = n_e + n_b$, and $n_b \ll n_e$. Show that there is no infinite gain peak near the ion-plasma frequency and that in the vicinity of this frequency the growing wave gain is nearly that for the electron–electron interaction by itself.

7.6 Interaction with a Circuit—Beam Loading in a Finite-Length Gap

We here again picture an electron beam traveling through a diode with walls at $z = 0$ and $z = d$, as in our previous problems. But now we allow for the presence of an impedance connected between the two walls of the diode other than a short circuit. Figure 7.6-1 shows a finite-diameter system of this type. Our one-dimensional theory would apply to such a finite system, if the gap length were short in comparison with its transverse dimensions. In the one-dimensional model that we will analyze we will consider an impedance per unit area connected across the gap, one element of which is that of the capacitance between the walls of the gap, assumed to be perfectly conducting plane surfaces permeable to the electrons of the beam. The object of our analysis will be to calculate the additional impedance introduced by the beam, as a function of beam current and voltage [10–13].

Our model will be that shown in Fig. 7.6-2 which includes a background of infinite-mass positive ions, as in the analysis in Section 7.4. We first wish to define the impedance of interest in terms of parameters that we can calculate. The ac current density J_1 that would flow in an external circuit is the total current, including both the displacement current $j\omega\varepsilon E_{z1}$ and the convection current density J_{e1} in the gap.

$$J_1 = J_{e1} + j\omega\varepsilon E_{z1} \tag{7.6.1}$$

It may readily be shown that J_1 is not a function of distance across the gap, although of course J_{e1} varies from zero at the injection side of the gap to a finite value at the exit. To see this, take the divergence of the fourth Maxwell equation and express the result for a system that varies only in z

$$\partial(J_{e1} + j\omega\varepsilon E_{z1})/\partial z = 0 \tag{7.6.2}$$

Integrating this equation we see that J_1 must be a constant in z although it may vary with time.

An even more stringent condition on J_1 results from simply expressing the z-component of the same Maxwell equation, as follows

$$(\nabla \times \mathbf{H})_z = J_1 = (\partial H_y/\partial x) - (\partial H_x/\partial y) = 0 \tag{7.6.3}$$

Now, it is noted that in a one-dimensional problem there are no variations in any quantity in either x or y, so the rhs of Eq. (7.6.3) is zero. So, in general, J_1 is zero in a one-dimensional problem. There are two ways in

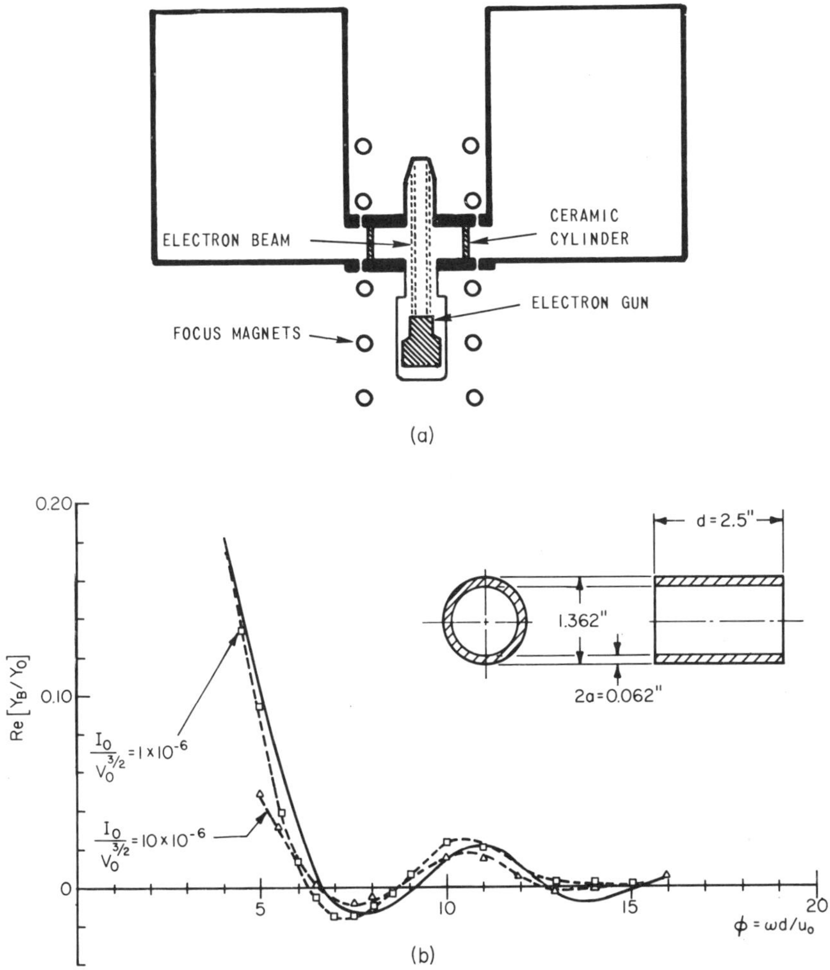

Fig. 7.6-1. An experimental electron diode.

(a) A thin hollow cylindrical electron beam is produced by an electron gun and transmitted through a diode to a collector. The beam is confined by a strong axial magnetic field. An external impedance is connected across the gap in the form of a resonant cavity. The Q of the cavity and its resonant frequency can be measured in the presence of the electron beam and the equivalent impedance of the beam can thereby be inferred.

(b) Measured beam admittance (dashed) as a function of ϕ for two values of beam perveance. A theoretical curve (solid) based on the one-dimensional theory presented in this section. [from D. A. Dunn and G. L. Sackman, Interaction of two beams with a cavity, *J. Electron. Contr.* **18**, 181–200 (1965)].

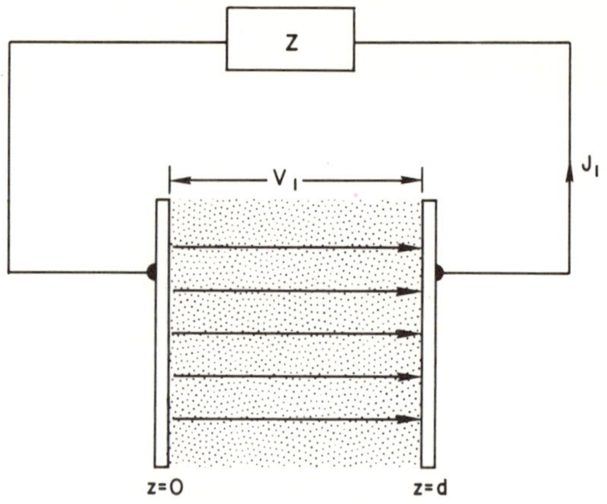

Fig. 7.6-2. A one-dimensional model of an electron diode with an external impedance Z. An electron beam travels at a constant dc velocity through the diode from $z = 0$ to $z = d$. A background of infinite-mass positive ions is postulated that maintains the dc velocity constant. An rf voltage V_1 exists across the diode, and the total ac current through the diode is J_1 which is also the current in the external circuit.

which J_1 can be zero. In the first, which is the case considered here, it is zero because the total current passing through the gap is exactly equal and opposite to the current passing through the external circuit lead. One can picture a return circuit at infinity, if desired. This means that the current J_1 is not zero in the gap, but must merely be constant, as required by Eq. (7.6.2). There is a second, and perhaps simpler, way for the total current to be zero and that is for the case of zero return current and zero gap current. This case is treated in the next section.

The ac voltage V_1 across the gap can be obtained from the field by integrating as if E_{z1} were a static field, for gaps short in comparison with a free-space wavelength, i.e., for conditions in which quasi-static theory applies, so

$$V_1 = \int_0^d E_{z1}\, dz \tag{7.6.4}$$

where d is the length of the gap. The impedance per unit area of the gap can then appropriately be defined as $Z = V_1/J_1$, and we can solve for Z from Eqs. (7.6.1) and (7.6.4), using the fact that J_1 is independent of z

$$Z = (d/j\omega\varepsilon)\left[1 - (1/J_1\, d) \int_0^d J_{e1}\, dz\right] \tag{7.6.5}$$

The two terms on the rhs of Eq. (7.6.5) can be identified as the reactance of the gap in the absence of the beam and a term due to the beam which we will define as Z_B, where

$$Z_B = -(1/j\omega\varepsilon J_1) \int_0^d J_{e1}\, dz \tag{7.6.6}$$

where Z_B is to be regarded as appearing in series with the cold gap reactance $(d/j\omega\varepsilon)$.

Evidently then to evaluate Z_B we must obtain an expression for $J_{e1}(z)$ and integrate over the gap length d. In Section 7.4 we obtained a differential equation for charge density as a function of z, Eq. (7.4.12). By some simple substitutions, using Eqs. (7.4.6), (7.4.9)–(7.4.11), and (7.6.1) we can derive an equivalent differential equation for J_{e1} in terms of J_1, as follows:

$$(\partial^2 J_{e1}/\partial z^2) + 2j\beta_e(\partial J_{e1}/\partial z) + (\beta_{pe}^2 - \beta_e{}^2)J_{e1} = \beta_{pe}^2 J_1 \tag{7.6.7}$$

where β_e and β_p are as defined in Eq. (7.4.12). Let us now rewrite Eq. (7.6.7) in the following normalized form:

$$(d^2 J_{e1}/d\zeta^2) + 2j(dJ_{e1}/d\zeta) - (1 - p^2)J_{e1} = p^2 J_1 \tag{7.6.8}$$

where

$\zeta = \beta_e z =$ a normalized length

$p = \omega_{pe}/\omega =$ ratio of plasma to applied frequency

The solution of this equation may be obtained by transform methods. For this problem, one may take advantage of the initial-value boundary conditions to apply the one-sided Laplace transform. If one assumes an unmodulated stream at the entrance to the gap, the ac current density and its derivative are zero at $z = 0$. The driving function J_1 is a step function in space and the equation can be transformed to an algebraic equation which can be solved for the transform I of the ac current density J_{e1}

$$I = p^2 J_1 \frac{1}{\Gamma(\Gamma - \Gamma_1)(\Gamma - \Gamma_2)} \tag{7.6.9}$$

where $\Gamma_{1,2} = -j(1 \pm p)$. The inverse transform is

$$J_{e1} = p^2 J_1 \left[\frac{1}{\Gamma_1 \Gamma_2} + \frac{\exp(\Gamma_1 \zeta)}{\Gamma_1(\Gamma_1 - \Gamma_2)} + \frac{\exp(\Gamma_2 \zeta)}{\Gamma_2(\Gamma_2 - \Gamma_1)}\right] \tag{7.6.10}$$

The beam impedance Z_B can be expressed in terms of the normalized parameters in the form

$$\left(\frac{j\omega\varepsilon}{d}\right)Z_B = -\frac{1}{\phi J_1}\int_0^\phi J_{e1}\,d\zeta \tag{7.6.11}$$

where $\phi = \beta_e d =$ the electronic transit angle.

The current J_{e1} from Eq. (7.6.10) may then be integrated to give the beam impedance as a function of the electronic transit angle ϕ and the parameter p:

$$\frac{j\omega\varepsilon}{d}Z_B = -p^2\left[\frac{1}{\Gamma_1\Gamma_2} + \frac{1}{\Gamma_1(\Gamma_1-\Gamma_2)}\cdot\frac{\exp(\Gamma_1\phi)-1}{\Gamma_1\phi} + \frac{1}{\Gamma_2(\Gamma_2-\Gamma_1)}\cdot\frac{\exp(\Gamma_2\phi)-1}{\Gamma_2\phi}\right] \tag{7.6.12}$$

The characteristic waves have wave numbers

$$\Gamma_{1,2} = -j(1\pm p) = -j\frac{(\omega\pm\omega_{pe})/u_{0e}}{(\omega/u_{0e})} \tag{7.6.13}$$

These two waves are of course the space-charge waves solved for previously with velocities slightly faster (lower sign) and slower (upper sign) than the beam velocity u_{0e}. Substituting the wave numbers from Eq. (7.6.13) into Eq. (7.6.12) and separating the real and imaginary parts puts the equation for beam impedance in the form:

$$Z_B = \frac{d}{\omega\varepsilon}\Bigg\{\underbrace{\frac{p^2}{j(1-p^2)}}_{(a)} - \underbrace{\frac{p}{2(1+p)}}_{(b)}\underbrace{\left[\frac{1-\cos(1+p)\phi}{(1+p)\phi} + j\frac{\sin(1+p)\phi}{(1+p)\phi}\right]}_{(c)} + \underbrace{\frac{p}{2(1-p)}}_{(d)}\underbrace{\left[\frac{1-\cos(1-p)\phi}{(1-p)\phi} + j\frac{\sin(1-p)\phi}{(1-p)\phi}\right]}_{(e)}\Bigg\} \tag{7.6.14}$$

For electron beams the parameter p is almost always less than unity, and normally less than 0.3. Assuming this is true, it can be seen that the expression for beam impedance is composed of three terms: a constant negative imaginary part (a) plus two terms of nearly equal magnitude and opposite sign which are associated with the slow and fast space-charge waves. These terms, (b) times (c) and (d) times (e), are in turn products of real coefficients:

$$\text{(b), (d)} = \frac{\pm p}{2(1\pm p)} \tag{7.6.15}$$

and trigonometric functions of a transit angle $(1 \pm p)\phi$

$$(\text{c}), (\text{e}) = \left[\frac{1 - \cos(1 \pm p)\phi}{(1 \pm p)\phi} + j\frac{\sin(1 \pm p)\phi}{(1 \pm p)\phi}\right] \tag{7.6.16}$$

The terms (c) and (e) are the gap-to-beam coupling coefficient for each wave. As a result of the $1/\phi$ factor, both terms shrink rapidly with increasing transit angle until at very large transit angles only the constant imaginary term (a) remains to represent the beam as a "dielectric substance" in the gap.

Equation (7.6.14) also displays the effect of the parameter p, which is the ratio of the plasma frequency to the operating frequency. The coefficients (b) and (d) are sketched in Fig. 7.6-3. For values of p near zero, they are

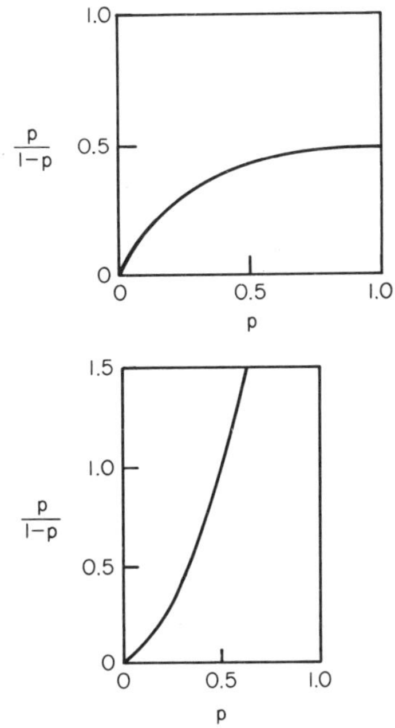

Fig. 7.6-3. The beam impedance Z_B contains terms proportional to the coefficients plotted here as a function of p. The upper curve is the negative of the coefficient of the slow space-charge wave and the lower curve is the coefficient of the fast space-charge wave. For p greater than 0.5 the fast-wave term predominates [from D. A. Dunn and G. L. Sackman, Interaction of two beams with a cavity, *J. Electron. Contr.* **18**, 181–200 (1965)].

nearly equal. However, the coefficient corresponding to the fast wave is positive and increases rapidly, having a pole at $p = 1$, whereas the coefficient corresponding to the slow wave is negative and bounded, with a value of $\frac{1}{2}$ at $p = 1$. From this it can be seen that negative resistance (negative real part of the beam impedance) due to the contribution of the slow wave soon becomes small compared to the positive contribution of the fast wave as the ratio of plasma frequency to operating frequency is increased.

It is customary to express the beam impedance as an admittance $Y_B = 1/Z_B$ and to normalize the rf beam admittance to the dc beam admittance Y_0 given by

$$Y_0 = J_0/V_0 \tag{7.6.17}$$

where

$$J_0 = \varrho_{0e}u_{0e} = \text{dc current density}$$
$$V_0 = m_e u_{0e}^2/2e = \text{dc beam voltage}$$

It is possible to put this dc admittance in terms of transit angles and the magnitude of the cold gap admittance $(d/\omega\varepsilon)$

$$Y_0 = 2(\omega\varepsilon/d)(\theta^2/\phi) \tag{7.6.18}$$

where $\theta = \beta_p d = p\phi$ = plasma transit angle.

After some algebra, the normalized beam admittance can be written as

$$\frac{Y_B}{Y_0} = \frac{\phi}{2\theta^2} \frac{(\omega\varepsilon/d)Z_B}{1 + j(\omega\varepsilon/d)Z_B} \tag{7.6.19}$$

This expression is plotted in the complex plane with ϕ and θ as parameters in Fig. 7.6-4. The curve of $\theta = 0$ corresponds to the normalized beam loading for negligible current density usually presented in the literature

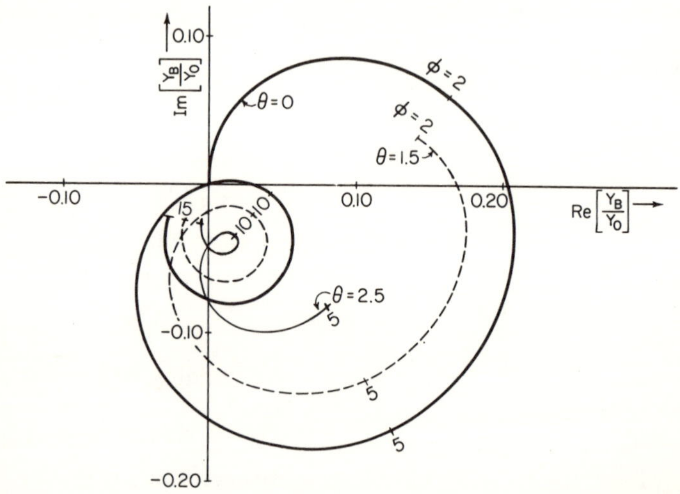

Fig. 7.6-4. The normalized beam admittance, Y_B/Y_0, as a function of the beam transit angle ϕ for several values of the plasma transit angle θ [from D. A. Dunn and G. L. Sackman, Interaction of two beams with a cavity, *J. Electron, Contr.* **18,** 181–200 (1965)].

as two curves that show the real and imaginary parts separately as functions of the electronic transit angle. It can be seen in this figure that the effect of increasing θ is to decrease the magnitude of the normalized beam admittance over the whole range of electronic transit angles.

The real part of the normalized beam admittance is plotted in Fig. 7.6-5 as a function of electronic transit angle for several values of plasma transit

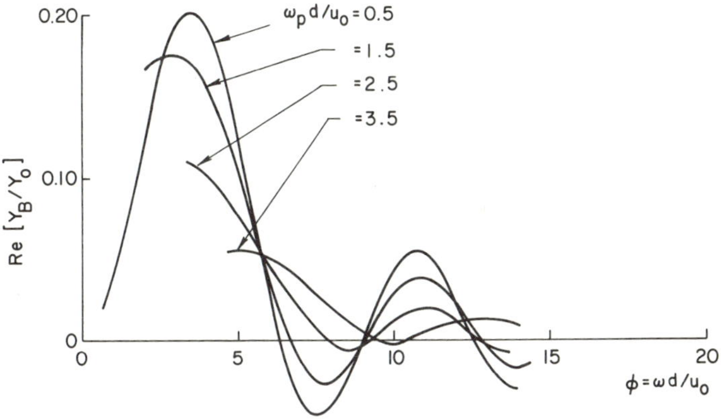

Fig. 7.6-5. The real part of the normalized beam admittance, Y_B/Y_0, as a function of ϕ for several values of $\theta = \omega_p d/u_0$ [from D. A. Dunn and G. L. Sackman, Interaction of two beams with a cavity, *J. Electron. Contr.* **18**, 181–200 (1965)].

angle. The curves each begin at values of ϕ corresponding to $p = 0.75$ for the particular value of θ. It is seen that the peak value of negative conductance in the first negative conductance region is decreasing in amplitude for increasing values of θ. To understand this effect it is convenient to look at the value of beam admittance in an unnormalized form.

The perveance is related to the plasma transit angle by the expression

$$\theta^2 = (\beta_{pe}\, d)^2 = \frac{1}{2\varepsilon(2e/m)^{1/2}} \left(\frac{J_0\, d^2}{V_0^{3/2}}\right) \tag{7.6.20}$$

where $J_0\, d^2/V_0^{3/2}$ is the perveance of a $d \times d$ square section in a one-dimensional beam. Therefore, if Eq. (7.6.19) is multiplied by the square of the plasma transit angle, the result is

$$\theta^2 \frac{Y_B}{Y_0} = \frac{d^2 Y_B}{2\varepsilon[(2e/m)V_0]^{1/2}} \tag{7.6.21}$$

which is the beam admittance as it would be measured in mhos, multiplied by a quantity that is independent of current or perveance. The product

$\theta^2(Y_B/Y_0)$ is thus a better quantity than Y_B/Y_0 for studying the effects of perveance and the real part of $\theta^2(Y_B/Y_0)$ is plotted in Fig. 7.6-6 as a function of ϕ for various θ. Using the new normalization, the beam loading at first positive maximum and first negative maximum is plotted as a function of the plasma transit angle in Fig. 7.6-7. Maximum negative gap loading is shown for a one-dimensional beam in this figure to be almost periodic in θ. Maximum positive beam conductance shown on the same plot for

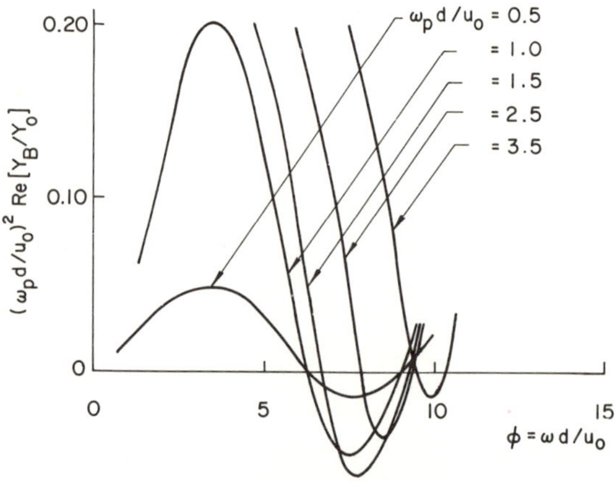

Fig. 7.6-6. The real part of the normalized beam admittance times $\theta^2 = (\omega_p d/u_0)^2$ as a function of ϕ for several values of θ [from D. A. Dunn and G. L. Sackman, Interaction of two beams with a cavity, *J. Electron. Contr.* **18**, 181–200 (1965)].

comparison increases monotonically with plasma transit angle. It is also much larger in magnitude than the negative conductance, which is shown ten times scale.

The usual expression for beam admittance for small space charge can be obtained directly from Eq. (7.6.19) as

$$\lim_{p\to 0} \frac{Y_B}{Y_0} = \left(\frac{1-\cos\phi}{\phi^2} - \frac{\sin\phi}{2\phi}\right) + j\left(\frac{\sin\phi}{\phi^2} - \frac{1+\cos\phi}{2\phi}\right) \tag{7.6.22}$$

This equation predicts that Y_B is proportional to dc current both at the positive and negative conductance peaks. As seen from Figs. 7.6-6 and 7.6-7 this is only true for the negative conductance peak for small currents, and a small current is one that yields $\theta < 2$.

The reason that we have made such a detailed study of the negative resistance portion of the characteristic is the possible application of this phenomenon in an oscillator called a "monotron" oscillator which is extremely simple and would consist of a beam passing through a single

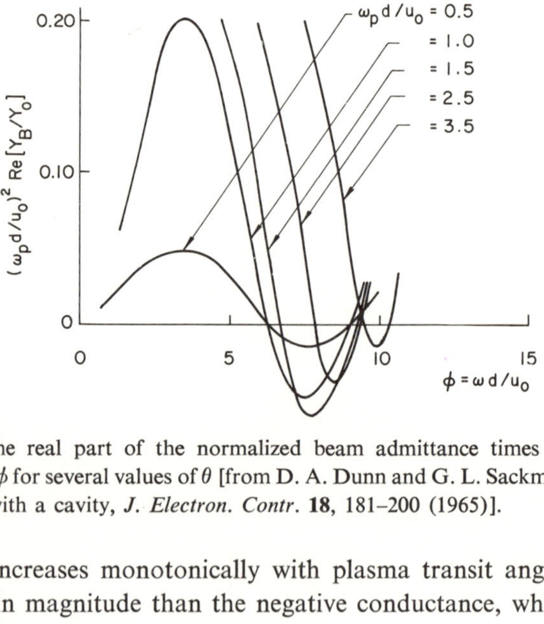

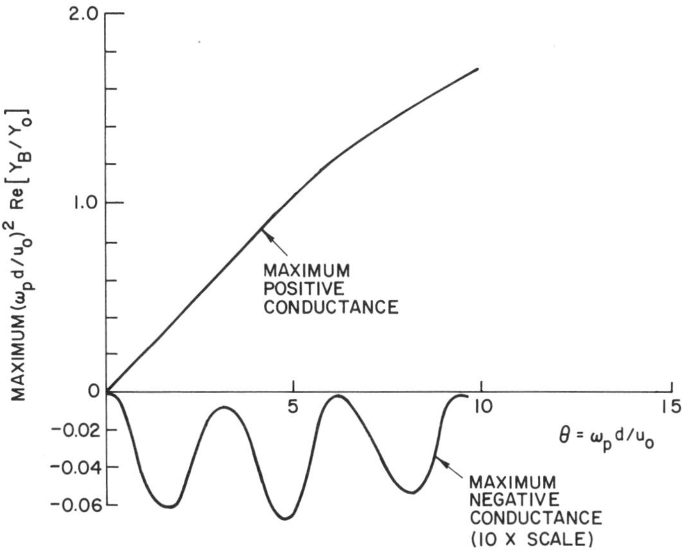

Fig. 7.6-7. The real part of the normalized beam admittance times θ^2 at the value of ϕ corresponding to the first positive and negative maxima, plotted as a function of θ [from D. A. Dunn and G. L. Sackman, Interaction of two beams with a cavity, *J. Electron. Contr.* **18**, 181–200 (1965)].

cavity. Because of its extreme simplicity it is potentially a useful device at extremely high frequencies where the device would be extremely small and only very simple devices can be built. The analysis we have been through here demonstrates that there is an optimum beam perveance for maximum negative beam conductance. However, when numerical values of the external equivalent resistance of typical simple resonant cavities using copper are compared with the numerical values of beam conductance available, it is found that this effect, even when optimized, is just barely capable of producing oscillations and is not practical for this purpose in this form. More complex cavity shapes using slow-wave structures can easily produce oscillations but are not easy to construct at high frequencies.

The positive resistance portion of the gap impedance characteristic is of importance primarily in estimating the power required to drive a short gap that is being used to modulate an electron beam. If the gap is of zero length, no power is required and the energy delivered from the electric field to the beam during half of the cycle is exactly returned during the other half of the cycle. For a finite length gap, Eq. (7.6.22) gives the resistance of the beam for small space charge, i.e., for small p, while Eq. (7.6.14) is correct for arbitrary values of p.

PROBLEM 7.6-1 MONOTRON OSCILLATOR

Consider a monotron oscillator at a relative low microwave frequency where gridded gaps can be used. The theory of this section will give a good approximation to the behavior of such a device. Suppose that you have a cavity with an $(R/Q) = 100$ and wish to operate with a beam perveance of 10×10^{-6} as required by an available electron gun. What is the loaded Q for start oscillation?

7.7 Interaction with a Distributed Circuit—the Traveling-Wave Tube

The traveling-wave tube is a device in which an electron beam is coupled along its length to a slow-wave structure such as a helix that propagates a mode with a strong electric field in the direction of electron motion. Figure 7.7-1 shows such a device. The basic principle of interaction in this type of device is similar to the two-stream interaction discussed in Section 7.5, except that the slow-wave structure takes the place of one stream. Bunching of electrons in the beam by the fields on the circuit results in a convection current in the beam that delivers energy back to the circuit. This is a distributed positive feedback system with a field- and convection-current amplitude that grow exponentially with distance along the device. As in the problems

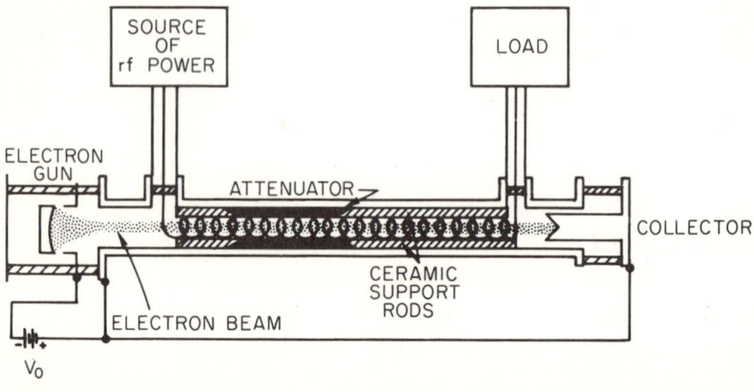

Fig. 7.7-1. A traveling-wave amplifier using a helix supported on ceramic rods as a slow-wave structure. A low VSWR match can be obtained between an external coaxial line and the helix over a very wide frequency range. Typical net amplifications of 30 dB are obtained. The attenuator made by impregnating the ceramic support rods with carbon is near the middle of the tube and prevents reflections from the output load from being rereflected at the input. Oscillations are thus avoided even when the output is short circuited.

of Chapter 4 involving slow-wave interaction with solids, energy can only be delivered from the electron beam to the circuit to provide this wave growth when the electron beam or moving medium is traveling faster than the growing wave.

As in the two-stream system discussed previously, a traveling-wave amplifier can easily be made to oscillate by allowing feedback to take place via a negatively traveling wave along the circuit. Such a negatively traveling wave is easily excited by any mismatch at the output end of the structure, so oscillations can be avoided only with great difficulty, if the structure has little attenuation. Ideally, a traveling-wave amplifier circuit would have zero attenuation for a positively traveling wave as required for high efficiency and very high attenuation for a negatively traveling wave, as required to avoid feedback and oscillation. Such a circuit can be approximated with the use of ferrite-loaded slow-wave structures in a dc magnetic field. However, interestingly enough, the introduction of loss into a circuit in the presence of an electron beam with a beam velocity in the positive direction turns out to introduce about three times as much attenuation in the negative direction as there is reduction in gain in the positive direction, as we shall show [14]. Because of this fact, most traveling-wave tubes are built using simple lossy attenuator sections near the center of the tube, to prevent oscillations.

In order to analyze this device we will formulate a one-dimensional model of this device, following the work of Olving [15]. Historically, this was not the way the device was first analyzed, but it has important advantages over the models first used in its degree of definiteness with respect to precisely what is being analyzed. The more complete theory of this device naturally includes the fact that the helix and the beam are round and the fact that there is some spacing or separation between beam and helix [14]. These facts limit the bandwidth of the device in comparison with the bandwidth of the model that we will analyze. However, the results of the theory presented here are otherwise essentially the same as those of the more complete theory.

Figure 7.7-2 is a drawing of the model that we will analyze. At first glance, the resemblance between this model and the device of Fig. 7.7-1 seems somewhat remote. We will first examine the properties of the circuit shown, without the beam, and show that it is very similar to the helix in its behavior. The beam equations for an infinite one-dimensional beam are those developed previously in this chapter. Coupling between beam and helix is extremely simple to put into this model in an unambiguous way, which is partly why this model is of interest.

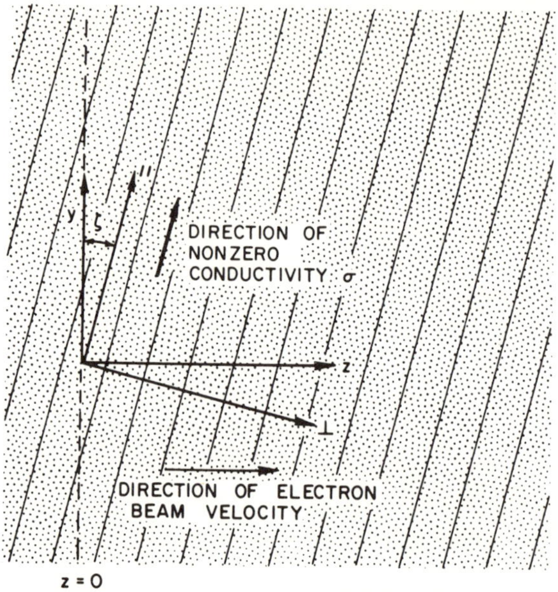

Fig. 7.7-2. A one-dimensional model of a traveling-wave amplifier. An infinite medium has permittivity ε and permeability μ that are uniform and isotropic, but has anisotropic conductivity that is σ in the $\|$ direction and zero in the $\perp$ direction. The medium is permeable to an electron beam traveling in the z-direction. There are no variations in any quantities in x or y and the $\|$ direction is at an angle ζ to the y-direction.

The circuit of Fig. 7.7-2 is a type of anisotropic spatially-uniform medium. We are interested in waves that have no variations in x or y and that propagate as $\exp[j(\omega t - kz)]$ in time and z. We let σ be the conductivity in the $\|$ direction at an angle ζ to the y-direction. The conductivity is zero in the $\perp$ direction at right angles to the $\|$ direction. Thus we have

$$J_\perp = 0, \qquad J_\| = \sigma E_\| \tag{7.7.1}$$

Since

$$E_\| = E_y \cos\zeta + E_z \sin\zeta \tag{7.7.2}$$

we have

$$\begin{aligned} J_x &= 0 \\ J_y &= J_\| \cos\zeta = \sigma(\cos\zeta)(E_y\cos\zeta + E_z\sin\zeta) \\ J_z &= J_\| \sin\zeta = \sigma(\sin\zeta)(E_y\cos\zeta + E_z\sin\zeta) \end{aligned} \tag{7.7.3}$$

If we now write out Maxwell's curl equations for a system of this type with no variations in x or y, using the values of J_x, J_y, and J_z from Eq. (7.7.3), we can show that there are two separate modes that can propagate in this anisotropic medium, one TEM and one TM with respect to z.

The TEM mode has field components E_x and H_y only and $J = 0$. It is unaffected by the anisotropic conductivity, because the electric field is purely in the $\perp$ direction. From Maxwell's equations it can be shown that

$$E_x/H_y = k/\omega\varepsilon = \omega\mu/k \tag{7.7.4}$$

so that

$$k^2 = \omega^2\mu\varepsilon \tag{7.7.5}$$

as in an isotropic nonconducting medium with permittivity ε and permeability μ. An electron beam constrained to move in the z-direction only would not couple to this mode at all, because the only electric field component is E_x.

The TM mode is of more interest. It has field components E_y, E_z, and H_x. From Maxwell's equations for this mode

$$\begin{aligned} E_y/H_x &= -\omega\mu/k \\ E_z/E_y &= -(\sigma \sin\zeta \cos\zeta)/(\sigma \sin^2\zeta + j\omega\varepsilon) \\ -jkH_x &= (\sigma \cos^2\zeta + j\omega\varepsilon)E_y + (\sigma \sin\zeta \cos\zeta)E_z \end{aligned} \tag{7.7.6}$$

These equations may be solved for the dispersion relation for $k(\omega)$ in the form

$$k^2 = \omega^2\mu\varepsilon(1 + j\Omega)/(\sin^2\zeta + j\Omega) \tag{7.7.7}$$

where $\Omega = \omega\varepsilon/\sigma$. In the limit of infinite conductivity σ, i.e., for $\Omega \to 0$,

$$k^2 = \omega^2\mu\varepsilon/\sin^2\zeta \tag{7.7.8}$$

The phase velocity of the wave v_p is then

$$v_p = \omega/k = (\sin\zeta)/(\mu\varepsilon)^{1/2} \tag{7.7.9}$$

and, if we have $\mu = \mu_0$ and $\varepsilon = \varepsilon_0$, the ratio (v_p/c) is given by

$$v_p/c = \sin\zeta \tag{7.7.10}$$

which is exactly the asymptotic value of (v_p/c) for a helix with pitch angle ζ in free space, i.e., this is the velocity of propagation along a helix in the axial direction at very high frequencies. Under these conditions the wave appears to travel at the velocity of light along the wire direction. Thus for

ζ very small, the z-directed velocity becomes very small and can be made synchronous with a low-velocity electron beam with a z-directed velocity. For the anisotropic medium of interest here, in the limit $\Omega = 0$, this mode, which is TM with respect to z, can also be viewed as TEM with respect to the $\|$ direction. From Eq. (7.7.6) it is evident that for $\Omega = 0$, E_z and E_y are in phase and

$$E_z/E_y = -\cot \zeta \tag{7.7.11}$$

so that $E_{\|} = 0$. However, note that although the wave is TEM with respect to the $\|$ direction and thus propagates at the velocity of light in this direction, it still has phase fronts that are normal to the z-direction. The electric field in the z-direction becomes much larger than the y-directed field for small ζ and low z-directed velocities, as desired for strong interaction with an electron beam constrained to move in the z-direction.

Let us now examine the equations governing the beam. It is more convenient in this problem to deal with an equation for current like Eq. (7.6.7) than an equation for charge like Eq. (7.4.12). Since our model for the beam is exactly the same as the beam model used in deriving these equations we can use Eq. (7.6.7), making the substitution $J_1 = 0$, as required in a one-dimensional model. Letting $J_1 = 0$ in Eq. (7.6.7) and Eq. (7.6.1), we have for the convection current in the beam J_{e1} in terms of the ac field E_{z1}

$$(\partial^2 J_{e1}/\partial z^2) + 2j\beta_e(\partial J_{e1}/\partial z) - \beta_e{}^2 J_{e1} = j\omega\varepsilon\beta_{pe}^2 E_{z1} \tag{7.7.12}$$

where we have used Eq. (7.6.1) to substitute E_{z1} for J_{e1} in the term proportional to β_{pe}^2. If we had not made this substitution, we would have had an equation equivalent to Eq. (7.4.14) from which we could have solved for the usual space charge waves as we did in Section 7.4. However, now we wish to model a beam in the presence of an electric field E_{z1} that includes not only the field due to space charge as in Section 7.4, but also the "circuit" field that can propagate through the anisotropic medium in the absence of the beam. This field E_{z1} must satisfy Maxwell's equations, just as it did for the circuit in the absence of the beam, but now the total z-directed ac current is the sum of the convection current J_{e1} and the current J_z from Eq. (7.7.3), i.e., total current for the coupled system comprising beam plus anisotropic medium is

$$\begin{aligned} J_{x1} &= 0 \\ J_{y1} &= \sigma \cos \zeta(E_{y1} \cos \zeta + E_{z1} \sin \zeta) \\ J_{z1} &= J_{e1} + \sigma \sin \zeta(E_{y1} \cos \zeta + E_{z1} \sin \zeta) \end{aligned} \tag{7.7.13}$$

where the subscript 1 has been used to indicate the ac part as before and to remind us that we are dealing with a small-signal theory, once we use Eq. (7.7.12). Note that there is no ambiguity with respect to this specification of the current, because the conduction current proportional to σ and the current J_{e1} occupy the same space. This step is somewhat less clear in other models. The currents J_{x1}, J_{y1}, and J_{z1} and the field E_{z1} must now satisfy Maxwell's equations as before, but now J_{e1} and E_{z1} must simultaneously satisfy Eq. (7.7.12). The field E_{z1} that now must satisfy Maxwell's equations for the coupled system is the self-consistent field that acts on the beam and exists in the medium in the presence of the beam. To solve for E_{z1} it is simplest first to convert Eq. (7.7.12) to an algebraic equation in terms of an unknown propagation constant k, letting J_{e1} and E_{z1} be proportional to $\exp(-jkz)$ as usual.

$$J_{e1}/E_{z1} = -j\omega\varepsilon\beta_{pe}^2/(k - \beta_e)^2 \tag{7.7.14}$$

We can use Eq. (7.7.14) to substitute for J_{e1} in Eq. (7.7.13), and we can then proceed to solve a set of equations like Eqs. (7.7.6) to obtain the dispersion relation for the coupled system which turns out to be

$$(k^2 - k_c{}^2) = \frac{\beta_{pe}^2}{(\beta_e - k)^2}\left[k^2 - k_c{}^2 - \frac{(k^2 - k_0{}^2)\sin^2\zeta}{\sin^2\zeta + j\Omega}\right] \tag{7.7.15}$$

where

$$k_0 = \omega(\mu\varepsilon)^{1/2}, \qquad k_c = \pm k_0\left(\frac{1 + j\Omega}{\sin^2\zeta + j\Omega}\right)^{1/2} = \text{cold circuit wave number}$$

Since we have coupled two systems, each having two possible waves, we now have a coupled system that propagates four waves. The four corresponding values of k are the solutions of the quartic of Eq. (7.7.15).

Let us first examine the approximate solutions to Eq. (7.7.15) obtained for $\Omega \to 0$ corresponding to high conductivity σ. As we have seen, for this condition

$$k_c = \pm k_0/\sin\zeta \tag{7.7.16}$$

and Eq. (7.7.15) reduces to

$$(\beta_e - k)^2(k^2 - k_c{}^2) = -\beta_{pe}^2 k_c{}^2\cos^2\zeta \tag{7.7.17}$$

which is still a quartic, but a simpler one than Eq. (7.7.15). Of the four waves corresponding to the four roots of this equation, three can be expected to be positively-traveling waves and only one a negatively traveling wave, because that was the situation for the uncoupled systems. We will use subscripts 1 through 4 to refer to the four solutions of Eq. (7.7.17) and the negatively traveling wave will be referred to as the fourth wave. Thus k_4 must have a value near the value of k_c corresponding to the negatively traveling wave, and k_1, k_2, and k_3 must be near the value of k_c for the positively traveling wave if the electron beam is "weak" and has a velocity near the circuit velocity. The beam can be regarded as "weak," if the rhs of Eq. (7.7.17) is small in comparison with the terms on the lhs. Certainly, if the rhs is zero the four solutions would be at β_e and k_c and, if the beam and circuit velocities are nearly equal, β_e is nearly equal to k_c. Thus, approximately

$$k_{1,2,3} \cong k_0/\sin\zeta \cong \beta_e, \qquad k_4 \cong -k_0/\sin\zeta \cong -\beta_e \tag{7.7.18}$$

A convenient way to solve for the deviations of the k's from $\pm\beta_e$ is to set

$$k = \beta_e - \delta \tag{7.7.19}$$

where δ can be complex. We can then put the value of k from Eq. (7.7.19) into Eq. (7.7.17) to obtain

$$\delta^2(\beta_e^2 - k_c^2 - 2\beta_e\delta + \delta^2) = -\beta_{pe}^2 k_c^2 \cos^2\zeta \tag{7.7.20}$$

Let us now consider the special case of beam velocity exactly equal to the positive unperturbed circuit velocity, i.e.,

$$\beta_e = k_0/\sin\zeta \tag{7.7.21}$$

so that

$$\delta^3(\delta - 2\beta_e) \cong -\beta_e^4 C^3 \tag{7.7.22}$$

where $C^3 = (\beta_{pe}^2 k_0^2 \cot^2\zeta)/\beta_e^4$.

The four values of δ, $\delta_{1,2,3,4}$, that are the roots of this equation yield the four values of k through Eq. (7.7.19). The negatively traveling wave must have a value of δ near $2\beta_e$. For δ_4 then it is approximately correct that

$$8\beta_e^3(\delta - 2\beta_e) \cong -\beta_e^4 C^3 \tag{7.7.23}$$

so that

$$\delta_4 \cong 2\beta_e - (\beta_e C^3/8), \qquad k_4 \cong -\beta_e[1 - (C^3/8)] \tag{7.7.24}$$

The three positively traveling waves must have values of δ near zero, so that for these roots, it is approximately correct that

$$-2\beta_e\delta^3 \cong -\beta_e^4 C^3 \tag{7.7.25}$$

so that

$$\delta^3 \cong \beta_e^3 C^3/2 \cong (\beta_e^3 C^3/2) \exp j2\pi \tag{7.7.26}$$

and thus the three values of δ are

$$\begin{aligned}
\delta_1 &\cong \frac{\beta_e C}{2^{1/3}} \exp\left(j\frac{2\pi}{3}\right) \cong \frac{\beta_e C}{2^{1/3}}\left(-\frac{1}{2} + j\frac{\sqrt{3}}{2}\right) \\
\delta_2 &\cong \frac{\beta_e C}{2^{1/3}} \exp\left(j\frac{4\pi}{3}\right) \cong \frac{\beta_e C}{2^{1/3}}\left(-\frac{1}{2} - j\frac{\sqrt{3}}{2}\right) \\
\delta_3 &\cong \frac{\beta_e C}{2^{1/3}} \exp\left(j\frac{6\pi}{3}\right) \cong \frac{\beta_e C}{2^{1/3}}
\end{aligned} \tag{7.7.27}$$

and the three values of k for positively traveling waves are

$$\begin{aligned}
k_1 &= \beta_e\left[1 + \frac{C}{2^{1/3}}\left(\frac{1}{2} - j\frac{\sqrt{3}}{2}\right)\right] \\
k_2 &= \beta_e\left[1 + \frac{C}{2^{1/3}}\left(\frac{1}{2} + j\frac{\sqrt{3}}{2}\right)\right] \\
k_3 &= \beta_e\left[1 - \frac{C}{2^{1/3}}\right]
\end{aligned} \tag{7.7.28}$$

A phasor diagram of the four wave numbers is shown in Fig. 7.7-3. Since the field and current amplitudes are proportional to $\exp(-jkz)$, wave

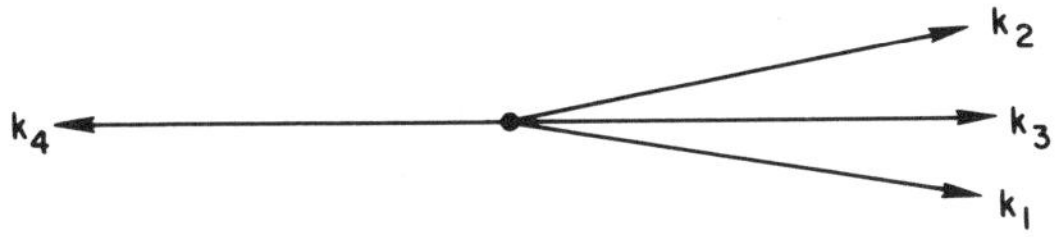

Fig. 7.7-3. Phasor diagram of the four waves that can exist in the one-dimensional model of the traveling-wave amplifier. Waves 1–3 are positively traveling waves and wave 4 is a negatively traveling wave. Waves 3 and 4 are constant-amplitude waves. Wave 1 is a decaying wave and wave 2 is a growing wave.

number 1 is a decaying wave and wave number 2 is a growing wave. The characteristic exponential wave growth thus occurs here in much the same manner as in the two-stream amplification process discussed previously. The rate of wave growth here is proportional to the parameter C which is typically of the order of 0.1 or somewhat less for a beam perveance of $(I/V^{3/2}) = 10^{-6}$. Note that the negatively traveling wave has only its phase constant altered by the presence of the beam and it does not become either a growing or a decaying wave in the presence of the beam.

Thus a traveling-wave amplifier which is perfectly matched at each end has a growing wave traveling in the positive z-direction and a constant-amplitude wave traveling in the negative z-direction. Waves numbers 1 and 3 which are positively traveling decaying- and constant-amplitude waves, respectively, will become small in comparison with the growing wave at a sufficiently great distance from the input and can thus be neglected in sufficiently long tubes. Just how great a distance is sufficiently great depends on the relative amount of each wave excited at the input. For typical simple excitation techniques the three positively traveling waves turn out to be excited about equally, just as was the case for the space-charge wave problem studied in Section 7.4. A typical case is illustrated in Fig. 7.7-4.

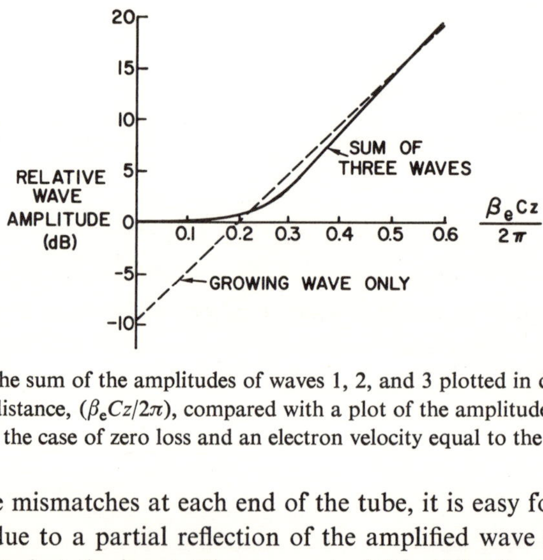

Fig. 7.7-4. The sum of the amplitudes of waves 1, 2, and 3 plotted in dB as a function of normalized distance, $(\beta_e Cz/2\pi)$, compared with a plot of the amplitude of the growing wave alone, for the case of zero loss and an electron velocity equal to the circuit velocity.

If there are mismatches at each end of the tube, it is easy for oscillations to build up due to a partial reflection of the amplified wave at the output being rereflected at the input. The way out of this difficulty is through the use of an attenuator, as mentioned above. To see how an attenuator works in a traveling-wave amplifier we return to Eq. (7.7.15) and consider the effects of small losses. To do this, let us consider the case of Ω small but finite. First, we find that the form of the cold circuit wave number for small Ω, taking $\Omega \ll \sin^2 \zeta$ and $\sin^2 \zeta \ll 1$, is

$$k_c \cong \pm \frac{k_0}{\sin \zeta}\left(1 - j\frac{\Omega}{2 \sin^2 \zeta}\right) \tag{7.7.29}$$

We have seen that the negatively traveling wave in the presence of the beam is not perturbed significantly by the beam, as seen from Eq. (7.7.24). In the presence of loss we find that the negatively-traveling wave has an attenuation constant given by the imaginary part of k_c in Eq. (7.7.29). We now wish to compare the effect of the same amount of loss on the positively traveling wave in the presence of the beam.

Using the same approximations used in Eq. (7.7.29), Eq. (7.7.15) can be written as

$$(k^2 - k_c^2) \cong \frac{\beta_{pe}^2}{(\beta_e - k)^2}\left[-\frac{k_0^2}{\sin^2\zeta}\left(1 - j\frac{\Omega}{\sin^2\zeta}\right)\right] \qquad (7.7.30)$$

Again taking the special case of beam velocity equal to unperturbed circuit velocity, i.e., $\beta_e = (k_0/\sin\zeta)$, and expressing k as in Eq. (7.7.19) we can express Eq. (7.7.30) as follows, for the positively-traveling waves:

$$\begin{aligned}\delta^3 &\cong \frac{\beta_{pe}k_0}{2\sin\zeta}\left(1 - j\frac{\Omega}{\sin^2\zeta}\right)\\ &\cong \frac{\beta_e^3C^3}{2}\left(1 - j\frac{\Omega}{\sin^2\zeta}\right)\\ &\cong \frac{\beta_e^3C^3}{2}\exp\left[j\left(2\pi - \frac{\Omega}{\sin^2\zeta}\right)\right]\end{aligned} \qquad (7.7.31)$$

So, if we examine the value of δ for the growing wave, δ_2, in the presence of finite loss, we find

$$\begin{aligned}\delta_2 &\cong \frac{\beta_e C}{2^{1/3}}\exp\left[j\left(\frac{4\pi}{3} - \frac{\Omega}{3\sin^2\zeta}\right)\right]\\ &\cong \frac{\beta_e C}{2^{1/3}}\left(-\frac{1}{2} - j\frac{\sqrt{3}}{2} + j\frac{\Omega}{6\sin^2\zeta} - \frac{\Omega}{2\sqrt{3}\sin^2\zeta}\right)\end{aligned} \qquad (7.7.32)$$

so that the value of k for this wave is

$$k_2 \cong \beta_e\left[1 + \frac{C}{2^{1/3}}\left(\frac{1}{2} + j\frac{\sqrt{3}}{2} - j\frac{\Omega}{6\sin^2\zeta}\right)\right] \qquad (7.7.33)$$

and thus, comparing Eqs. (7.7.29) and (7.7.33), the growth rate of this positively-traveling wave is reduced by exactly one-third the amount that the negatively-traveling wave is attenuated. The circuit with a beam present is therefore unsymmetrical in its response to the presence of loss, which implies that a lossy amplifier with net gain can be built that is "short-circuit stable." This means that the device will not oscillate, even if both ends are short circuited, and yet the device when reasonably well matched will give net gain from input to output. This idea is further explained in Fig. 7.7-5.

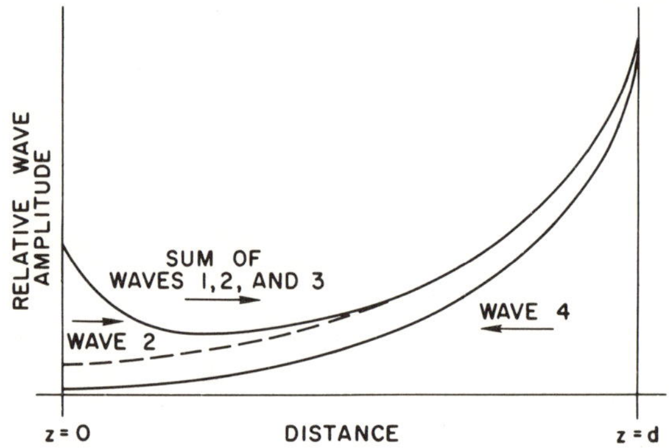

Fig. 7.7-5. Relative wave amplitudes as a function of distance in a traveling-wave amplifier with uniformly distributed loss over its entire length, for the case of a short-circuited output.

The effect of the attenuation is to cause wave 4 to decay to a sufficiently small amplitude at the input to prevent oscillation. Wave 2 grows at a more rapid rate than wave 4 decays. Waves 1 and 3 are both decaying waves in the presence of loss so the total forward wave amplitude first decays and then grows. At the output, because of the short circuit, wave 4 has the same amplitude as the sum of the three forward waves.

Another basic feature of the traveling-wave amplifier that we will not explore in detail is the requirement that beam and wave velocities be approximately in synchronism. The case that we have studied above is that of exact synchronism, and for this case amplification is observed. The range of beam velocities over which amplification is observed is, as a fraction of the beam velocity, about equal to the parameter C. Since C^3 is proportional to beam current, a low beam current results in a narrow velocity bandwidth.

Figure 7.7-6 is a photograph of a traveling-wave amplifier designed for high efficiency operation in space applications.

Problem 7.7-1 Wave Velocity in a Traveling-Wave Amplifier

For the case of beam velocity equal to the unperturbed circuit velocity show that the beam velocity is greater than the velocity of the growing wave.

Problem 7.7-2 Wave Excitation in a Traveling-Wave Amplifier

Show that the three positively traveling waves are excited equally for the case in the text of zero loss and exact synchronism of beam and unperturbed circuit velocity, if ac velocity and current are equal to zero at $z = 0$. Assume the output is perfectly matched so the fourth wave has zero amplitude.

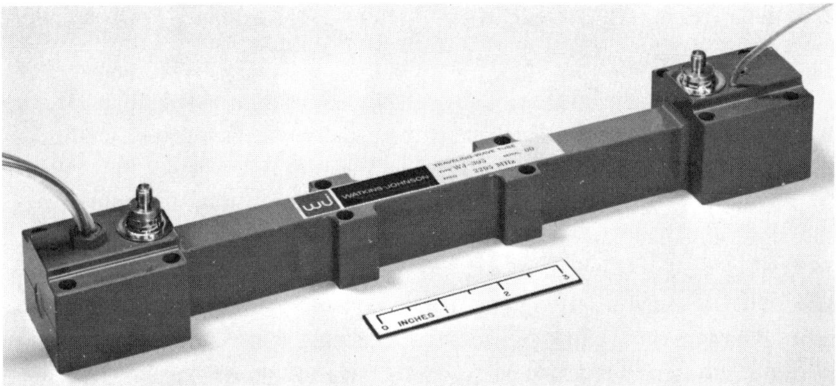

(a)

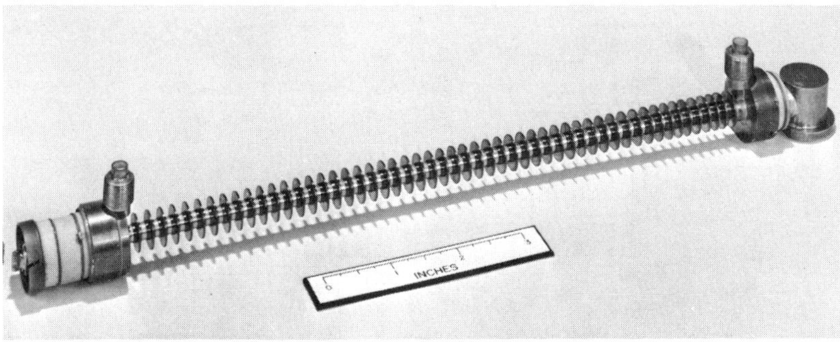

(b)

Fig. 7.7-6. A traveling-wave amplifier designed for use in an earth satellite. The tube provides 100 W output at 2.3 GHz at an efficiency of 48%. (a) Photograph of tube in housing which encloses periodic permanent magnet focusing system. (b) Photograph of tube prior to installation of permanent magnets. Discs are of magnetic material to carry flux into a small radius. [Courtesy Watkins Johnson Co.]

PROBLEM 7.7-3 NET GAIN OF A TRAVELING-WAVE AMPLIFIER

For a traveling-wave amplifier with the following properties, use the model in the text to calculate the net gain, assuming that the power at the input is equally divided among waves 1, 2, and 3. Assume both ends of the tube are perfectly matched, i.e., VSWR = 1.0.

$$\text{beam voltage} = 10{,}000 \text{ V}, \qquad \text{beam current} = 0.1 \text{ A}$$

$$\text{length} = 12 \text{ in.}, \qquad \text{frequency} = 3 \text{ GHz}$$

$$\varepsilon = \varepsilon_0, \qquad \mu = \mu_0, \qquad \Omega = 0.1$$

7.8 Interaction with a Distributed Circuit—the Resistance-Wall Amplifier

In the preceding section we analyzed the interaction between an electron beam and a propagating circuit such as a helix with a phase velocity near the velocity of the beam. Here we treat an even simpler device in which the beam interacts with a circuit which supports only a decaying wave, as shown in Fig. 7.8-1. The fact that such a device could be an amplifier at all is very odd and when it was first proposed by Birdsall, Brewer, and Haeff [16] it came as a complete surprise to those doing research in this field. Prior to their work the idea of passing a beam through a resistive medium had been suggested as a way of attenuating the noise on the beam in order to produce a low-noise amplifier. In fact the precise opposite of attenuation occurs, and beam noise is amplified by this procedure.

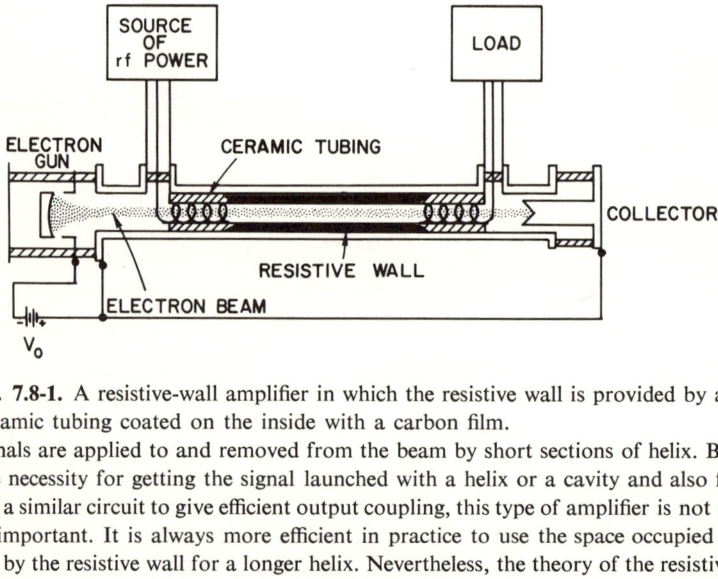

Fig. 7.8-1. A resistive-wall amplifier in which the resistive wall is provided by a piece of ceramic tubing coated on the inside with a carbon film.

Signals are applied to and removed from the beam by short sections of helix. Because of the necessity for getting the signal launched with a helix or a cavity and also for the use of a similar circuit to give efficient output coupling, this type of amplifier is not practically important. It is always more efficient in practice to use the space occupied in this figure by the resistive wall for a longer helix. Nevertheless, the theory of the resistive wall has been important in understanding other devices, especially in connection with the attenuator region of a conventional traveling-wave amplifier.

We could approach the analysis of this device by simply increasing the attenuation in the preceding analysis of the traveling-wave amplifier. We would find a steadily decreasing amplification and a steadily increasing velocity bandwidth as the amount of loss increased. Finally we would reach the point at which it would no longer be of importance that the distributed circuit have a wave-slowing property in the absence of loss at all. In other words, we could replace the helix plus attenuator with simply the attenuator, as shown in Fig. 7.8-1.

In our model of the preceding section, this means that we could simply set $\zeta = \pi/2$ and allow conduction in the same direction as the direction of beam travel. We could also allow for conduction at right angles to the beam direction and consider the case of an isotropic medium. Because we have already analyzed the anisotropic medium let us here simply continue that analysis for the configuration shown in Fig. 7.8-2 in which we set $\zeta = \pi/2$ and consider the effects of large loss.

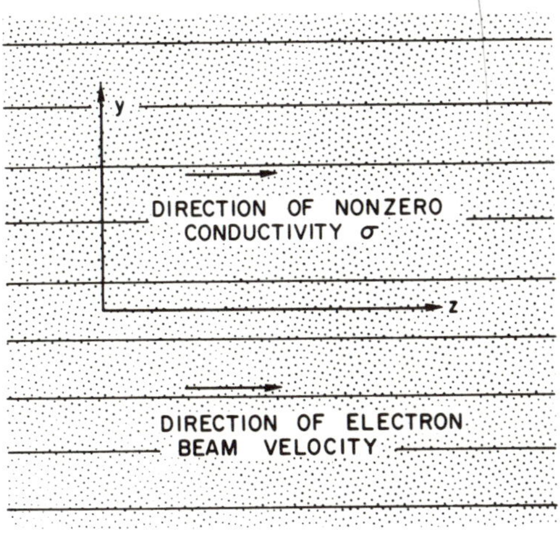

Fig. 7.8-2. A one-dimensional model of the resistive-wall portion of the resistive-wall amplifier of Fig. 7.8-1. An infinite medium has permittivity ε and permeability μ that are uniform and isotropic, but has an anisotropic conductivity that is σ in the z-direction and zero in the x- and y-directions. The medium is permeable to an electron beam traveling in z-direction. There are no variations in any quantities in the x- or y-directions.

We first examine the cold circuit behavior, setting $\sin \zeta = 1$ in Eq. (7.7.7). Evidently for this value of ζ

$$k^2 = \omega^2 \mu \varepsilon = k_0^2 \tag{7.8.1}$$

and the unperturbed circuit wave is unattenuated, without regard to the value of Ω. We also see from Eq. (7.7.6) that $(E_z/E_y) = 0$ which indicates that this mode is TEM with respect to z. From our previous experience we would expect that a beam constrained to move only in the z-direction would not interact with such a wave. This is correct as we now will see. Note that if we had analyzed an isotropic lossy medium, the unperturbed circuit wave would have also been a TEM wave, but an attenuated wave rather than a constant amplitude wave. It also would have had zero E_z and hence also would have had zero coupling to the beam.

The behavior of the coupled system can be found from Eq. (7.7.15) by setting $\sin \zeta = 1$. We find that this equation reduces to

$$(k^2 - k_0^2) = \frac{\beta_{pe}^2}{(\beta_e - k)^2}\left[1 - \frac{1}{1 + j\Omega}\right](k^2 - k_0^2) \qquad (7.8.2)$$

Evidently two solutions are those corresponding to the roots of Eq. (7.8.1) and these two waves are not coupled to the space-charge waves of the beam, because the factor $(k^2 - k_0^2)$ appears on both sides of Eq. (7.8.2). Canceling it out we obtain the dispersion relation for the waves of interest:

$$(\beta_e - k)^2 = \beta_{pe}^2\left[1 - \frac{1}{1 + j\Omega}\right] \qquad (7.8.3)$$

This is a quadratic in k which has two solutions which reduce to the usual space charge waves for $\Omega = 0$. For finite Ω we can solve Eq. (7.8.3) for k as follows:

$$k = \beta_e \pm \beta_{pe}[1 + (1/j\Omega)]^{-1/2} \qquad (7.8.4)$$

Evidently k has one solution with a positive imaginary part representing a growing waves. Clearly these two waves are perturbed forms of the usual space-charge waves that are a result of passing the beam through a medium with a finite conductivity. This idea can be generalized to consider media with arbitrary properties [17]. From this point of view the problem treated previously of a beam traveling through a plasma can be shown to be equivalent to that of a beam traveling through a medium with a negative permittivity. We will return to this generalized point of view after examining Eq. (7.8.4) in more detail.

Equation (7.8.4) can be written in the form

$$k = \beta_e \pm \beta_{pe}(p + jq) \qquad (7.8.5)$$

where $p + jq = [1 + (1/j\Omega)]^{-1/2}$. Figure 7.8-3 is a plot of p and q as a function of Ω. The quantities p and q specify the phase and amplification of the wave, respectively. Maximum gain occurs for maximum q which is seen to occur for

$$\Omega = 1/\sqrt{3}, \qquad q = 1/\sqrt{8} \qquad (7.8.6)$$

The value of k at maximum gain is seen, from Fig. 7.8-3, to be

$$k = \beta_e \pm \beta_{pe}(0.63 + j0.353) \qquad (7.8.7)$$

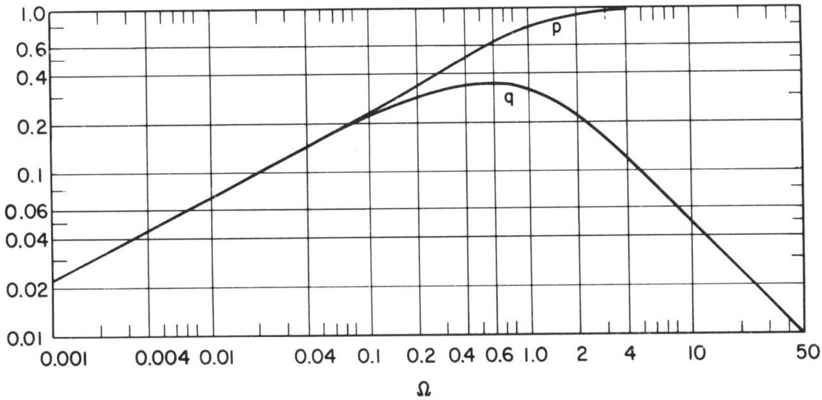

Fig. 7.8-3. The gain factor q and phase factor p as a function of $\Omega = \omega\varepsilon/\sigma$ for the one-dimensional model of the resistive-wall amplifier [from C. K. Birdsall, G. R. Brewer, and A. V. Haeff, The resistive wall amplifier, *Proc. IRE* **41**, 865–875 (1953)].

This equation turns out to predict a gain that is less than the gain of a comparable low loss traveling-wave tube for ordinary currents and voltages. This is as we might expect, since this device is equivalent to a traveling-wave tube with heavy loss.

Let us now look at this problem from a somewhat more general point of view. Instead of a resistive medium, let us consider a medium with arbitrary values of complex permittivity ε_c. As we have seen, the waves of interest here are simply space charge waves in a medium other than free space. The values of k in this case can be derived by simply rederiving the space-charge wave equations, substituting ε_c for ε_0 throughout. The result can obviously be written

$$k = \beta_e \pm \beta_{pe}[\varepsilon_r]^{-1/2} \tag{7.8.8}$$

where $\varepsilon_c = \varepsilon_0\varepsilon_r$. Evidently for the resistive medium

$$\varepsilon_r = 1 + (1/j\Omega) \tag{7.8.9}$$

For arbitrary values of ε_r we can obtain arbitrary values of p and q in Eq. (7.8.5) and hence arbitrary values of gain and phase change. We will consider several special cases.

First, consider the case of an ordinary lossless dielectric for which ε_r is a positive purely real number. From Eq. (7.8.8) it is evident that in this case k is purely real and there is no amplification.

Next, consider the case of a negative dielectric medium for which ε_r

is a negative purely real number. Here it is evident that k takes the form

$$k = \beta_e \pm j\beta_{pe}[|\,\varepsilon_r\,|]^{-1/2} \tag{7.8.10}$$

which is seen to be of exactly the form of Eq. (7.5.6) for the beam traveling through a background of mobile positive ions. Such a medium can thus be characterized as having a negative permittivity.

For complex values of ε_r, the analysis is equivalent to what we have been through for the resistive medium. Note that a negative resistance would not behave in any significantly different way from what we have already observed for a positive resistance. Similarly for purely imaginary ε_r, k is always complex and the only effect of changing the sign of ε_r is to reverse the rôles of the two waves, but there is always a growing wave.

PROBLEM 7.8-1 RESISTIVE-MEDIUM AMPLIFIER

Compare the gains of a resistive-medium amplifier and a double-stream amplifier with equal values of β_{pe} for both beams. The double-stream amplifier has a value of β_{pe} for the two beams together exactly equal to the β_{pe} for the resistive-medium amplifier.

PROBLEM 7.8-2 RESISTIVE-MEDIUM AMPLIFIER

Compare the gains of a resistive-medium amplifier and a traveling-wave amplifier with zero attenuation, for the case where the β_{pe} of each device is the same and for equal values of β_e. Assume that the traveling-wave amplifier is operating at exact synchronism and that the resistive-medium amplifier is operating at maximum gain.

7.9 The Magnetic Pinch Effect

In the preceding sections of this chapter we have used a fluid model to represent various electron and ion beam configurations in which only axial perturbations of the beam were allowed. In this section we consider radial perturbations in a perfectly compressible fluid model of a high energy electron beam neutralized by positive ions. Magnetic effects are assumed to be dominant in this beam as a result of the neutralization of space-charge forces, so this problem is analogous to the problem treated by the Lagrangian technique in Section 6.8.

The magnetic pinch effect is caused by the mutual attraction of current elements in a current-carrying conductor such as a wire or a plasma column. The effect occurs in all pressure ranges. For example, in a solid wire the effect exists as a force, but there is no motion unless the wire is melted or vaporized. Exploding wires result from a sudden application of a very high current which heats and vaporizes the wire while the current is flowing. The pinch effect carries the material of the wire together to a maximum density condition after which the material bounces back and "explodes." Similar phenomena can occur in high current arcs in plasmas, and a class of fusion machines makes use of this effect to attempt to obtain, on a transient basis, the very high density and temperature necessary for thermonuclear fusion [18]. The difficulty with these machines is that there are too many ways that the arc can do something other than maintain perfect axial symmetry and shrink down to the desired small, straight, high-density current-carrying column. Two other basic ways that the column can behave are the result of the so-called sausage and hose instabilities. If the column starts to shrink slightly more rapidly at one point along its length than at adjacent points, its tendency is to shrink more rapidly in the smaller region than in the larger regions, thus enhancing any random perturbations of diameter. This type of instability can be stabilized by the application of an axial dc magnetic field. However, then the column can readily develop the hose instability, in which, instead of shrinking, the column forms itself into a helix which rapidly grows in diameter [19]. We shall not attempt to deal with problems of this complexity, but it is important to realize that, lurking just beyond the elementary one-dimensional pinch phenomenon discussed here, is a large class of interesting and complex instabilities based on the mutual attraction of current elements in two-dimensional configurations.

We shall treat only the case of a current-carrying fluid that is infinitely compressible, i.e., we neglect the effects of collisions entirely, as is appropriate for low pressure, highly ionized media. Our analysis would thus be incorrect for a mercury column carrying a current. In an incompressible fluid like mercury, sausage-type instabilities are enhanced even beyond what occurs in compressible fluids, because, if one squeezes on an incompressible fluid at one point, adjacent points are immediately pushed out by mechanical forces. In our fluid, this additional effect is not present.

Figure 7.9-1 is a drawing of an experimental apparatus that readily exhibits the effect that we will analyze here. A high-energy electron beam is

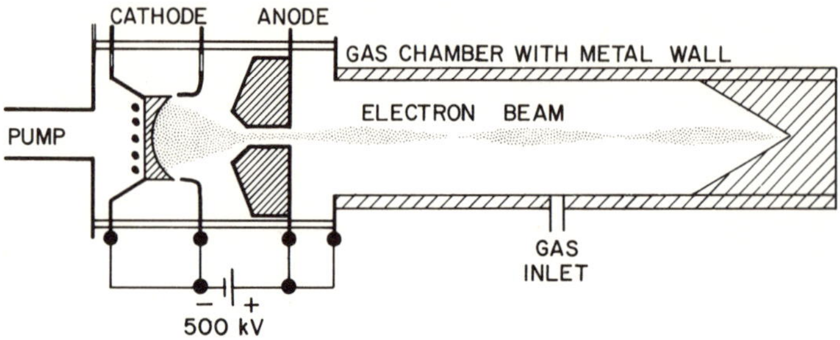

Fig. 7.9-1. A high-voltage (such as 500 kV) electron beam is injected into a chamber containing a gas at low pressure (such as 10^{-3} Torr). The electron gun region is differentially pumped with respect to the gas chamber to maintain a sufficiently high vacuum (such as 10^{-6} Torr) to avoid breakdown between cathode and anode. The beam diameter varies as it travels, resulting in a beam profile that has a series of minimum diameter points repeating at fixed intervals, giving it the appearance of a sausage.

injected into a neutral gas, ionizing the gas and thereby making enough ions available to neutralize the dc space charge of the electron beam. The details of how this neutralization takes place and just how many ions and low-energy electrons accumulate in the beam is a function of the temperature of the low-energy electrons as well as of other factors [20]. For our purposes, we are only interested in the fact that the beam indeed does become sufficiently neutralized that the electrostatic forces tending to spread the beam become small in comparison with the magnetic forces that we wish to consider here. In a physical neutralized beam, there will also be electrostatic forces that will tend to pinch the beam. For high voltage, high current beams the magnetic pinch effect is the dominant one. For low voltages, the electrostatic effect is dominant [21].

The model that we treat is shown in Fig. 7.9-2. It is an axially symmetric model in which only radial motion is allowed [21, 22]. For this cylindrical configuration we can use the Biot–Savart law to find the magnetic field H_θ within the beam of outer radius r_c, as a function of r, as follows. The magnetic field H_θ at r_c for a beam current I_b is

$$I_b = \oint \mathbf{H} \cdot dl = H_\theta \cdot 2\pi r_c \tag{7.9.1}$$

The field at r is then

$$H_\theta = \frac{I_b}{2\pi r_c}\left(\frac{r}{r_c}\right) \tag{7.9.2}$$

The radial inward force F_r on an electron at radius $r_e < r_c$ is that due to

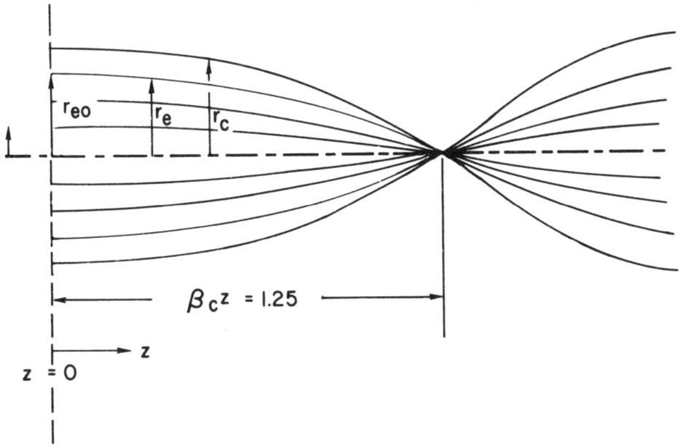

Fig. 7.9-2. A one-dimensional model of the beam of Fig. 7.9-1. The radius of an arbitrary electron is r_e and at $z = 0$, $r_e = r_{e0}$. The outer radius of the beam is r_c. The velocity of all particles is in the z-direction at the plane $z = 0$ [from A. S. Halsted and D. A. Dunn, Electrostatic and magnetic pinch effects in beam-generated plasmas, *J. Appl. Phys.* **37**, 1810–1820 (1966)].

the Lorentz force

$$F_r = -\frac{eu_b}{\mu_0 H_\theta} = -\left[\frac{e\mu_0}{2\pi r_c{}^2}\left(\frac{2e}{m_e}\right)^{1/2}\right]\left[\frac{I_b}{V_b^{3/2}}\right]V_b{}^2 r_e \qquad (7.9.3)$$

where μ_0 is the permeability of free space, e is charge of the electron, m_e is mass of the electron, u_b is beam velocity, V_b is beam voltage, and $u_b{}^2 = (2eV_b/m_e)$. It has been written in this form, because the beam perveance $(I_b/V_b^{3/2})$ is a quantity determined by electron gun design with a typical value of 10^{-7} to 10^{-6}, and is typically fixed for a given experiment. This equation is obviously based on a nonrelativistic calculation of the beam velocity as a function of energy. Note that the force on the outermost beam electrons, at $r_e = r_c$, is inversely proportional to beam radius. The beam is thus more strongly pinched as it becomes smaller.

Making the assumption of relatively low energy, i.e., $V_b < 300$ kV or so, so that Eq. (7.9.3) applies, and looking at the path of the outer beam electrons, we set $r_c = r_e$ in Eq. (7.9.3) and write it in the form

$$m_e\frac{d^2 r_e}{dt^2} = -\left[\frac{\mu_0}{4\pi r_{e0}^2}\left(\frac{2e}{m_e}\right)^{1/2}\right]\left[\frac{I_b}{V_b^{3/2}}\right]\frac{2er_{e0}^2 V_b{}^2}{r_e} \qquad (7.9.4)$$

where r_{e0} is the initial value of r_e. This equation gives us the behavior in time of a particular electron. If we move with an electron, this equation describes its behavior.

Let us now change from a frame of reference following the electron to the rest frame where we can observe the beam profile. Evidently we can use a Galilean transformation to a frame moving at the beam velocity in the negative z-direction, so that

$$\frac{d^2}{dz^2} = \left(\frac{2eV_b}{m_e}\right)\frac{d^2}{dt^2} \tag{7.9.5}$$

We can then rewrite Eq. (7.9.4) in the form

$$\frac{d^2}{dz^2}\left(\frac{r_e}{r_{e0}}\right) = -\beta_c{}^2\left(\frac{r_{e0}}{r_e}\right) \tag{7.9.6}$$

where

$$\beta_c{}^2 = \left(\frac{\mu_0}{4\pi r_{e0}^2}\right)\left(\frac{2e}{m_e}\right)^{1/2}\left(\frac{I_b}{V_b^{3/2}}\right)V_b \tag{7.9.7}$$

The solution to Eq. (7.9.6) can be obtained in terms of the error function in the form

$$\beta_c z = 2^{1/2}\int_0^{[\ln(r_{e0}/r_e)]^{1/2}} \exp(-u^2)\, du \tag{7.9.8}$$

A plot of Eq. (7.9.8) is given in Fig. 7.9-3. The value of the integral in Eq. (7.9.8) at the first crossover when $r_e = 0$ is $(\pi/2)^{1/2}$, so the first focus point occurs at $\beta_c z = (\pi/2)^{1/2} = 1.25$ as shown. Two other curves, parabolic and sinusoidal in shape, are shown for comparison.

Let us now review what we have found. Our prediction is that a neutralized electron beam will pinch itself to a focus, then expand, and repetitively

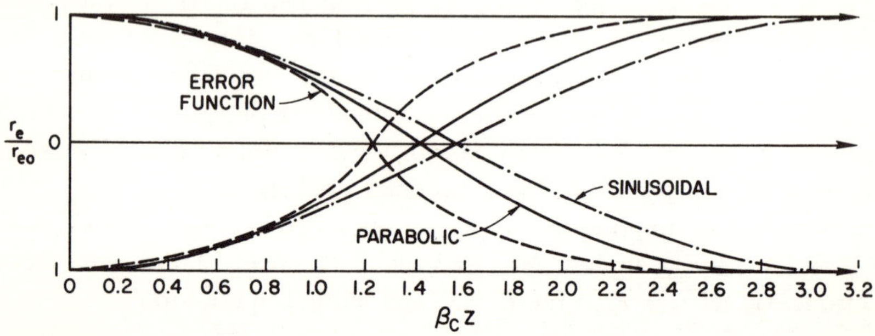

Fig. 7.9-3. Beam radius r_c as a function of normalized axial distance $\beta_c z$ for an electron beam experiencing the magnetic pinch effect is the curve marked error function. This curve results from a magnetic field at the outer beam radius r_c that increases inversely with radius. If the field were independent of radius, the parabolic curve would be obtained. If the field increased linearly with radius, the sinusoidal curve would be obtained [from A. S. Halsted and D. A. Dunn, Electrostatic and magnetic pinch effects in beam-generated plasmas, *J. Appl. Phys.* **37**, 1810–1820 (1966)].

come to a focus again, as a result of the magnetic pinch effect. This is precisely what has been observed [23] in an experiment similar to that of Fig. 7.9-1, except that it was performed at a higher beam energy than is suggested there and hence made use of a more complex electron source than that indicated in the figure.

Our derivation differs from that usually carried out for the sausage instability in a plasma column. There are three major differences between the regime in which our theory applies and that of the typical plasma column. First, the drift velocity of the current carriers here is enormously greater than in an arc or other similar discharge where 10 volts is a typical carrier energy. As a result, for the same current, our plasma has a much lower particle density than an arc. Second, the random radial velocity of the carriers is insignificant percentagewise in our case, but can be greater than the drift velocity in an arc. Third, collisions with neutrals have an insignificant effect in our case, but strongly effect the behavior of an arc, which can approach the incompressible-fluid behavior of a liquid-mercury column. Because of these differences, our analysis has been somewhat simpler than the analysis for a higher-pressure, lower-voltage discharge. The spatial period of the instability in our case is fixed by the drift velocity and beam-current density, whereas the sausage instability in a low-voltage arc has a wavelength that depends on a number of other factors such as the pressure and the temperature.

The fundamental cause of the oscillations in both systems is the same, i.e. the magnetic pinch effect. And there are a number of similarities in the detailed mechanisms involved, such as the fact that the inward force increases with decreasing radius, thereby enhancing any perturbation in the radius of an initially uniform plasma column. It should be noted, however, that the force that is responsible for these perturbations could also result in similar perturbation patterns under some circumstances, even if it increased with increasing radius, as suggested by the other curves in Fig. 7.9-3. Note that the planar model of Section 6.8 led to parabolic trajectories appropriate to a constant force, independent of radius, obtained in that model.

REFERENCES

1. G. W. Sutton and A. Sherman, "Engineering Magnetohydrodynamics." McGraw-Hill, New York, 1965.
2. W. F. Hughes and F. J. Young, "The Electromagnetodynamics of Fluids." Wiley, New York, 1966.

3. B. Salzberg and A. V. Haeff, Effects of space charge in the grid–anode region of vacuum tubes, *RCA Rev.* **2**, 336–374 (1938).
4. C. E. Fay, A. L. Samuel, and W. Shockley, On the theory of space charge between parallel plane electrodes, *Bell Syst. Tech. J.* **17**, 49–79 (1938).
5. H. F. Ivey, Space charge limited currents, *Advan. Electron. Electron Phys.* **6**, 137–256 (1954).
6. D. A. Dunn, V. Hamza, and J. A. Jolly, Computer designs for beam pentodes, *IEEE Trans. Electron Devices* **ED-12**, 343–351 (1965).
7. Ramo-Wooldridge Staff, Electrostatic propulsion, *Proc. IRE* **48**, 447–491 (1960).
8. H. Derfler, Nonexistence of quiescent plasma states in ion propulsion, *Phys. Fluids* **7**, 1625–1637 (1964).
9. W. W. Harman, "Fundamentals of Electronic Motion." McGraw-Hill, New York, 1953.
10. J. Marcum, Interchange of energy between an electron beam and an oscillating electric field, *J. Appl. Phys.* **17**, 4–11 (1946).
11. M. Chodorow and C. Susskind, "Fundamentals of Microwave Electronics." McGraw-Hill, New York, 1964.
12. D. A. Dunn and G. L. Sackman, Interaction of two beams with a cavity, *J. Electron. Contr.* **18**, 181–200 (1965).
13. A. H. W. Beck, "Thermionic Valves." Cambridge Univ. Press, London and New York, 1953.
14. J. R. Pierce, "Traveling Wave Tubes." Van Nostrand-Reinhold, Princeton, New Jersey, 1950.
15. S. Olving, "A new method for space charge wave interaction studies," Rept. 37, Res. Lab. Electron., Chalmers Univ. Techn., Goteborg, Sweden, 1956.
16. C. K. Birdsall, G. R. Brewer, and A. V. Haeff, The resistive wall amplifier, *Proc. IRE* **41**, 865–875 (1953).
17. C. K. Birdsall and J. R. Whinnery, Waves in an electron stream with general admittance walls, *J. Appl. Phys.* **24**, 315–323 (1953).
18. D. J. Rose and M. Clark, Jr., "Plasmas and Controlled Fusion." Wiley, New York, 1961.
19. H. Alfven and C. G. Falthammer, "Cosmical Electrodynamics." Oxford Univ. Press, London and New York, 1963.
20. D. A. Dunn and S. A. Self, Static theory of density and potential distribution in a beam-generated plasma, *J. Appl. Phys.* **35**, 113–122 (1964).
21. A. S. Halsted and D. A. Dunn, Electrostatic and magnetic pinch effects in beam-generated plasmas, *J. Appl. Phys.* **37**, 1810–1820 (1966).
22. J. D. Lawson, Perveance and the Bennett pinch relation in partially neutralized electron beams, *J. Electron. Contr.* **5**, 146–151 (1958).
23. S. E. Graybill and S. V. Nablo, Observations of magnetically self-focusing electron streams, *Appl. Phys. Lett.* **8**, 18–22 (1966).

Author Index

Numbers in parentheses are reference numbers and indicate that an author's work is referred to, although his name is not cited in the text. Numbers in italics show the page on which the complete reference is listed.

Subject Index